STRATEGIES FOR BUSINESS AND TECHNICAL WRITING

Sixth Edition

KEVIN J. HARTY

La Salle University

PEARSON
Longman

New York San Francisco Boston
London Toronto Sydney Tokyo Singapore Madrid
Mexico City Munich Paris Cape Town Hong Kong Montreal

Acquisitions Editor: Lauren A. Finn
Editorial Assistant: Bristol Maryott
Marketing Manager: Tom DeMarco
Senior Cover Designer/Manager: Nancy Danahy
Production Manager: Ellen MacElree
Project Coordination and Electronic Page Makeup: Nesbitt Graphics, Inc.
Cover Photo: PhotoDisc, Inc./Getty Images; Digital Vision/Getty Images, Inc.; Artville Images.
Manufacturing Manager: Mary Fischer
Printer and Binder: Courier
Cover Printer: Courier

For permission to use copyrighted material, grateful acknowledgment is made to the copyright holders on pp. 381–383, which are hereby made part of this copyright page.

Library of Congress Cataloging-in-Publication Data

Harty, Kevin J.
  Strategies for business and technical writing / Kevin J. Harty.—6th ed.
    p. cm.
  Includes bibliographical references and index.
  ISBN-13: 978-0-205-56206-0
  ISBN-10:   0-205-56206-X
 1. Commercial correspondence. 2. Business report writing. 3. Technical writing. I. Title.

  HF5721.H37 2008
  808'.066651—dc22

                                                              2007043464

Please visit our website at http://www.ablongman.com

ISBN-13: 978-0-205-56206-0
ISBN-10:   0-205-56206-X

1 2 3 4 5 6 7 8 9 10—CRS—10 09 08 07

In memory of John Keenan
and
John Christopher Kleis
—both good friends, generous colleagues, skilled writers,
and exacting, but always supportive, editors

Contents

Preface

This sixth edition of *Strategies for Business and Technical Writing* represents a thorough reworking of the previous edition. The sixth edition retains 26 of the 32 selections from the fifth edition and adds six new selections. In revising *Strategies* for this new edition, I have been guided by the same principle that informed the selection of materials for inclusion in the five previous editions: *Strategies* should present the best advice from the best sources about important issues in business, technical, and other kinds of professional writing.

The changes I have made in preparing this sixth edition broaden the coverage in *Strategies* of key issues in business and technical writing, and increase the usefulness of *Strategies* as both a textbook and a reference. The six new selections in this sixth edition of *Strategies* reflect the ever-changing demands writers in the world of work face to meet the needs of their many audiences.

In particular, this sixth edition of *Strategies:*

- expands the previous coverage of the problems posed by legalese and reemphasizes the need for using Plain English in a variety of kinds of documents;
- includes a new selection that offers extensive advice on how to avoid international miscommunication;
- contains a completely revised and expanded discussion of e-mail and other forms of electronic communication;
- expands the collection's previous discussions of persuasive writing to include suggestions on how writers might match style with purpose better to achieve their goals;
- provides a completely revised and expanded section on the job search that now offers job candidates advice from corporate recruiters and headhunters as well as from university placement officers.

This sixth edition of *Strategies* can, like its predecessors, be used in three ways. *Strategies* can serve as a supplement to a business or technical writing textbook. *Strategies* can also be used as a supplement to a composition textbook or a writing handbook. And *Strategies* can be used on its own as a textbook or as a reference source.

I have designed *Strategies* to appeal to practical-minded instructors, students, and women and men already working in business and technical occupations and the professions. The contributors to this volume write with the kind of purpose and understanding that come only from a career dedicated to improving the effectiveness of business and technical communication. As a result, the selections in *Strategies* not only teach professional writing but are also fine examples of the genre.

In selecting essays for this sixth edition of *Strategies,* I have been guided by the belief that all courses in business and technical writing should firmly implant one idea in the minds of students (and of those in the world of work). *All* successful writing consists of clear and effective prose no matter in which medium it is composed or sent. However, students should recognize that business and technical writing differs in important ways from the expository prose usually taught in first-year or advanced composition courses. Because business and technical writers communicate with multiple audiences and with a variety of intentions, learning to write well for the world of work may require multiple adjustments in technique. Therefore, in addition to touching on such broad problems as style, jargon, and diction, *Strategies* includes comprehensive discussions of specific forms of writing—for example, different kinds of resumes, letters of application, business letters, memos, e-mail, reports, and proposals. Students learn the specific techniques used by successful men and women in business, industry, and technical professions who communicate information as part of their jobs.

The essays in this sixth edition of *Strategies* have, like those in the five previous editions, survived the most rigorous scrutiny—that of my students in seminars and courses in business and technical writing at La Salle University, Temple University, Rhode Island College, Centenary College of Louisiana, the Federal Reserve Bank of Philadelphia, Kroll Lindquist Avey, Philadelphia Telco Credit Union, the Board of Pensions of the United Presbyterian Church (USA), First Pennsylvania Bank and Fidelity Bank (now both Wachovia Bank), Blue Cross of Greater Philadelphia, and Philadelphia Newspapers, Inc.

For permission to reprint the selections included in this sixth edition of *Strategies,* I am happy to thank the authors, agents, and editors who handled my permissions requests with efficiency and good cheer. At La Salle University, I owe a continuing debt to my many good friends and colleagues, especially to Joe Brogan, Jim Butler, Madeleine Dean, Gabe Fagan, Bob Fallon, Francine Lottier, Emery Mollenhauer, and Lynne Texter. I am especially grateful to the staff of the La Salle University library, in particular to Stephen Breedlove and Nancy Tresnan in Interlibrary Loan Services.

Finally, I am indebted to the editorial and production staff at Pearson Longman, especially to my editor, Lauren Finn, and to the reviewers whom she recruited to critique the previous edition of this book: Daniel Ding, Ferris State University; Lou Ann Karabel, Indiana University Northwest; Timothy Miller, Millersville University; Rob Rector, Delaware Technical and Community College; and Adina Sanchez-Garcia, University of Miami.

Kevin J. Harty
Philadelphia
January 2008

Introduction

Writing consumes a substantial portion of the working day for almost all college-educated workers.[1]

Ability to express ideas cogently and goals persuasively—in plain English—is the most important skill to leadership. I know of no greater obstacle to the progress of good ideas and good people than the inability to compose a plain English sentence.[2]

Whether you are planning a career in business, in industry, or in some technical field, or whether you already have such a career, you may be ill-prepared for what is perhaps the most difficult part of your job. Although you may be an excellent accountant, salesperson, manager, engineer, or scientist, all of your training may not help you when you sit down to write. If this is the case, you are not alone in being frustrated by the writing demands of your job. Nevertheless, writing is essential to most technical and professional occupations; the ability to communicate effectively both in person and on paper can help you advance in your profession. Good writing can mean the difference between winning or losing a major sale, a pleased customer, or a challenging new position with a higher salary.

The selections in this book will help you make your letters, reports, memos, and other professional documents more effective. The selections represent some of the best advice veteran teachers and practitioners of business and technical writing have to offer. None of the selections, however, offers a quick cure for writing ills. There is no such cure. Like any worthwhile skill, good professional writing requires time, practice, and, most of all, discipline.

To begin with, we need to realize that there are only two kinds of professional writing:

1. clear, effective writing that meets the combined needs of the reader and the writer, and
2. bad writing.

Bad writing is unclear and ineffective; it wastes time and money. Even worse, it ignores its readers, usually raising more questions than answers.

---

[1] Paul V. Anderson, "What Survey Research Tells Us about Writing at Work," in *Writing in Nonacademic Settings,* ed. Lee Odell and Dixie Goswarmi (New York: Guilford, 1985), p. 30.

[2] The late John D. deButts, former Chairman, AT&T. Quoted as the epigraph to the third edition of the Bell Laboratories' *Editorial Style Guide* (Whippany, NJ: 1979).

To be effective writers, each of us needs to develop a variety of strategies for the different writing situations we face. We must also respect writing as something more than words typed or written on the page or fed into a dictaphone or word processor. Writing takes thought and planning.

THREE KEY QUESTIONS

Business and technical writing situations present us with a variety of challenges. One of the ways to meet these challenges is to begin by asking ourselves these three key questions:

1. Exactly who is my audience?
2. What is the most important thing I want to tell my audience?
3. What is the best way of making sure my audience understands what I have to say?

Exactly Who Is My Audience?

Whatever we write on the job, somebody—across the hall, across the city, across the country, or even across the world—will eventually read what we have written. Most likely, our readers will receive our message in their own offices or homes. Since we won't be on hand to explain anything that is unclear or that doesn't make sense, our message will have to speak for itself.

Whenever we sit down to write, we become experts. That is in large part why we are the writers and not the readers. Our readers depend on us for clear, effective writing, so that they may share in our expertise. Although this point seems obvious, business, industrial, and technical practice shows that the single most common cause of bad writing is ignoring the reader. There are any number of ways to ignore our audience: we can talk over their heads; we can talk down to them; we can beat around the bush; we can leave out important details—and, as a result, we can undermine our credibility.

When you next sit down to write, therefore, it is important to ask yourself what you know about your audience—you must determine your audience's situation and anticipate your audience's potential reaction to your message. To paraphrase the golden rule, make sure you are prepared to write unto others the way you would have them write unto you.

What Is the Most Important Thing I Want to Tell My Audience?

Whether you write one page or one hundred pages, you should be able to condense your most important idea into about 25 words. If you can't, you aren't ready to write, since you haven't figured out what you are trying to say. You may not be able to explain every reason for a decision or a proposal in 25 words, but 25 words should be enough for a statement of the decision or proposal itself. For instance, you may have several reasons for recommending that your company replace its Brand X word processors with Brand Y word processors. You can make the recommendation in 22 words flat:

I recommend that we immediately replace all our Brand X word processors with new Brand Y units for the reasons presented below.

In short, you can cut through all the bull and come right out with your recommendation. Then, in the body of your memo or report, you can present the relevant facts. Busy people dislike having their time wasted, and they hate surprises. So, in professional writing, the bottom line should also be the top line.

What Is the Best Way of Making Sure My Audience Understands What I Have to Say?

Different writing situations require different strategies. After answering the first two questions, you may decide that you shouldn't write at all—that you should phone instead or make a personal visit. But once you do commit yourself to writing, you need to determine how to put the particular document together. You will need to use one strategy for good news and a different one for bad news, one for giving information and another for changing someone's mind, one for accepting a job offer and another for turning the offer down. But in each case you will need to get your main idea out as soon as possible in as few words as possible, and then later defend or elaborate on that idea as clearly and as effectively as possible.

To illustrate the importance of answering these three key questions properly, let's look at a brief excerpt from a booklet formerly used by the New York State Income Tax Bureau:

> The return for the period before the change of residence must include all items of income, gain, loss or deduction accrued to the taxpayer up to the time of his change of residence. This includes any amounts not otherwise includable in the return because of an election to report income on an installment basis.
>
> Stated another way, the return for the period prior to the change of residence must be made on the accrual basis whether or not that is the taxpayer's established method of reporting. However, in the case of a taxpayer changing from nonresident to resident status, these accruals need not be made with respect to items derived from or connected with New York sources.

How did you respond to this information? If you were a taxpayer in need of quick and concise information to solve a particular problem, you would surely be disappointed and probably even annoyed. Tax publications and forms need not be complicated; since ordinary people normally read the publications and fill out these forms, they should be written for the average taxpayer, not for a trained accountant. In fact, all professional writing should be reader-directed.

Furthermore, the result of such convoluted writing is that these two paragraphs do not fulfill their purpose; they do not tell taxpayers how to adjust their returns when changing residence. Although the grammar is correct, the message is certainly not effective. Among other faults, the writer forgot to ask, "Who is my audience?" Unfamiliar words or usages such as *includable, accrual basis,* and *election* appear without explanation. If taxpayers cannot understand such language, they will complete their returns incorrectly and cheat either themselves or the state out of money. In the end, these two paragraphs—and others like them in the tax booklet—will have missed their mark, and unnecessary complications for both taxpayers and the state will be the result.

GUIDELINES FOR EFFECTIVE WRITING

To make sure you answer the three key questions correctly, keep these guidelines in mind:

- *Put your main idea up front.* Then provide whatever background or evidence your reader needs to understand your main idea.
- *Avoid using technical terms with nontechnical readers.* If you are an accountant writing to another accountant, use all the technical terminology you need. But if you are an accountant writing to a layperson, use ordinary language that the layperson will understand. If you must use technical terms, define them carefully.
- *Limit the length and complexity of your sentences.* Whenever possible, try to include only one main idea in each sentence. More mechanically, try also to limit your sentences to 25 to 35 words (or about 2½ typed lines). Generally speaking, the shorter the sentences, the fewer the difficulties you and your reader will get into. To ensure that your sentences are easily understood, move the main subject and verb to the front of the sentence and cut out any unnecessary words.
- *Limit the length and content of your paragraph.* Nothing intimidates a reader more than a half-page-long block of dense prose. Rather than using one paragraph to discuss every aspect of a topic, use a separate paragraph for each aspect of your topic. In that way, you can limit your paragraphs to five or six sentences. If you really need a longer paragraph, use indentation or enumeration so that your subpoints stand out from the pack.
- *Give your reader all necessary information in a clear and unmistakable way.* If your reader needs to let you know something by June 12, tell the reader so plainly. If you want your reader to check back with you for additional information, make sure you provide your e-mail address or telephone number, and your travel or vacation schedule if you are going to be out of the office. If you are attaching or enclosing anything, specify exactly what. If your message is a follow-up to some previous correspondence or to a recurring problem, make sure you give the reader this information too.

The selections that follow suggest many ways in which you can make your professional writing more effective. They also suggest ways you can make the process more enjoyable. If you view your professional writing as a chore, its effectiveness will be limited. But the more you get into the habit of asking the three key questions before you write your memos, letters, reports, and other documents, the easier, the more effective—and the more enjoyable—writing will become for you.

Part 1

Process as Well as Product

The old adage tells us to think before we speak. We should also think before we write. Actually, we should think before we write, while we write, and after we write. Thinking is what the writing process is all about. A process approach to writing is simply a writing strategy that asks writers to make sure they carefully plan each written message, rather than simply letting words fall where they may in the hope they will somehow organize themselves effectively and make sense to their intended audience.

People in business and industry are, however, trained to think in terms of finished products, bottom lines, and deadlines, so at first they may react with skepticism—if not with antagonism—to a process that requires them to slow down and think, especially when it is 4:30 P.M. and they need to get a memo to the boss by 5:00 P.M. Sometimes, sheer momentum and adrenaline will actually help, and a writer will produce an acceptable, even an effective, document. More times than not, though, the document will fail to achieve its purpose. The writer who balked at spending an extra ten minutes thinking a document through ends up spending an additional hour writing a follow-up document to clarify what should have been clear in the first document or, even worse, to mend fences because the original document created new problems.

In the introduction, I suggested an approach to the writing process that involved having writers ask themselves three key questions:

1. Exactly who is the audience?
2. What is the most important thing they want to tell this audience?
3. What is the best way of making sure this audience understands what they have to say?

These three key questions are simply one of any number of approaches to the writing process. The essays that follow in this section of *Strategies* offer other approaches and additional insights into a process approach to writing.

In the first essay, John Keenan offers a five-part approach to the writing process that asks writers to consider purpose, audience, format, evidence, and organization. Michael Adelstein follows Keenan with another five-part approach to writing that lays down a schedule for any writing task: 15% worrying, 10% planning, 25% writing, 45% revising, and 5% proofreading. The number of steps involved is, of course, unimportant. What is important is that there be a conscious effort to understand that, in business and technical writing, the process is as important as the goal. Keenan's approach is more open-ended, allowing writers considerable freedom in the amount of time they spend on the different steps in the writing process. While allowing for some variation among writing situations, Adelstein suggests a more regimented approach that will appeal to other writers.

Peter Elbow takes a different tack than Keenan or Adelstein by simply dividing writing time in half. Using what Elbow calls "the direct writing process," writers spend the first half of their available time "fast writing" in an attempt to get their ideas on the page before worrying about organization, correctness, or precision. Elbow suggests that these latter issues are best left to the second step in his approach to the writing process, revision. Revision is the key to successful writing, and Linda Flower and John Ackerman suggest ways in which the process of revision can help writers produce more effective documents.

In an often reprinted and adapted essay, John S. Harris next examines the writing process from a different perspective. The fault in poor quality documents may lie as much with the managers who assign them as with those who produce them. To counter problems in the management of writing, Harris offers a project worksheet designed to facilitate efficient writing management.

What follows in this first section of *Strategies* are five different discussions of the idea that it is important to take a process approach to any writing assignment in the many worlds of work. Writing is hard work. Writing takes planning. Throughout the remaining essays collected in this entire book, this message will be repeated in a variety of ways by experts with a variety of backgrounds and from a variety of fields, such as business, science, technology, industry, and the academy. Think before you write, while you write, and after you write—and only then send your message.

John Keenan

Using PAFEO Planning

A former technical writer for SmithKline, the late John Keenan was Professor of English at La Salle University in Philadelphia and coauthor with Kevin J. Harty of Writing for Business and Industry: Process and Product *(Macmillan, 1987).*

Surveys tell us that "lack of clarity" is the most frequently cited weakness when executives evaluate their employees' writing. But since lack of clarity results from various bad habits, it is often used as a convenient catchall criticism. Pressed to explain what they meant by lack of clarity, readers would probably come up with comments such as these:

"I had to read the damn thing three times before I got what he was driving at. At least I *think* I got his point; I'm not sure."

"I couldn't follow her line of thought."

"I'm just too busy to wade through a lot of self-serving explanation and justification. If he has something to say, why didn't he just say it without all that buildup?"

Most unclear writing results from unclear thinking. But it is also true that habits of clear writing help one to think clearly. When a writer gets thoughts in order and defines the purpose and goal, what he wants to say may no longer be exactly the same as it was originally. Sometimes the writer may even decide it is better *not* to write. Whatever the case, the work done *before* the first draft in the prewriting process and *after* the draft in the revising process is often the difference between successful communication and time-wasting confusion. . . .

THINK PAFEO

Some years ago I made up a nonsense word to help my students remember the important steps in constructing a piece of writing. It seemed to stick in their minds. In fact, I keep running into former students who can't remember my name, but remember PAFEO because it proved to be so useful. Here are the magic ingredients:

P stands for purpose
A stands for audience
F stands for format
E stands for evidence
O stands for organization.

Put together, they spell PAFEO, and it means the world to me!

PURPOSE

Put the question to yourself, "Why am I writing this?" ("Because I have to," is not a sufficient answer.) This writing you're going to do must aim at accomplishing something; you must be seeking a particular response from your reader. Suppose, for example, you're being a good citizen and writing to your congressman on a matter of interest to you. Your letter may encompass several purposes. You are writing to inform him of your thinking and perhaps to persuade him to vote for or against a certain bill. In explaining your position, you may find it useful to narrate a personal experience or to describe a situation vividly enough to engage his emotions. Your letter may therefore include the four main kinds of writing—exposition, argument, narration, and description—but one of them is likely to be primary. Your other purposes would be subordinated to your effort to persuade the congressman to vote your way. So your precise purpose might be stated: "I am writing to Congressman Green to persuade him to vote for increased social security benefits provided by House Bill 5883."

Write it out in a sentence just that way. Pin it down. And make it as precise as you can. You will save yourself a good deal of grief in the writing process by getting as clear a focus as you can on your purpose. You'll know better what to include and what to omit. The many choices that combine to form the writing process will be made easier because you have taken the time and the thought to determine the exact reason you have for writing this communication.

When you're writing something longer than a letter, like a report, you can make things easier for yourself and your reader by making a clear statement of the central idea you're trying to develop. This is your thesis. To be most helpful, it must be a complete sentence, not just the subject of a sentence. "Disappointing sales" may be the subject you are writing about, but it is not a thesis. Stating it that way won't help you organize the information in your report. But if you really say something specific about "disappointing sales," you will have a thesis that will provide a framework for development. For example:

> Disappointing sales may be attributed to insufficient advertising, poor selection of merchandise, and inadequate staffing during peak shopping hours.

The frame for the rest of the report now exists. By the time you have finished explaining how each cause contributes to the result (disappointing sales), the report is finished. Not only finished, but clear; it follows an orderly pattern. The arguments, details, illustrations, statistics, or quotations are relevant because the thesis has given you a way of keeping your purpose clearly in focus.

Of course the ideal method is to get your purpose and your thesis down before you write a first draft, but I can tell you from experience that it doesn't always work that way. Sometimes the only way to capture a thesis statement is to sneak up on it by means of a meandering first draft. When you've finished that chore, read the whole thing over and try to ambush the thesis by saying, "OK, now just what am I trying to say?" Then *say it* as quickly and concisely as you can—out loud, so you can hear how it sounds. Then try to write it down in the most precise and concise way you can. Even if you don't capture the exact thesis statement you're looking for, there's a good chance that you will have clarified your purpose sufficiently to make your next revision better.

AUDIENCE

Most of the writing the average person does is aimed at a specific reader: the congressman, the boss, a business subordinate, a customer, a supplier. Having a limited audience can be an advantage. It enables the writer to analyze the reader and shape the writing so that it effectively achieves the purpose with that particular reader or group of readers.

Here are some questions that will help you communicate with the reader:

1. How much background do I need to give this reader, considering his or her position, attitude toward the subject, and experience with this subject?
2. What does the reader need to know, and how can I best give him or her this information?
3. How is my credibility with this reader? Must I build it gradually as I proceed, or can I assume that he or she will accept certain judgments based on my interpretations?
4. Is the reader likely to agree or disagree with my position? What tone would be most appropriate in view of this agreement or disagreement?

The last question, I think, is very important. It implies that the writer will try to see the reader's point of view, will bend every effort to look at the subject the way the reader will probably look at it. That isn't easy to do. Doing it takes both imagination and some understanding of psychology. But it is worth the considerable effort it involves; it is a gateway to true communication.

Let me briefly suggest a point here that is easily overlooked in the search for better techniques of communication. It is simply this: *better writing depends on better reading.* Technique isn't everything. One cannot learn to be imaginative and understanding in ten easy lessons, but reading imaginative literature—fiction, poetry, drama—is one way of developing the ability to put yourself in someone else's place so you can see how the problem looks from that person's angle. The writer who limits reading to his or her own technical field is building walls around the imagination. The unique value of imaginative literature is the escape that it allows from the prison of the self into the experience and emotions of persons of different backgrounds. That ability to understand why a reader is likely to think and act in a certain way is constantly useful to the writer in the struggle to communicate.

FORMAT

Having thought about your purpose and your audience, consider carefully the appropriate format for this particular communication you are writing. Of course the range of choice may be limited by company procedures, but even the usual business formats allow some room for the writer to use ingenuity and intelligence.

The key that unlocks this ingenuity is making the format as well as the words work toward achieving your purpose. You can call attention to important points by the way you arrange your material on the page.

Writers sometimes neglect this opportunity, turning the draft over to the typist without taking the trouble to plan the finished page. Remember the reader, his or her desk piled high with things to be read, and try to make your ideas as clear as possible on the page as the reader glances at it.

External signals such as headings, underlining, and numerals are ways of giving the reader a quick preview. Use them when they are appropriate to your material.

Don't be afraid of using white space as a way of drawing attention to key ideas.

To see how experts use format for clarity and emphasis, take a look at some of the sales letters all of us receive daily. Note how these persuasive messages draw the reader's attention to each advantage of the product. You may not want to go that far, but the principle of using the format to save the reader's time is worth your careful consideration.

Overburdened executives are always looking for shortcuts through the sea of paper. Many admit that they cannot possibly read every report or memo. They scan, they skip, they look for summaries. A good format identifies the main points quickly and gives an idea of the organizational structure and content. If the format looks logical and interesting, readers may be lured into spending more time on your ideas. You've taken the trouble to separate the important from the unimportant, thus saving them time. The clear format holds promise of a clear analysis, and what boss can be too busy for that?

EVIDENCE

Unless you're an authority on your subject, your opinions carry only as much weight as the evidence you can marshal to support them. The more evidence you can collect before writing, the easier the writing task will be. Evidence consists of the facts and information you gather in three ways: (1) through careful observation, (2) through intelligent fieldwork (talking to the appropriate persons), and (3) through . . . research.

When you follow the evidence where it leads and form a hypothesis, you are using inductive reasoning, the scientific method. If your evidence is adequate, representative, and related to the issue, your conclusion must still be considered a probability, not a certainty, since you can never possess or weigh *all* the evidence. At some point you make the "inductive leap" and conclude that the weight of evidence points to a theory as a probability. The probability is likely to be strong if you are aware of the rules of evidence.

The Rules of Evidence

Rule 1. Look at the Evidence and Follow Where It Leads.

The trick here is not to let your own bias seduce you into selecting only the evidence you agree with. If you aren't careful, you unconsciously start forcing the evidence to fit the design that seems to be emerging. When fact A and fact B both point toward the same conclusion, there is always the temptation to *make* fact C fit. Biographer Marchette Chute's warning is worth heeding: ". . . you will never succeed in getting at the truth if you think you know, ahead of time, what the truth ought to be."

A reliable generalization ought to be based on a number of verifiable, relevant facts—the more the better. Logicians tell us that evidence supporting a generalization must be (1) known or available, (2) sufficient, (3) relevant, and (4) representative. Let's see how these apply to the generalization, "Cigarette smoking is hazardous to your health."

The evidence accumulated in animal studies and in comparative studies of smokers and nonsmokers has been published in medical journals: it is therefore known and available. Is it sufficient? The government thinks so; the tobacco industry does not. Which is more likely to have a bias that might make it difficult to follow where the evidence leads?

Tobacco industry spokesmen argue that animal studies are not necessarily relevant to human beings and point to individuals of advanced age who have been smoking since childhood as living refutations of the supposed evidence. But the number of cases studied is now in the thousands, and studies have been done with representative samplings. The individuals who have smoked without apparent damage to their health are real enough, but can they be said to be numerous enough to be representative of the usual effects of tobacco on humans? "Proof" beyond doubt lies only in mathematics, but the evidence in this example is the kind on which a sound generalization can be based.

Rule 2. Look for the Simplest Explanation that Accounts for All the Evidence.

When the lights go out, the sudden darkness might be taken as evidence of a power failure. But a quick investigation turns up other evidence that must be accounted for: the streetlights are still on; the refrigerator is still functioning. So a simpler explanation may exist, and a check of the circuit breakers or fuse box would be appropriate.

Rule 3. Look at All Likely Alternatives.

Likely alternatives in the example just discussed would include such things as burned-out bulbs, loose plugs, and defective outlets.

Rule 4. Beware of Absolute Statements.

In the complexity of the real world, it is seldom possible to marshal sufficient evidence to permit an absolute generalization. Be wary of writing general statements using words like *all, never* or *always*. Sometimes these words are implied rather than stated, as in this example:

> Jogging is good exercise for both men and women.
>
> [Because it is unqualified, the statement means *all* men and women. Since jogging is not good for people with certain health problems, this statement would be better if it said *most* or *many* men and women.]

Still, caution is always necessary. Induction has its limitations, and a hypothesis is best considered a probability subject to change on the basis of new evidence.

The other kind of reasoning we do is called deduction. Instead of starting with particulars and arriving at a generalization, deduction starts with a general premise or set of premises and works toward the conclusion necessarily implied by them. If the premise is true, it follows that the conclusion must be true. This logical relationship is called an inference. A fallacy is an erroneous or unjustified inference. When you are reasoning deductively, you can get into trouble if your premise is faulty or if the route from premise to conclusion contains a fallacy. Keep the following two basic principles in mind, and watch out for the common fallacies that can act as booby traps along the path from premise to conclusion.

Two Principles for Sound Deduction

1. The ideas must be true; that is, they must be based on facts that are known, sufficient, relevant, and representative.
2. The two ideas must have a strong logical connection.

If your inductive reasoning has been sound enough to take care of the first principle, let me warn you that the second is not so simple as it sounds. We have all been exposed to so much propaganda and advertising that distort this principle that we may have trouble recognizing the weak or illogical connection. Nothing is easier than to tumble into one of the commoner logic fallacies. As a matter of fact, no barroom argument would be complete without one of these bits of twisted logic.

Twisted Logic

Begging the question. Trying to prove a point by repeating it in different words.

> Women are the weaker sex because they are not as strong as men.
> Our company is more successful because we outsell the competition.

Non sequitur. The conclusion does not follow logically.

> I had my best sales year; the company's stock should be a big gainer this year.
> I was shortchanged at the supermarket yesterday. You can't trust these young cashiers at all.

Post hoc. Because an event happened first, it is presumed to be the cause of the second event.

> Elect the Democrats and you get war; elect the Republicans and you get a depression.

Oversimplification. Treating truth as an either-or proposition, without any degrees in between.

> Better dead than Red.
> America—love it or leave it.
> If we do not establish our sales leadership in New York, we might as well close up our East Coast outlets.

False analogy. Because of one or two similarities, two different things are assumed to be *entirely* similar.

> You can lead a horse to water but you can't make him drink; there is no sense in forcing kids to go to school.
> Giving the Canal back to Panama would be like giving the Great Lakes back to the Indians.

Seen in raw form, as in the examples, these fallacies seem easy to avoid. Yet all of us fall into them with dismaying regularity. A writer must be on guard against them whenever he or she attempts to draw conclusions from data.

ORGANIZATION

Experienced writers often use index cards when they collect and organize information. Cards can be easily arranged and rearranged. By arranging them in piles, you can create an organizational plan. Here's how you play the game:

1. You can write only one point on each card—one fact, one observation, one opinion, one statistic, one whatever.
2. Arrange the cards into piles, putting all closely related points together. All evidence related to marketing goes in one pile, all evidence related to product development goes in another pile, and so on.
3. Now you can move the piles around, putting them in sequence. What kind of sequence? Consider one of the following commonly used principles:

   Chronological. From past to present to future.
   Background, present status, prospects
   Spatial. By location.
   New England territory, Middle Atlantic, Southern
   Logical. Depends on the topic.
   Classification and division according to a consistent principle (divisions of a company classified according to function)
   Cause and effect (useful in troubleshooting manuals)
   Problem-analysis-solution. Description of problem, why it exists, what to do about it.
   Order of importance. From least important to most important or from most important to least important.

   The choice of sequence will depend largely on the logic of the subject matter and the needs of your audience.
4. Go through each pile and arrange the cards in an understandable sequence. Which points need to precede others in order to present a clear picture?
5. That's it! You now have an outline that is both orderly and flexible. You can add, subtract, or rearrange whenever necessary.

SUMMARY

Think PAFEO. Use this word to remind you to clarify your purpose and analyze your audience before you write. When you have a clear sense of purpose, create a thesis that will act as a frame for your ideas.

Choose a format for your communication that will help the reader identify the main points quickly.

Collect your evidence before you write by observing, interviewing, and doing library research. Use the index card system to help you organize your evidence. Keep in mind the rules of evidence before you draw conclusions. Be alert to the common logic fallacies that can easily undermine the deductive process.

Try the file card system as a way of organizing your material. Write one point on each card and then place closely related points together in a pile. Place the piles in sequence according to a principle of organization you select. Some useful organizing principles include: chronological, spatial, logical, problem-analysis-solution, and order of importance. The sequence you choose will depend on the logic of the subject matter and the needs of your audience.

Michael E. Adelstein

The Writing Process

Michael E. Adelstein was formerly Professor of English at the University of Kentucky and coauthor with W. Keats Sparrow of Business Communications *(Harcourt Brace Jovanovich, 1983).*

THE WRITING PROCESS

Their prevalent belief in the myths of genius, inspiration, and correctness prevents many people from writing effectively. If these individuals could realize that they can learn to write, that they cannot wait to be inspired, and that they must focus on other aspects besides correctness, then they can overcome many obstacles confronting them. But they should be aware that writing, like many other forms of work, is a process. To accomplish it well, people should plan on completing each one of its five stages. The time allocations may vary with the deadline, subject, and purpose, but the work schedule should generally follow this pattern:

1. Worrying—15%
2. Planning—10%
3. Writing—25%
4. Revising—45%
5. Proofreading—5%

Note that only 25% of the time should be spent in writing; the rest—75%—should be spent in getting ready for the task and perfecting the initial effort. Observe also that more time is spent in revising than in any other stage, including writing.

Stage One—Worrying (15%)

Worrying is a more appropriate term for the first operation than *thinking* because it suggests more precisely what you must do. When you receive a writing assignment, you must avoid blocking it out of your mind until you're ready to write;

instead, allow it to simmer while you try to cook up some ideas. As you stew about the subject while brushing your teeth, taking a shower, getting dressed, walking to class, or preparing for bed, jot down any pertinent thoughts. You need only note them on scrap paper or the back of an envelope. But if you fail to write them down, you'll forget them.

Another way to relieve your worrying is to read about the subject or discuss it with friends. The chances are that others have wrestled with similar problems and written their ideas. Why not benefit from their experience? But if you think that there is nothing in print about the subject or a related one—either because your assignment is restricted, localized, or topical—then you should at least discuss it with friends.

Let's say, for example, that as a member of the student government on your campus, you've been asked to investigate the possibility of opening a student-operated bookstore to combat high textbook prices. During free hours, you might drop into the library to read about the retail book business and, if possible, about college bookstores. If your student government is affiliated with the National Student Association, you might write to it for information about policies and practices on other campuses. In addition, you might start talking about the problem to the manager of your college bookstore, owners of other bookstores in your community, student transfers from other schools, managers of college bookstores in nearby communities, and some of your professors. Because you might be concerned about the public relations repercussions of a student bookstore, you should propose the venture to college officials. At this point, you will have heard many different ideas. To discuss some of them, bring up the subject of a student bookstore with friends. As you talk, you will be clarifying and organizing your thoughts.

The same technique can be applied to less involved subjects, such as a description of an ideal summer job, or an explanation of a technical matter like input-output analysis. By worrying about the subject as you proceed through your daily routine, by striving to get ideas from books and people or both, and by clarifying your thoughts in conversation with friends, you will be ready for the next step. Remember: you can proceed only if you have some ideas. You must have something to write about.

Stage Two—Planning (10%)

Planning is another term for organizing or for the task that students dread so—*outlining*. You should already have had some experience with this process. Like many students, unfortunately, you may have learned only how to outline a paper after writing it. This practice is a waste of time. If you want to operate efficiently, then it follows that you will have to specialize in each of the writing stages in turn. When you force yourself simultaneously to conceive of ideas, to arrange them in logical order, and to express them in words, you cannot perform all of these complex activities as effectively as if you had concentrated on each one separately. By dividing the writing process into five stages, and by focusing on each one individually, the odds are that the result will be much better than you had thought possible. Of course you will hear about people who never plan their writing but who do well anyway, just as you may know someone who skis or plays golf well without having had a lesson. There are always some naturals who can flaunt the rules. But perhaps these people could have significantly improved their writing, golf, or skiing if they had received some formal instruction and proceeded in the prescribed manner.

An efficient person plans his work. Many executives take a few minutes before leaving the office or retiring at night to jot down problems to attend to the following day. Many vacationers list tasks to do before leaving home: stop the newspaper, turn down the refrigerator, inform the mailman, cut off the hot water, check the car, mail the mortgage payment, notify the police, and the like. Of course, you can go on vacation without planning—just as you can write without planning—but the chances are that the more carefully you organize, the better the results will be.

In writing, planning consists mainly in examining all your ideas, eliminating the irrelevant ones, and arranging the others in a clear, logical order. Whether you write a formal outline or merely jot down your ideas on scrap paper is up to you. The outline is for your benefit; follow the procedure that helps you the most. But realize that the harder you work on perfecting your outline, the easier the writing will be. As you write, therefore, you can concentrate on formulating sentences instead of also being concerned with thinking of ideas and trying to organize them. For the present, realize the importance of planning. . . .

Stage Three—Writing (25%)

When you know what you want to say and have planned how to say it, then you are ready to write. This third step in the writing process is self-explanatory. You need only dash away your thoughts, using pen, pencil, or typewriter, whichever you prefer. Let your mind flow along the outline. Don't pause to check spelling, worry about punctuation, or search for exact words. If new thoughts pop into your mind, as they often do, check them with your outline, and work them into your paper if they are relevant. Otherwise discard them. Keep going until either the end or fatigue halts you. And don't worry if you write slowly: many people do.

Stage Four—Revising (45%)

Few authors are so talented that they can express themselves clearly and effectively in a first draft. Most know that they must revise their papers extensively. So they roll up their sleeves and begin slashing away, cutting out excess verbiage, tearing into fuzzy sentences, stabbing at structural weaknesses, and knifing into obfuscation. Revision is painful: removing pet phrases and savory sentences is like getting rid of cherished possessions. Just as we dislike to discard old magazines, books, shoes, and clothes, so we hate to get rid of phrases and sentences. But no matter how distasteful the process may be, professional writers know that revising is crucial; it is usually the difference between a mediocre paper and an excellent one.

Few students revise their work well; they fail to realize the importance of this process, lack the necessary zeal, and are unaware of how to proceed. Once you understand why you seem to be naturally adverse to revision, you may overcome your resistance.

Because writing requires concentrated thought, it is about the most enervating work that humans perform. Consequently, upon completing a first draft, we are so relieved at having produced something that we tidy it up slightly, copy it quickly, and get rid of it immediately. After all, why bother with it anymore? But there's the rub. The experienced writer knows that he has to sweat through it again and again, looking for trouble, willing to rework cumbersome passages, and striving always to find a better word, a more felicitous phrase, and a smoother sentence. John Galbraith, for example, regularly writes five drafts, the last one reserved for inserting "that note of spontaneity" that ironically makes his work appear natural and effortless. The record for revision is probably held by Ernest Hemingway, who wrote the last page of *A Farewell to Arms* thirty-nine times before he was satisfied!

To revise effectively, you must not only change your attitude, but also your perspective. Usually we reread a paper from our own viewpoint, feeling proud of our accomplishment. We admire our cleverness, enjoy our graceful flights, and glow at our fanciful turns. How easy it is to delude ourselves! The experienced writer reads his draft objectively, looking at it from the standpoint of his reader. To accomplish this, he gets away from the paper for a while, usually leaving it until the following morning. You may not be able to budget your time this ideally, but you can put the paper aside while you visit a friend, grab a bite to eat, or phone someone. Unless you divorce yourself from the paper, you will probably be under its spell; that is, you will reread it with your mind rather than with your eyes. You will see only what you think is on the page instead of what is actually there. And you will not be able to transport yourself from your role of writer to that of critic.

Only by attacking your paper from the viewpoint of another person can you revise it effectively. You must be anxious to find fault and you must be honest with yourself. If you cannot or will not realize the weaknesses in your writing, then you cannot correct them. This textbook should open your eyes to many things that can go wrong, but above all you must *want* to find them. If you blind yourself to the bad, then your work will never be good. You must convince yourself that good writing depends on good rewriting. This point cannot be repeated too frequently or emphasized too much. Tolstoy revised *War and Peace*—one of the world's longest and greatest novels—five times. The brilliant French stylist Flaubert struggled for hours, even days, trying to perfect a single sentence. But question your professors about their own writing. Many—like me—revise papers three, four, or five times before submitting them for publication. You may not have this opportunity, your deadlines in school and business may not permit this luxury; but you can develop a respect for revision and you can devote more time and effort to this crucial stage in the writing process. . . .

Stage Five—Proofreading (5%)

Worrying, planning, writing, and revising are not sufficient. The final task, although not as taxing as the others, is just as vital. Proofreading is like a quick check in the mirror before leaving for a date. A sloppy appearance can spoil a favorable acceptance of you or your paper. If, through your fault or your typist's, words are missing or repeated, letters are transposed, or sentences juxtaposed, a reader may be perturbed enough to ignore or resist your ideas or information. Poor proofreading of an application letter can cost you a job. Such penalties may seem unfair, but we all react in this fashion. What would you think of a doctor with dirty hands? His carelessness would not only disgust you but would also raise questions about his professional competence. Similarly, carelessness in your writing antagonizes readers and raises questions in their minds about your competence. The merit of a person or paper should not depend on appearances, but frequently it does. Being aware of this possibility, you should keep yourself and your writing well groomed.

Poor proofreading results from failure to realize its importance, from inadequate time, and from improper effort. Unless you are convinced that scrutinizing your final copy is important, you cannot proofread effectively. If you realize the significance of this fifth step in the writing process, then you will not only have the incentive to work hard at proofreading but also will allow time for it in planning.

Like most things, proofreading takes time. We usually run out of time for things that we don't care about. But there's always time for what we enjoy or what is important to us. So with proofreading; lack of time is seldom a legitimate excuse for doing it badly.

But lack of technique is. To proofread well, you must forget the ideas in a paper and focus on words. Slow down your reading speed, stare hard at the black print, and search for trouble rather than trying to finish quickly. . . . Some professional proofreaders start with the last sentence and read backward to the first. Whatever technique you adopt, work painstakingly, so that a few careless errors will not spoil your efforts. If you realize the importance of proofreading, view it as one of the stages in the writing process, and labor at it conscientiously, your paper will reflect your care and concern. . . .

Each Stage Is Important

These then are the five stages in the writing process: worrying, planning, writing, revising, and proofreading. Each is vital. You need to become proficient at each of them, although, at some later point in your life, an editor or efficient secretary may relieve you of some proofreading chores. But in your career as a student and in your early business years, you will be on your own to guide your paper through the five stages. Since it is your paper, no one else will be as interested in it as you. It's your offspring—to nourish, to cherish, to coddle, to bring up, and finally, to turn over to someone else. You will be proud of it only if you have done your best throughout its growth and development.

Peter Elbow

The Direct Writing Process for Getting Words on Paper

Peter Elbow is a former member of the faculty at the University of Massachusetts at Amherst. He is also the author of Writing without Teachers *(Oxford, 1973), an innovative introduction to the writing process.*

The direct writing process is most useful if you don't have much time or if you have plenty to say about your topic. It's a kind of let's-get-this-thing-over-with writing process. I think of it for tasks like memos, reports, somewhat difficult letters, or essays where I don't want to engage in much new thinking. It's also a good approach if you are inexperienced or nervous about writing because it is simple. . . .

Unfortunately, its most common use will be for those situations that aren't supposed to happen but do: when you have to write something you *don't* yet understand, but you also don't have much time. The direct writing process may not always lead to a satisfactory piece of writing when you are in this fix, but it's the best approach I know.

The process is very simple. Just divide your available time in half. The first half is for fast writing without worrying about organization, language, correctness, or precision. The second half is for revising.

Start off by thinking carefully about the audience (if there is one) and the purpose for this piece of writing. Doing so may help you figure out exactly what you need to say. But if it doesn't, then let yourself put them out of mind. You may find that you get the most benefit from ignoring your audience and purpose at this early stage of the writing process. . . .

In any event spend the first half of your time making yourself write down everything you can think of that might belong or pertain to your writing task: incidents that come to mind for your story, images for your poem, ideas and facts for your essay or report. Write fast. Don't waste any time or energy on how to organize it, what to start with, paragraphing, wording, spelling, grammar, or any other matters of presentation. Just get things down helter-skelter. If you can't find the right word just leave a blank. If you can't say it the way you want to say it, say it the wrong way. (If it makes you feel better, put a wavy line under those wrong bits to remind you to fix them.)[1]

I'm not saying you must never pause in this writing. No need to make this a frantic process. Sometimes it is very fruitful to pause and return in your mind to some productive feeling or idea that you've lost. But don't stop to worry or criticize or correct what you've already written.

While doing this helter-skelter writing, don't allow too much digression. Follow your pencil where it leads, but when you suddenly realize, "Hey, this has nothing to do with what I want to write about," just stop, drop the whole thing, skip a line or two, and get yourself back onto some aspect of the topic or theme.

Similarly, don't allow too much repetition. As you write quickly, you may sometimes find yourself coming back to something you've already treated. Perhaps you are saying it better or in a better context the second or third time. But once you realize you've done it before, stop and go on to something else.

When you are trying to put down everything quickly, it often happens that a new or tangentially related thought comes to mind while you are just in the middle of some train of thought. Sometimes two or three new thoughts crowd in on you. This can be confusing: you don't want to interrupt what you are on, but you fear you'll forget the intruding thoughts if you don't write them down. I've found it helpful to note them without spending much time on them. I stop right at the moment they arrive—wherever I am in my writing—and jot down a couple of words or phrases to remind me of them, and then I continue on with what I am writing. Sometimes I jot the reminder on a separate piece of paper. When I write at the typewriter I often just put the reminder in caps inside double parentheses ((LIKE THIS)) in the middle of my sentence. Or I simply start a new line LIKE THIS
and then start another new line to continue my old train of thought. But sometimes the intruding idea seems so important or fragile that I really want to go to work on it right away so I don't lose it. If so, I drop what I'm engaged in and start working on the new item. I know I can later recapture the original thought because I've already written part of it. The important point here is that what you produce during this first half of the writing cycle can be very fragmented and incoherent without any damage at all.

There is a small detail about the physical process of writing down words that I have found important. Gradually I have learned not to stop and cross out something I've just written when I change my mind. I just leave it there and

---

[1] An excerpt from a letter giving me feedback on an earlier draft: "I tried the direct writing process. Though it sounds simple enough, I . . . see now that in the past I've often interrupted the flow of writing by spending disproportionate time on spelling, punctuation, etc. I can spend hours on an opening paragraph stroking the words to death; then, if there's a deadline, to rush through the remainder." (Joanne Turpin, 7/24/78.)

write my new word or phrase on a new line. So my page is likely to have lots of passages that look like

Many of my pages
Still I don't mean that you should stop and rewrite every passage till you are happy with it.
This kind of appearance.

What is involved here is developing an increased tolerance for letting mistakes show. If you find yourself crumpling up your sheet of paper and throwing it away and starting with a new one every time you change your mind, you are really saying, "I must destroy all evidence of mistakes." Not quite so extreme is the person who scribbles over every mistake so avidly that not even the tail of the *y* is visible. Stopping to cross out mistakes doesn't just waste psychic energy, it distracts you from full concentration on what you are trying to say.

What's more, I've found that leaving mistakes uncrossed out somehow makes it easier for me to revise. When I cross out all my mistakes I end up with a *draft*. And a draft is hard to revise because it is a complete whole. But when I leave my first choices there littering my page along with some second and third choices, I don't have a draft, I just have a succession of ingredients. Often it is easier to whip that succession of ingredients into something usable than, as it were, to *undo* that completed draft and turn it into a better draft. It turns out I can just trundle through that pile of ingredients, slash out some words and sections, rearrange some bits, and end up with something quite usable. And quite often I discover in retrospect that my original "mistaken phrase" is really better than what I replaced it with: more lively or closer to what I end up saying.

* * *

If you only have half an hour to write a memo, you have now forced yourself in fifteen minutes to cram down every hunch, insight, and train of thought that you think might belong in it. If you have only this evening to write a substantial report or paper, it is now 10:30 P.M., you have used up two or two and a half hours putting down as much as you can, and you only have two or more hours to give to this thing. You must stop your raw writing now, even if you feel frustrated at not having written enough or figured out yet exactly what you mean to say. If you started out with no real understanding of your topic, you certainly won't feel satisfied with what is probably a complete mess at this point. You'll just have to accept the fact that of course you will do a poor job compared to what you could have done if you'd started yesterday. But what's more to the point now is to recognize that you'll do an even crummier job if you steal any of your revising time for more raw writing. Besides, you will have an opportunity during the revising process to figure out what you want to say—what all these ingredients add up to—and to add a few missing pieces. It's important to note that when I talk about revising . . . I mean something much more substantial than just tidying up your sentences.

So if your total time is half gone, stop now no matter how frustrated you are and change to the revising process. That means changing gears into an entirely different consciousness. You must transform yourself from a fast-and-loose-thinking person who is open to every whim and feeling into a ruthless, tough-minded, rigorously logical editor. Since you are working under time pressure, you will probably use quick revising or cut-and-paste revising. . . .

* * *

Direct writing and quick revising are probably good processes to start with if you have an especially hard time writing. They help you prove to yourself that you *can* get things written quickly and acceptably. The results may not be the very best you can do, but they work, they get you by. Once you've proved you can get the job done you will be more willing to use other processes for getting words down on paper and for revising—processes that make greater demands on your time and energy and emotions. And if writing is usually a great struggle, you have probably been thrown off balance many times by getting into too much chaos. The direct writing process is a way to allow a limited amount of chaos to occur in a very controlled fashion.

It's easiest to explain the direct writing process in terms of pragmatic writing: you are in a hurry, you know most of what you want to say, you aren't trying for much creativity or brilliance. But I also want to stress that the direct writing process can work well for very important pieces of writing and ones where you haven't yet worked out your thinking at all. But one condition is crucial: you must be confident that you'll have no trouble finding lots to say once you start writing. . . .

As I wrote many parts of this book [i.e., *Writing with Power*], for example, I didn't have my thinking clear or worked out by any means, I couldn't have made an outline at gunpoint, and I cared deeply about the results. But I knew that there was lots of *stuff* there swirling around in my head ready to go down on paper. I used the direct process. I just wrote down everything that came to mind and went on to revise.

But if you want to use the direct writing process for important pieces of writing, you need plenty of time. You probably won't be able to get them the way you want them with just quick revising. You'll need thorough revising or revising with feedback. . . . For important writing I invariably spend more time revising than I do getting my thoughts down on paper the first time.

MAIN STEPS IN THE DIRECT WRITING PROCESS

- If you have a deadline, divide your total available time: half for raw writing, half for revising.
- Bring to mind your audience and purpose in writing but then go on to ignore them if that helps your raw writing.

- Write down as quickly as you can everything you can think of that pertains to your topic or theme.
- Don't let yourself repeat or digress or get lost, but don't worry about the order of what you write, the wording, or about crossing out what you decide is wrong.
- Make sure you stop when your time is half gone and change to revising, even if you are not done.
- The direct writing process is most helpful when you don't have difficulty coming up with material or when you are working under a tight deadline.

Linda Flower and John Ackerman

Evaluating and Testing as You Revise

Both Linda Flower, a member of the faculty at Carnegie Mellon University, and John Ackerman, a member of the faculty at the University of Utah, are nationally recognized authorities on the teaching of writing.

IMAGINING A READER'S RESPONSE

Even if you have a model or text to work from, as a writer you will work hard to construct a text: generating and discarding ideas, trying to figure out your point, sketching out alternative organizations, and then trying to signal that point and structure to your reader. But the story does not end there because your various readers have to work equally hard to construct a meaning based on your text. That is, readers want to construct *in their own minds* a coherent text, with a hierarchical organization (like an issue tree) based on key points. And they need to see a purpose for reading. Readers want to know *Why read this? What is the point?* and *How is all of this connected?*—and they want to find answers as quickly as possible.

Readers begin to predict the structure of a text and its meaning as soon as they face a page. They will look for cues that the text might offer about point, purpose, and structure, but if they do not find them, they will go ahead and construct their own version of the text *and assume that is what you intended*. It may be helpful to imagine your readers as needing to *write* your text for themselves. Thus, their own goals and interests will strongly influence what they look for in a text and the meaning they make out of it. Your goal, then, as a writer is to do the best job you can to make sure a reader's process through your text and his or her understanding matches or comes close to what you intended.

Once your ideas are down in a draft of some form, **revision** lets you anticipate how readers will respond and adjust the text to get the response you hope for. **Local revision** involves editing, correcting spelling and grammar, and making local improvements in wording or sentence style. **Global revision** involves looking at the big picture, but that does not mean throwing the draft out and starting again, and it may not even take a lot of time. Global revision, however, does mean looking at the text as a whole—thinking globally about the major rhetorical decisions and plans you made—now that the draft is complete. Global revisions alter the focus, organization, argument, or detail to improve the overall text.

Revision begins with a tricky reading process, a close reading of your own text. You know from your own experience in school that readers can read for different purposes: to skim a text to prepare for class discussion and then to read more carefully for an examination. In business, because writing usually augments a transaction between people, readers look for information but they also look for how well a piece of writing is adapted to their needs. The key to your success as a reviser is your ability to read your own writing critically for different features and purposes. At the simplest level, revision is *re-vision*—the process of stepping back from one's text and seeing it anew. . . .

STRATEGY 1: LOOK FOR WRITER-BASED PROSE

Why is it that even experienced writers typically choose to draft and revise rather than write a final text in one pass? Suppose you put time into planning your paper; you followed a good model; you thought about your reader as you chose what to say. Why should your first draft not do the job? One reason is because first drafts often contain large sections of **writer-based prose.** Writer-based prose appears when writers are essentially talking to themselves, talking through a problem, exploring their own knowledge, or trying to get their ideas out. In fact, producing writer-based prose is often a smart problem-solving strategy. Instead of getting blocked or spending hours staring at a blank page trying to write a perfect text the first time through, writers can literally walk through their own memory, talking out on paper what they know. Other concerns, including what the reader needs to hear, are put on temporary hold. Writer-based prose is, in fact, a very effective strategy for searching your memory and for dealing with difficult topics or lots of information. Writer-based prose may also be the best a writer can do in a new situation. A writer facing an unfamiliar task may produce writer-based prose, uncertain of what readers expect.

Whether it is a strategy or consequence, the downside of this strategy is that the text it produces is typically focused on the writer's thinking—not the readers' questions or needs—and it often comes out organized as a narrative or a river of connections. A number of studies of writers new to organizations demonstrate the reasoning behind writer-based prose and its ill-effect. New employees write consciously and unconsciously to report their discovery process or to survey all they know. And their readers impatiently wade through the river looking for the specific ideas and information they need.

Here [see Figure 1] is the first draft of a progress report written by four students in an organizational psychology course who were doing a consulting project with a local organization, the Oskaloosa Brewing Company. As a reader, put yourself first in the position of Professor Charns. He is reading the report to answer three questions: As analysts, what assumptions and decisions did these students make in setting up their study? Why did they make them? And where

---

Draft 1

Group Progress Report

(1)   Work began on our project with the initial group decision to evaluate the Oskaloosa Brewing Company. Oskaloosa Brewing Company is a regionally located brewery manufacturing several different types of beer, notably River City and Brough Cream Ale. This beer is marketed under various names in Pennsylvania and other neighboring states. As a group, we decided to analyze this organization because two of our group members had had frequent customer contact with the sales department. Also, we were aware that Oskaloosa Brewing had been losing money for the past five years, and we felt we might be able to find some obvious problems in its organizational structure.

(2)   Our first meeting, held February 17th, was with the head of the sales department, Jim Tucker. Generally, he gave us an outline of the organization, from president to worker, and discussed the various departments that we might ultimately decide to analyze. The two that seemed the most promising and more applicable to the project were the sales and production departments. After a few group meetings and discussions with the personnel manager, Susan Harris, and our advisor, Professor Charns, we felt it best suited our needs and Oskaloosa Brewing's needs to evaluate their bottling department.

(3)   During the next week we had a discussion with the superintendent of production, Henry Holt, and made plans for interviewing the supervisors and line workers. Also, we had a tour of the bottling department that gave us a first-hand look at the production process. Before beginning our interviewing, our group met several times to formulate appropriate questions to use in interviewing, for both the supervisors and the workers. We also had a meeting with Professor Charns to discuss this matter.

(4)   The next step was the actual interviewing process. During the weeks of March 14–18 and March 21–25, our group met several times at Oskaloosa Brewing and interviewed ten supervisors and twelve workers. Finally, during this past week, we have had several group meetings to discuss our findings and the potential problem areas within the bottling department. Also, we have spent time organizing the writing of our progress report.

1 of 2

FIGURE I   The Oskaloosa Brewing Progress Memo, Draft I

(5)   The bottling and packaging division is located in a separate building, adjacent to the brewery, where the beer is actually manufactured. From the brewery the beer is piped into one of five lines (four bottling lines and one canning line) in the bottling house, where the bottles are filled, crowned, pasteurized, labeled, packaged in cases, and either shipped out or stored in the warehouse. The head of this operation, and others, is production manager Phil Smith. Next in line under him in direct control of the bottling house is the superintendent of bottling and packaging, Henry Holt. In addition, there are a total of ten supervisors who report directly to Henry Holt and who oversee the daily operations and coordinate and direct the twenty to thirty union workers who operate the lines.

(6)   During production, each supervisor fills out a data sheet to explain what was actually produced during each hour. This form also includes the exact time when a breakdown occurred, what it was caused by, and when production was resumed. Some supervisors' positions are production-staff-oriented. One takes care of supplying the raw material (bottles, caps, labels, and boxes) for production. Another is responsible for the union workers' assignments each day.

These workers are not all permanently assigned to a production-line position. Workers called "floaters" are used, filling in for a sick worker or helping out after a breakdown.

(7)   The union employees are generally older than 35, some in their late fifties. Most have been with the company many years and are accustomed to having more workers per a slower moving line. . . .

2 of 2

FIGURE I   *(Continued)*

are they in the project now? Then take on the role of the client, the company vice president who follows their progress and wants to know: O.K. What is the problem (i.e., how did they define it)? And what did they conclude? Would this draft answer these questions for either of its intended readers?

Put yourself in the shoes of the professor. What would you look for in a progress report? According to Charns, he used this report to evaluate the group's progress: Were they on schedule; were they on task; did he need to intervene? However, he didn't need a blow-by-blow story to do that. As an evaluator he wanted to see whether they knew how to analyze an organization: Were they making good decisions (that is, decisions they could justify in this report); had they made any discoveries about this company? His needs as a reader, then, reflected his dual role as a teacher (supervisor) and an evaluator.

When we showed this draft to a manager (with comparable experience and responsibility to the Oskaloosa VP), we got a very different response. Here was a reader looking quickly for the information she wanted and building an image of the writers' business savvy based on their text. Here is part of her response as she read and thought aloud (the student text she reads is underlined):

> Work begun on our project with the initial group decision to evaluate . . . *OK.* *Our project What project is this? I must have a dozen "projects" I keep tabs on. And who is this group?* This beer is marketed . . . *blah, blah, I'm tempted to skim. This must be a student project. But why am I reading about the fact somebody bought a lot of beer for their frat? Maybe the next paragraph.*
> Our first meeting . . . *Ok, they saw Jim and Susan,* . . . *looked at bottling,* . . . *wrote their paper. And now they are telling me where my packaging division is located! This is like a shaggy plant tour story. They are just wasting my time. And I suppose I should say that I am also forming an image of them as rather naive, sort of bumbling around the plant, interrupting my staff with questions. I mean, what are they after? Do they have any idea of what they are doing?*

Now put yourself in the shoes of the professor. What would you look for in a progress report? How would you evaluate this group as decision makers? Have they learned anything about analyzing organizations? How would you evaluate their progress at this point? Have they made any discoveries, or are they just going through the steps?

Fortunately, this is not the draft the writers turned in to their professor or the company manager. The revised draft [as shown in Figure 2] was written after a short conference with a writing instructor who instead of offering advice, asked the writers to predict what each of their readers would be looking for. It took ten minutes to step back from their draft and to rethink it from the perspective of their professor and the Oskaloosa manager. They found that they needed global revision—a revision that kept the substance of their writer-based first draft, but transformed it into reader-based text. As you compare these two drafts, notice the narrative and survey organization in the first draft and the "I did it" focus that are often a tip-off to writer-based prose, and how they improved the second draft.

How would you characterize the differences between drafts one and two? One clear difference is the use of a conventional memo/report format to focus the reader's attention. But beyond the visual display of information, the writers moved away from narrative organization, an "I" focus, and a survey form or "textbook" pattern of organization. Watch for these three patterns as you revise.

Narrative Organization

The first four paragraphs of the first draft are organized as a narrative, starting with the phrase "Work began. . . ." We are given a story of the writers' discovery process. Notice how all of the facts are presented in terms of when they were discovered, not in terms of their implications or logical connections. The writers want to tell us what happened and when; the reader, on the other hand, wants to ask "why?" and "so what?"

A narrative organization is tempting to write because it is a prefabricated order and easy to generate. All of us walk around with stories in our head, and chronology is a common rhetorical move. Instead of creating a **hierarchical** organization among ideas or worrying about a reader, the writer can simply

Draft 2

MEMORANDUM

TO: Professor Martin Charns

FROM: Nancy Lowenberg, Todd Scott, Rosemary Nisson,
           Larry Vollen

DATE: March 31, 1987

RE: Progress Report: The Oskaloosa Brewing Company

Why Oskaloosa Brewing?

Oskaloosa Brewing Company is a regionally located brewery manufacturing
several different types of beer, notably River City and Brough Cream Ale. As
a group, we decided to analyze this organization because two of our group
members have frequent contact with the sales department. Also, we were
aware that Oskaloosa Brewing had been losing money for the past five years
and we felt we might be able to find some obvious problems in its organizational structure.

Initial Steps: Where to Concentrate?

After several interviews with top management and a group discussion, we felt
it best suited our needs, and Oskaloosa Brewing's needs, to evaluate the production department. Our first meeting, held February 17, was with the head
of the sales department, Jim Tucker. He gave us an outline of the organization and described the two major departments, sales and production. He
indicated that there were more obvious problems in the production department, a belief also suggested by Susan Harris, the personnel manager.

Next Step

The next step involved a familiarization with the plant and its employees. First,
we toured the plant to gain an understanding of the brewing and bottling
processes. Next, during the weeks of March 14–18 and March 21–25, we interviewed ten supervisors and twelve workers. Finally, during the past week we
had group meetings to exchange information and discuss potential problems.

The Production Process

Knowledge of the actual production process is imperative in understanding
the effects of various problems on efficient production. Therefore, we have
included a brief summary of this process.

The bottling and packaging division is located in a separate building, adjacent
to the brewery, where the beer is actually manufactured. From the brewery
the beer is piped into one of five lines (four bottling lines and one canning
line) in the bottling house, where the bottles are filled, crowned, pasteurized,
labeled, packaged in cases, and either shipped out or stored in the warehouse.

Problems

Through extensive interviews with supervisors and union employees, we have
recognized four apparent problems within the bottling house operations.
The first is that the employees' goals do not match those of the company. . . .
This is especially apparent in the union employees, whose loyalty lies with the
union instead of the company. This attitude is well-founded, as the union
ensures them of job security and benefits. . . .

FIGURE 2   The Oskaloosa Brewing Progress Memo, Draft 2

remember his or her own discovery process and write a story. Remember that in a hierarchical structure, such as an issue tree, the ideas at the top of the structure work as the organizing concepts that include other ideas. The alternative is often a string of ideas simply linked by association or by the order in which the writer thought about them. Papers that start out, "In studying the reasons for the current decline in our return customers, ..." are often a dead giveaway. They tell us we are going to watch the writer's mind at work and follow him or her through the process of thinking out conclusions. Following one's own associations makes the text easier to write. But another reason new employees are tempted to write narrative reports is that they were often rewarded for narratives at some point in their career *as students*. They fail to realize that in business the reader is someone who expects to *use* this text (not check off whether or not they did the assignment).

A narrative pattern, of course, has the virtue of any form of drama—it keeps you in suspense by withholding closure. But this drama is an effective strategy only if the audience is willing to wait that long for the point. Most professional and academic readers are impatient, and they tend to interpret such narrative, step-by-step structures either as wandering and confused (Is there a point?) or as a form of hedging. Narrative structures may be read as veiled attempts to hide what really happened or the writers' actual position. Although a progress report naturally involves narrative, how has Draft 2 been able to *use* the narrative to answer readers' questions?

The "I" Focus

The second feature of Draft 1 is that it is a discovery story starring the writers. Its drama, such as it is, is squarely focused on the writer: "I did/I thought/I felt. . . ." Of the fourteen sentences in the first three paragraphs, ten are grammatically focused on the writer's thoughts and actions rather than on the issues. For example: "Work began . . . ," "We decided . . . ," Also, "we were aware . . . and we felt. . . ." Generally speaking, the reader is more interested in issues and ideas than in the fact that the writer thought them.

In pointing out the "I" focus in Draft 1, we are not saying that writers cannot refer to themselves or begin a sentence with "I," as many learned in school. Sometimes a specific reference to oneself is exactly the information a reader needs, and a reader may respond to the honesty and directness. Use "I" or "we" to make a claim or when it is an important piece of information, not just as a convenient way to start a sentence. In Draft 2, the students are clearly present as people doing the research, but the focus is on the information the reader wants to hear.

Survey Form or Textbook Organization

In the fifth paragraph of Draft 1 the writers begin to organize their material in a new way. Instead of a narrative, we are given a survey of what the writers observed. Here, the raw facts of the bottling process dictated the organization of the paragraph. Yet the client-reader already knows this, and the professor probably does not care. In the language of computer science we could say the writers are performing a "memory dump": printing out information in the exact form in which they stored it in memory. Notice how in the revised version the writers try to use their observations to understand production problems.

The problem with a survey or "textbook" pattern is that it ignores the reader's need for a different organization of the information. Suppose, for example, you are writing to model airplane builders about wind resistance. The information you need comes out of a physics text, but that text is organized around the field of physics; it starts with subatomic particles and works up from there. To meet the needs of your reader, you have to adapt that knowledge, not lift it intact from the text. Sometimes writers can simply survey their knowledge, but generally the writer's main task is to use knowledge rather than reprint it.

To sum up, in Draft 2 of the Oskaloosa report, the writers made a real attempt to write for their readers. Among other things, the report is now organized around major questions readers might have, it uses headings to display the overall organization of the report, and it makes better use of topic sentences that tell the reader what each paragraph contains and why to read it. Most important, it focuses more on the crucial information the reader wants to obtain.

Obviously this version could still be improved. But it shows the writers attempting to transform writer-based prose into reader-based prose and change their narrative and survey pattern into a more issue-centered top-to-bottom organization.

STRATEGY 2: TEST YOUR TEXT FOR A READER-BASED STRUCTURE

Reading your text for writer-based prose lets you spot places where you were still exploring ideas or talking to yourself—places that probably call for some sort of global or structural revision to make this text a reader-based document. But that does not tell you how to revise. **Reader-based prose** foregrounds and makes explicit the information a reader needs or expects to find. Reader-based prose tries to anticipate and support an active reader, one who probably will use your writing for some specific end. What do you want your reader to see, think, or do? How will your reader respond? One important approach to global revision is to look at your text as a conversation with the reader in which you set up some initial agreements and expectations and then fulfill your promises. We offer three ways to test for a reader-based structure: *testing your drafts against your initial plans, using clues that reveal this plan to a reader,* and *keeping the promises you made in your writing.*

Does Your Text Reflect Your Plans?

Texts have a way of running off by themselves. The more text you produce, the more convinced you are that your prose is complete, readable, even entertaining. This is natural given the commitment it takes to write anything, but because texts often drift away from your intentions, you need a way to keep your text honest or to realize that you have come up with a better approach. So instead of reading your text "as written" and just going with the flow, start by setting up an image of your purpose and plans in your mind's eye, then test your text against that image. Can you find any evidence of your plans in the text?

To get a good image of your plan, return to . . . your planning notes to review your plans consciously. In your mind or on paper, restate the plans *to do* and *say* that you produced prior to your current draft, and find the exact places in your text where your writing satisfies (or departs from) your plans to reach a reader. You could set the mental exercise up as a checklist [Figure 3].

Holding your draft up against the backdrop of your own plans can help you notice how well the two fit together. Did you have important goals, or good points in your notes that have just not appeared in your text yet? Did you simply forget ideas? Or did you find—as many writers do—that the act of writing was itself an inventive, generative process? Your plans may have changed as a result of writing. If that happened, what should you do? At this point, inexperienced writers often abandon their old plans and follow wherever the text seems to be taking them. However, experienced writers make another move. They go back into planning and look for possible ways to consolidate their new ideas with other old plans. They try to build a new plan that makes use of the good parts

---

My plans for writing, to do and say were...

  The important Goals and Purposes I gave myself...
  My Key Points...
  How I wanted my Reader to respond (and what other responses I anticipated)...
  My choice of Text Conventions...
  How I planned to make all of this work together...

As a checklist, I could evaluate my writing this way:

Plans *to do* and say	Draft	Text Reference
Goals...	✓	-------------------------
Key points...	✓	-------------------------
Intended Response...	✓	-------------------------
Text Conventions, Models...	✓	-------------------------
Making it all work together...	????	*oops, gotta work on this!!!*

FIGURE 3 Checking Your Text Against Your Plans

of both ideas. They use this round of planning to guide their global revisions consciously.

This strategy is obviously one you will use more than once—a kind of in-process evaluation that lets you keep checking in with your goals for writing and checking your text against the big picture of what you want it to do.

Cues That Reveal Your Plans to Your Reader

Maybe you are satisfied that you have indeed <u>defined</u> a real and shared <u>problem</u> for your reader, and <u>compared</u> some <u>alternative</u> ways to respond to it, <u>supported</u> them with <u>examples</u>, while <u>proposing</u> your favored course of <u>action</u>. But will the reader recognize all of those rhetorical moves? It is possible that you "talked about" this information, but did not make your good rhetorical plan to define, compare, support and propose fully apparent or explicit.

In most pieces of writing there are two conversations going on. One is the information that the reader needs or expects to find: the recommendations that you want to make, the results of your study, the idea you are proposing, the specifics of a solution. As we shall see, this conversation is held together, usually, by a strong chain of topics and an appropriate rhetorical pattern. But the second conversation is explicitly between the writer and the reader, announcing what the reader will find, in what order, and reminding the reader of where he or she is in the text.

This second conversation is often called **metadiscourse.** *Meta-* is a Greek prefix that means "along with" or "among," and the basic strategy is to include explicit statements and cues to the reader that announce and reinforce your intentions along with your content. There are two main ways writers give such cues. One is to talk directly to the reader, inserting metacomments that preview what will come, remind, predict, or summarize. This lets the writer step back, make the plan of the text more visible, and direct traffic, by telling the reader, "In the next section I will argue that. . . ." The other kind of cues work more like traffic signals—they are the conventional words and phrases that signal transitions, or logic, or the structure of ideas. Readers often expect to find these metacomments and signalling cues in some standard places. Some common places to insert cues to the reader include:

Title, title page
Table of contents
Abstract
Introduction or first paragraphs
Headings
The beginning and end of paragraphs (i.e., topic sentences)
Entire paragraphs in between long sections

So review your text first to see if you have used enough cues to make your plan clear, and second to see if you have included cues in places readers expect to see some guidance on what to look for and how to read this document [Figure 4].

Cues that signal your plan and guide the reader:

Cues that lead the reader forward

To show addition: *To show time:*

Again	Moreover,	At length	And then
And	Nor,	Immediately thereafter,	Later,
And then,	Too,	Soon,	Previously,
Besides	Next,	After a few hours,	Formerly,
Equally important,	First, second, etc.	Afterwards,	First, second, etc.
Finally	Lastly,	Finally,	Next, etc.
Further,	What's more,	Then	
Furthermore,			

Cues that make the reader stop and compare

But	Notwithstanding,	Although
Yet,	On the other hand,	Although this is true,
And yet,	On the contrary,	While this is true,
However,	After all,	Conversely,
Still	For all that,	Simultaneously,
Nevertheless,	In contrast,	Meanwhile
Nonetheless,	At the same time,	In the meantime,

Cues that develop and summarize

To give examples: *To emphasize:*   *To repeat:*   *To signal a relationship:*

For instance,	Obviously,	In brief,	Finally
For example,	In fact,	In short,	Because
To demonstrate,	As a matter of fact,	As I have said,	Yet
To illustrate,	Indeed,	As I have noted,	For instance,
As an illustration,	In any case,	In other words,	
	In any event,	That is,	

To introduce conclusions: *To summarize:*

Hence,	In brief,
Therefore,	On the whole,
Accordingly,	Summing up,
Consequently,	To conclude,
Thus,	In conclusion,
As a result,	

FIGURE 4   Giving Cues to the Reader

Metacomment cues that announce and reinforce your intentions

To ask a question about your topic or the argument unfolding:

What series of events led to event? . . .
To answer that question . . .

To preview what will come:

In the next section, we will see how this formula applies . . .
The third paragraph will reveal how . . .

To summarize what has been said thus far:

In the preceding pages, I've described . . .
Thus far, I have argued . . .

To comment on your writing and thinking as it unfolds:

I haven't mentioned yet that . . .
I'm talking about . . .
My main point is . . .

FIGURE 4   *(Continued)*

There is an endless variety of sentences and phrases that can be invented and inserted to announce and reinforce your main ideas and the progression that you want your reader to follow. To show you the power of metadiscourse and how it plays out in a text, here is an excerpt from a shareowner's letter with the metadiscourse highlighted [as shown in Figure 5 on page 38]. We offer the original paragraph numbers with the cues underlined and the rhetorical purpose of the cues as we read them. . . .

Did You Keep Your Promises?

Your text started out with the best of intentions—a strong rhetorical plan and cues that keep your reader on track. The next test is to see if you followed through on the promises that you made. Read your text as if you were outlining its key points and promises and then look back to see if you have delivered the necessary detail. For instance:

- Your problem/purpose statement promises four main points and an extended example: Does each paragraph keep that promise? By referring back to your announced plan and delivering four main (i.e., well-developed) points in the same order that you promised?
- Your topic sentence in the seventh paragraph promises the two key instances that support a legal precedent: Does the paragraph deliver them in the order and detail necessary?

Chairman's Letter
Rhetorical Purpose/Cues

1		Often the greatest opportunities..........................
4	ties to history	Our strong 1984 results speak for themselves ......
5		Pacific Telesis Group earned $829 million............
6	forces question	What happened? How did the corporation that many observers predicted would be the biggest loser in the AT&T breakup turn out to be one of the biggest winners?
7	links date with solution	To answer that question, you have to go back to 1980, when we developed..........................
8	signals logic	Our employees mobilized to make it work. And work it did. More specifically:..............
8c	signals order & emphasis	Third, and very important, we've built a relationship with the California Public Utilities Commission ....
	previews, emphasizes	a subject I'll return to later on in this letter.............

The Future of Your Investment

34	previews	A corporation, particularly one as new as ours, must operate with a clearly articulated and widely understood vision of its future....................
37	summarizes	In the preceding pages I've described to you our strategies for deploying technology, marketing technology, and diversifying into new lines ....
38	summarizes, emphasizes	I've talked about our determination ....
	signals logic, emphasis uses authority	But there is another very significant factor I haven't mentioned. I'm talking about the people who work for the Pacific Telesis Group

FIGURE 5   Metadiscourse in the Chairman's Letter

John S. Harris

The Project Worksheet for Efficient Writing Management

John S. Harris is Professor Emeritus of English at Brigham Young University. The author of several books on technical writing, he was the founding president of the Association of Teachers of Technical Writing.

When employees produce poor quality documents, the fault may lie not with their incompetence or lack of devotion to the job, but with inadequate and unclear assignments from their project managers or publications managers. The Project Worksheet is designed to help managers give initial writing assignments, effectively manage documents during production, and evaluate finished products. The following pages describe the Worksheet and suggest ways publications managers can use it effectively.

THE PROJECT WORKSHEET

The Project Worksheet consists of a series of questions that help writers and managers consider the factors involved in planning a document. I developed the Worksheet years ago, and faculty in the technical writing program at Brigham Young University have successfully used it to teach technical writing for some years. After graduation, many of our students continue to use it in industry to plan documents and make document assignments.

What the Writer Must Consider

The manager and writer must determine the document's intended readers, its purpose, scope, form, length, graphic aids, and sources of information.[1] They must know such things before the writing begins. Failure to consider them results in inefficient document production and poor-quality documents.

Experienced writers in an organization usually answer these planning questions intuitively or subconsciously. Or they get the answers by direct conference, by study of past documents, or through the grapevine. They then efficiently plan and write good documents.

But less-experienced writers learn the answers to the questions only by the inefficient and expensive trial-and-error method of submitting draft documents and then repeatedly amending them as management requires.

Managers—whether of engineering projects, scientific studies, or computer software documentation—can reduce the trial-and-error writing cycle by providing complete answers for such planning questions during a writing assignment conference. But managers cannot provide the answers to the planning questions without first knowing the questions and their consequences. The Project Worksheet in Figure 1 guides both writers and managers in considering the important factors in planning documents. It can, of course, be adapted for special situations. Managers may want to reformat it to allow more space for answers, and some may wish to put it into a word processor as a template. Let us first examine the Worksheet questions and then consider procedures for using the Worksheet for document management.

Housekeeping Headings

Headings such as assignment date, due date, writer, and assignment authority are self-explanatory. Still, listing them together permits the use of the Worksheet as a tracking document for projects. The *document subject* is the file subject. The *document title* may be the same, or it may be something less prosaic—especially if the document is going out of house. Thus a document subject could be *Dental Hygiene,* and the document title could be *Your Teeth: An Owner's Manual.*

Primary Reader

No factor is more important for the writer than full understanding of the nature and needs of the document's primary reader. Experienced writers know this, of course. But even they do not always know everything that management knows and believes about the readers of the projected document. Too frequently both managers and writers assume they understand who the readers will be, but they may have two quite different audiences in mind. Using the Worksheet, managers and writers can reach an understanding of this and other questions.

---

[1]Although this . . . [essay] explains how managers can use the Project Worksheet for managing documents, the books listed under Selected Readings explain how writers can use the Worksheet and other methods to design technical documents for various audiences and purposes.

PROJECT WORKSHEET

Subject _____ Title _____

Assigned by _____ Approved _____ Assignment date _____

Writer _____ Agreed to _____ Due date _____

Primary Reader:

      Technical level (education, experience, etc.):
      Position (title or organizational relationship):
      Attitude toward subject:
      Other factors:

Secondary Readers (others who may read the document):

      Technical level:
      Position:
      Attitude toward subject:
      Other factors:

Reader's Purpose:

      What should the reader know after reading?
      What should the reader be able to do after reading?
      What attitude should the reader have after reading?
      How will the reader access the material?

Writer's Purpose:

      Intellectual purpose:
      Career or monetary purpose:

Logistics:

      Sources (lab reports, library research, etc.):
      Physical size limits (if any):
      Form or medium prescribed or desirable:
      Available aids (graphics, etc.):
      Means of production:
      Outline (Preliminary):

Distribution and Disposition:

FIGURE I   Project Worksheet

Reader's Technical Level

By carefully considering the technical level of the reader, the manager and writer can decide the proper level of explanation needed. How much information does the reader already have? Are definitions of *photo-grammetry* or *perihelion* or *debentures* needed? Considering such things can help avoid losing the reader in a maze of technical jargon, or alternatively avoid alienating a reader with overly simplistic explanations. In the trade-off between those two extremes, erring on the side of simplicity is nearly always better.

Reader's Organizational Relationship

The writer must also consider whether the document will go to the executive tower or to the shop floor. Since writers often prepare documents for someone else's signature, the manager—and the writer—must consider the organizational relationship of the reader *to the document's signer.* Thus, if the document will be signed by the boss, the writer must—for the task—carefully assume the persona of the boss. Is the signer addressing superiors, subordinates, or peers? Such situations raise delicate questions, and the manager must recognize the entailed problems and pass on an understanding of those problems to the writer.

Reader's Attitude

Similarly, the attitude of the reader must be considered. Is the reader hostile to the new manufacturing procedure because it will result in downsizing her department? Or is the reader skeptical about the efficacy of new cutting-edge technology because it has not yet been proved? Or is the reader torpid and apathetic and needing a stiff jolt to see the promise—or the threat—of the subject of the document?

Other Factors

Often even more subtle factors about the reader must be considered. Some of these may overlap the preceding questions, but they may also affect the kind of document needed. Is the reader a white, conservative Republican woman of sixty, raised on an Iowa farm, or a young Afro-American male from the South Bronx, or a highly educated but condescending academic? Is the reader an Ivy-League MBA obsessed with this year's bottom line, a high school dropout production-line worker who is fearful that increased automation will eliminate jobs, a displaced homemaker with an acute case of computer anxiety, or an immigrant with limited understanding of English? Such demographic factors may require delicate adjustments in the document.

Secondary Reader

Often a secondary reader must also be considered. The company annual report may be read by both stockholders and the CEO. An advertising brochure intended for customers may also be read by the competition. An environmental impact statement may be read by both the Douglas Fir Plywood Association and the Sierra Club. A report intended for Level 2 management may have to get past a Level 3 gatekeeper. Such situations require that the writer juggle the factors of two or more audiences at once, and this requires some skill. If the readers differ in attitude, the writer must watch out for red-flag statements that may cause a secondary reader to charge. If the differences are in technical level, the secondary reader's need for more elementary explanations or more technical information can often be taken care of in footnotes, glossaries, sidebars, or appendices.

The Reader's Purpose

Organizational documents have purposes as varied as securing approval for a new manufacturing process, justifying spending public funds for a reclamation project, instructing a software purchase on how to use a spreadsheet, or easing the worries of environmentalists about the effect of a new highway near a prime trout-fishing stream. Four basic questions about the reader's purposes need consideration:

- What should the reader know after the reading?
- What should the reader be able to do after reading?
- What attitude should the reader have after reading?
- How will the reader access the information?

What Should the Reader Know after Reading?

What factual information does the reader need to obtain from the document? Does the reader need to know the percentage of radial keratotomy patients who can pass a 20/40 eye examination without glasses after the operation? The percentage of impurities remaining in palladium after the macrocycle solvent extraction process? The change in numbers of predator kills of sheep after reintroduction of wolves to the grazing range? Often the specific information the reader is to gain from the document can be presented in lists, outlines, tables, or graphs.

What Should the Reader Be Able to Do after Reading?

What the reader can *do* after reading may be quite different from what the reader *knows*. Knowing the names for the stages of mitosis is different from being able to recognize telophase through a microscope. Knowing the stoichiometric ratio for combustion of a gasoline/air mixture is different from being able to adjust a fuel-injection system to achieve the ideal mixture.

What a reader can do after reading may depend on technical skills. It may also depend on the reader's capability to make executive decisions based on the information contained in the document. Thus the reader's needs probably extend beyond gaining general information to applying specific knowledge. A reader may use the information on color changes in a chemical solution to perform titration in the assay laboratory. Or a reader may use the information on the size of natural gas reserves of the overthrust belt to decide the feasibility of building a 24-inch pipeline from Wyoming to California.

What Attitude Should the Reader Have after Reading?

Earlier the Worksheet asked about the reader's attitude before reading. Here the Worksheet asks about the attitude the writer and manager want the reader to have *after* reading the document. Do they want the reader to believe and feel that the proposed SDI weapons system is an effective deterrent to foreign aggression? Or do they want the reader to believe that it is a horrendously expensive and impractical pork-barrel boondoggle? Though such attitudes in technical documents may be based on hard data, they are nonetheless attitudes—sometimes emotionally charged attitudes—and the writer must realize that shaping those attitudes is sometimes part of the job, even a moral responsibility. The technical writer uses different tools for shaping attitudes than the politician or advertiser does, and may have a higher regard for truth, but like the politician or advertiser, the technical writer may be an attitude shaper.

And one other attitude deserves consideration: the attitude that the reader should have toward the preparer of the document. The writer wants the reader to think that the document was prepared by a credible and meticulous professional who cares about the subject, the needs of the reader, and the needs of his or her employer. Though this point may seem obvious, writers do not always automatically consider it.

How Will the Reader Access the Information?

We read novels, murder mysteries, and perhaps newspaper editorials beginning-to-end, but almost everything else, we read in some other fashion. We do not read dictionaries, phone books, repair manuals, computer documentation, or encyclopedias beginning-to-end. We often read textbooks and journal articles in a nonlinear fashion too. Sometimes the text is well designed to help us get the critical information quickly. Or, exasperatingly, the text may bury critical information in the middle of a full-page paragraph. Whatever way the text is designed, readers scan, or skip around trying to find what they want without reading every blessed word.

Realizing what information the reader wants, the writer or manager can design documents so that the information is easy to find. Information accessibility relies on such devices as

- Headings
- Outlines
- Indexes
- Graphics
- Underlining
- Sidebars
- Glosses
- Varied typefaces
- Cross-references
- Bulleted lists (like this).

USA Today uses many of these devices. The *New York Times* uses fewer. *USA Today* lacks the substance of the *Times*, but it is easier to get the news from it quickly. A similar comparison could be made between *Popular Science* and *Scientific American*. The differences in format reflect how the writers and editors anticipate the publications will be read.

The writer and manager must carefully consider whether the reader will read beginning-to-end or access pieces of data individually. The basic consideration should be the importance of the information being conveyed. The more important the information, the more accessible it should be. The document manager must help the writer anticipate *all* the ways that *all the readers* will want to access the information and require the writer to provide the machinery that will allow that access. Most writers of technical and scientific documents should pay much more attention to accessibility. Few readers of technical documents read them beginning-to-end.

Writer's Purpose

The writer's and reader's purposes for a document differ just as the buyer's and seller's purposes differ. Both have their agendas. The writer—or the manager—may want to have a project approved. The target reader may want to know if the project is feasible.

The writer probably has potential salary, prestige, and promotion purposes for writing, but may also be writing because the task is challenging or intellectually interesting. The writer and manager should decide in advance whether the project is routine and can be done rather perfunctorily, or whether it will affect the safety of the user, the prosperity of the company, the security of the nation, the preservation of the environment—or the continued employment of the writer. These questions help the writer and manager determine the priorities of the task.

Logistics

The logistics are the nuts and bolts things—such as where the information comes from, the size of the projected document, and what form the document will take. These too must be considered in advance.

Sources of Information

Will the document be based on observation, personal opinion, government reports, a literature search, engineers' notes, customer interviews, public opinion polls, laboratory tests, compilations of vendors' brochures, or what? Frequently, lead time may be required, so early consideration of sources is wise to allow tests to be run, surveys to be conducted, or publications to be ordered.

Size of Document Expected

A two-page memo will not do if a twenty-page report is expected, and a twenty-page report will not do if a two-page memo is expected. Again, the question should be resolved between the writer and the manager before the writing begins.

Form or Medium Prescribed or Desirable

The form or medium of the eventual document should also be considered in advance. Should the document follow the company style sheet? Would a film, a wall chart, a wallet card, or a video be a more useful medium? *Such choices should be made according to subject, purpose, reader (user), and cost.*

Graphic Aids Available

Would photographs or art work be useful? Are facilities available to prepare them? Is there time? Are funds available? Again, these must be considered early.

Means of Document Production

The means of production of the document also need to be considered early. Will the document be laid out on a Macintosh with PageMaker software? Will it be photocopied or multilithed? How will the means of production affect the legibility and credibility of the document, and how much will it cost?

Outline

Even at the early planning stages, the writer and manager should consider the topics to be treated, their order, and their proportion of the expected document. Such an outline can, of course, be revised during preparation.

Distribution and Disposition

How many copies will be needed? Where do they go? What happens to them? The manager and writer should consider these questions—and their many consequences—in advance. Will the document be read by the competition? Will it come back to haunt everybody ten years later? Will it be subpoenaed in court? The wise consider such things.

PROJECT PLANNING CONFERENCE

The manager and the writer must come to an agreement about a document's audience, purpose, and scope, and the logistics of its development, production, and distribution. Such negotiation is most efficient in a planning conference. Usually the manager calls the conference, but the negotiation can take a variety of forms, depending on the situation and the experience of the writers.

With an inexperienced writer, the manager may say, "Willoughby, I want you to write a proposal for my signature recommending adoption of the hyperbaric procedure. It should be addressed to Allardyce, the plant manger. He is a chemical engineer with twenty years' experience on the job. But it will also be read by Sung. She is the company comptroller and has ultimate approval power on expenditures. She is pretty conservative, but her attention to cash-flow probably saved the company during the last recession."

With a more experienced writer, the manager can more inductively and democratically ask, "Kim, we need a brochure about our prefabricated forms for concrete. What ways do you see to make a brochure that will meet the needs of all of its readers?" Obviously, here the manager respects the writer's knowledge and expects useful suggestions.

Whatever the method, the manager and writer should agree upon the answers, write them down—Yes, and *sign off—each keeping a copy.* Such a signed-off Worksheet is then a kind of contract agreed to by both parties, and both can feel more secure with it. The manager is now more secure knowing that the assignment is clearly understood, and the writer is more secure knowing what is expected.

POSTCONFERENCE MANAGEMENT

Obviously other factors may arise during the writing. The manager may receive new data, or the writer may see new ways to deal with the problems. Depending on their impact on the document, such things may require additional conferences and a renegotiating of the specifications.

As needed, the manager may discreetly ask about the progress of the document, or may even ask to see sections in draft. However, if the initial assignment has been done carefully, a responsible and skilled writer should be able to produce a sound document with little intermediate prodding.

The writer properly attaches the Project Worksheet to the draft form of the document when submitting it. Then the manager checks to see if it fulfills the assignment. If it does, the manager approves it for production.

Quite often, however, problems appear. Of these, failures to match the assignment are the most obvious, and the manager can ask for a rewrite to make the document match the specifications. The most common problem seems to be a failure to consider adequately the technical levels of all the readers.

Or it may become clear that the specifications need changing. If for good reasons they do need changing, the manager can change them and ask that the new specifications be followed, but the manager who has agreed to the original specifications is less likely to make such changes capriciously or arbitrarily. If lessening the power to make arbitrary changes seems to take the fun out of being boss, then the manager should recognize that the resultant empowerment of the writer is not only enlightened and trendy, but also pragmatically effective.

OTHER APPLICATIONS

If the writer's manager does not make such detailed assignments, the writer obviously can still use the Project Worksheet for planning. Or the writer can use the Worksheet to initiate a conference and negotiate with the manager. We have found that managers are often impressed when writers request such conferences and demonstrate through their questions—the Worksheet questions—that they have a clear understanding of the situation.

The Worksheet is also effective in negotiating contracts in freelance writing assignments, or as an assignment device between editor and writer. In the university, a professor can make writing assignments following the Worksheet questions. Or a sharp student can use the questions to clarify a vaguely given assignment. And many problems of master's theses and doctoral dissertations would be avoided if candidates and supervising professors negotiated the specifications for theses and dissertations through the Worksheet.

In much technical writing, the most important work is the planning done before the first words of the document are written. If managers will use the Project Worksheet approach to document management, their writers will work more efficiently and produce better documents.

SELECTED READINGS

Anderson, P. V. (ed.), *Teaching Technical Writing: Teaching Audience Analysis and Adaptation*, ATTW Anthology No. 1, Association of Teachers of Technical Writing, St. Paul, Minnesota, 1980.

Caernarven-Smith, P., *Audience Analysis and Response,* Firman Technical Publications, Pembroke, Massachusetts, 1983.

Mathes, J. C., and D. Stevenson, *Designing Technical Reports: Writing for Audiences in Organizations* (2nd Edition), Macmillan, New York, 1991.

Pearsall, T., *Audience Analysis for Technical Writing,* Glencoe Press, Beverly Hills, California, 1969.

Souther, J. W., and M. L. White, *Technical Report Writing* (2nd Edition), John Wiley & Sons, New York, 1977.

Spilka, R., Orality and Literacy in the Workplace: Process- and Text-Based Strategies for Multiple-Audience Adaptation. *Journal of Business and Technical Communication,* 4:1, pp. 44–67, 1990.

Part 2

Problems with Language

In Lewis Carroll's *Through the Looking Glass,* Humpty Dumpty and Alice have the following exchange:

> "When *I* use a word," Humpty Dumpty said, in a rather scornful tone, "it means just what I choose it to mean—nothing more or less."
>
> "The question is," said Alice, "whether you *can* make words mean so many different things."

After reading some examples of business and technical writing, it is easy to suspect that Humpty Dumpty has been promoted to project manager—or even company president.

It is not always clear why business and technical writers get themselves into the problems with language that they do. At times, they mistakenly assume their readers will know what they are talking about, or that their goal is to impress rather than inform their readers. At other times, they simply imitate their superiors or model their documents after those in the files, as if the filing cabinet were a repository for sacred writings handed down from on high.

To solve problems with language, business and technical writers must remember their readers and then be guided by two principles:

- they should write everything as clearly, straightforwardly, and simply as possible; and
- they should write in a manner or style with which they would be comfortable speaking.

These two principles will help writers avoid problems with jargon, gobbledy-gook, legalese, and sexist and biased language. They will also serve as a guide in selecting the appropriate style for different writing situations.

Jargon is simply technical language unique to a profession or occupation. It only becomes a problem when writers use it in writing to an audience without the necessary background or training to understand it. A systems engineer writing to another systems engineer could rightly assume that the following passage would make sense to his or her reader:

> If you are using a CONFIG.SYS file to modify version 2.0 of the Personal Computer DOS, you may want to create a CONFIG.SYS file to use with the program instead of the AUTOEXEC.BAT file.

Unfortunately, this passage comes from an introductory computer manual.

Jargon in the extreme becomes gobbledygook, mindless gibberish akin to double-talk and characterized by pretentiousness. For example, the host of a popular television game show tells contestants who are losing that they are in "a deficit situation"; ordinary No. 2 pencils become, on the requisition forms of some bureaucrats, "writing implements, standard issue"; politicians wary of angering voters call tax increases "revenue enhancements" and "misspeak themselves" rather than lie when they are caught making false claims or leveling inaccurate charges against their opponents; "turn out the lights when you leave" becomes "illumination is required to be extinguished upon vacation of the premises"; the threat that "anyone caught smoking on the premises will be dealt with accordingly" may at first cause someone to think twice before lighting up, but, in the final analysis, we may well wonder just what the threat is really supposed to mean.

Gobbledygook also includes words and phrases that writers mistakenly think sound the way business and technical writing should sound. They may be in vogue or out of date, but either way, they have unfortunately come to be viewed by some writers as examples of good professional writing practice. People no longer meet for discussions—they "interface"—and every issue about which they interface has "parameters," "viable" and otherwise. Some of these same people see nothing wrong with beginning their letters as follows:

> Pursuant to yours of the twelfth, enclosed herewith please find our check in the amount of $24.95.

Subsequently in the same letter, they "are pleased to advise that," and they go on to refer to data "hereunder discussed" or "above referenced."

Legalese is an overreliance on legal terminology—or legal *sounding* terminology—when Plain English will serve the reader and the writer as well.

Sexist or biased language is linguistic discrimination usually, although not exclusively, against women and groups that are perceived to be minorities. Just as all managers are not men, so all secretaries are not women. Common sense, sensitivity to audience and language, and fairness can help writers avoid sexism and bias.

Problems in any of these areas can be exacerbated in international business settings. As the world grows smaller thanks to technical advances in communication, audiences grow larger and much more complex. An American automobile manufacturer learned the hard way that marketing a model called "Nova" to customers who spoke Spanish was a potentially no-win situation since, in Spanish, *nova* can mean "doesn't go or run."

The essays in this section of *Strategies* address further the issues touched upon in my introductory comments here. In a now classic, and often-reprinted, piece, Stuart Chase sends up gobbledygook, the linguistic overkill that is all too common in the writing of government officials, bureaucrats, lawyers, and, sadly, even academics (who especially should know better). William Zinsser, an author and teacher whose works on writing are a model for practicing what he preaches, follows Chase with an essay that shows how it is possible to write for the world of work and still sound like a human being. Alan Siegel and Mark Mathewson weigh in with comments about legalese. Siegel offers a brief history of the Plain English movement in his account of the work his company has done to help clients reduce legalese in the documents they produce. Mathewson, writing from his position as Director of Publications for the Illinois State Bar Association, addresses other legal professionals (and anyone else prone to write "like a lawyer") who balk at abandoning the legalese they too readily adopt in the documents they produce for their clients and the courts.

Guidelines developed at the University of Wisconsin then offer business and technical writers some suggestions on how to avoid language that stereotypes or denigrates men as well as women. Finally, Gwyneth Olofsson, who heads an international training and consulting firm based in Sweden, discusses almost two dozen letters drawn from the global workplace in which international communication very quickly turns into international miscommunication.

Stuart Chase

Gobbledygook

Stuart Chase worked for many years as a consultant to various government agencies; his other books include The Tyranny of Words *(1938) and* Democracy Under Pressure *(1945).*

Said Franklin Roosevelt, in one of his early presidential speeches: "I see one-third of a nation ill-housed, ill-clad, ill-nourished." Translated into standard bureaucratic prose his statement would read:

> It is evident that a substantial number of persons within the Continental boundaries of the United States have inadequate financial resources with which to purchase the products of agricultural communities and industrial establishments. It would appear that for a considerable segment of the population, possibly as much as 33.3333* of the total, there are inadequate housing facilities, and an equally significant proportion is deprived of the proper types of clothing and nutriment.
> *Not carried beyond four places.

This rousing satire on gobbledygook—or talk among the bureaucrats—is adapted from a report[1] prepared by the Federal Security Agency in an attempt to break out of the verbal squirrel cage. "Gobbledygook" was coined by an exasperated Congressman, Maury Maverick of Texas, and means using two, or three, or ten words in the place of one, or using a five-syllable word where a single syllable would suffice. Maverick was censuring the forbidding prose of executive departments in Washington, but the term has now spread to windy and pretentious language in general.

"Gobbledygook" itself is a good example of the way a language grows. There was no word for the event before Maverick's invention; one had to say: "You know, that terrible, involved, polysyllabic language those government people use down in Washington." Now one word takes the place of a dozen.

---

[1]This and succeeding quotations from FSA report by special permission of the author, Milton Hall.

A British member of Parliament, A. P. Herbert, also exasperated with bureaucratic jargon, translated Nelson's immortal phrase, "England expects every man to do his duty":

> England anticipates that, as regards the current emergency, personnel will face up to the issues, and exercise appropriately the functions allocated to their respective occupational groups.

A New Zealand official made the following report after surveying a plot of ground for an athletic field:[2]

> It is obvious from the difference in elevation with relation to the short depth of the property that the contour is such as to preclude any reasonable developmental potential for active recreation.

Seems the plot was too steep.

An office manager sent this memo to his chief:

> Verbal contact with Mr. Blank regarding the attached notification of promotion has elicited the attached representation intimating that he prefers to decline the assignment.

Seems Mr. Blank didn't want the job.

> A doctor testified at an English trial that one of the parties was suffering from "circumorbital haematoma."

Seems the party had a black eye.

> In August 1952 the U.S. Department of Agriculture put out a pamphlet entitled: "Cultural and Pathogenic Variability in Single-Condial and Hyphaltip Isolates of Hemlin-Thosporium Turcicum Pass."

Seems it was about corn leaf disease.

On reaching the top of the Finsteraarhorn in 1845, M. Dollfus-Ausset, when he got his breath, exclaimed:

> The soul communes in the infinite with those icy peaks which seem to have their roots in the bowels of eternity.

Seems he enjoyed the view.

A government department announced:

> Voucherable expenditures necessary to provide adequate dental treatment required as adjunct to medical treatment being rendered a pay patient in in-patient status may be incurred as required at the expense of the Public Health Service.

Seems you can charge your dentist bill to the Public Health Service. Or can you?

---

[2]This item and the next two are from the piece on gobbledygook by W. E. Farbstein, *New York Times,* March 29, 1953.

LEGAL TALK

Gobbledygook not only flourishes in government bureaus but grows wild and lush in the law, the universities, and sometimes among the literati. Mr. Micawber was a master of gobbledygook, which he hoped would improve his fortunes. It is almost always found in offices too big for face-to-face talk. Gobbledygook can be defined as squandering words, packing a message with excess baggage and so introducing semantic "noise." Or it can be scrambling words in a message so that meaning does not come through. The directions on cans, bottles, and packages for putting the contents to use are often a good illustration. Gobbledygook must not be confused with double talk, however, for the intentions of the sender are usually honest.

I offer you a round fruit and say, "Have an orange." Not so an expert in legal phraseology, as parodied by editors of *Labor:*

> I hereby give and convey to you, all and singular, my estate and interests, right, title, claim and advantages of and in said orange, together with all rind, juice, pulp and pits, and all rights and advantages therein . . . anything hereinbefore or hereinafter or in any other deed or deeds, instrument or instruments of whatever nature or kind whatsoever, to the contrary, in any wise, notwithstanding.

The state of Ohio, after five years of work, has redrafted its legal code in modern English, eliminating 4,500 sections and doubtless a blizzard of "where-ases" and "hereinafters." Legal terms of necessity must be closely tied to their referents, but the early solons tried to do this the hard way, by adding synonyms. They hoped to trap the physical event in a net of words, but instead they created a mumbo-jumbo beyond the power of the layman, and even many a lawyer, to translate. Legal talk is studded with tautologies, such as "cease and desist," "give and convey," "irrelevant, incompetent, and immaterial." Furthermore, legal jargon is a dead language; it is not spoken and it is not growing. An official of one of the big insurance companies calls their branch of it "bafflegab." Here is a sample from his collection.[3]

> One-half to his mother, if living, if not to his father, and one-half to his mother-in-law, if living, if not to his mother, if living, if not to his father. Thereafter payment is to be made in a single sum to his brothers. On the one-half payable to his mother, if living, if not to his father, he does not bring in his mother-in-law as the next payee to receive, although on the one-half to his mother-in-law, he does bring in the mother or father.

You apply for an insurance policy, pass the tests, and instead of a straightforward "here is your policy," you receive something like this:

> This policy is issued in consideration of the application therefor, copy of which application is attached hereto and made part hereof, and of the payment for said insurance on the life of the above-named insured.

---

[3]Interview with Clifford B. Reeves by Sylvia F. Porter, *New York Evening Post,* March 14, 1952.

ACADEMIC TALK

The pedagogues may be less repetitious than the lawyers, but many use even longer words. It is a symbol of their calling to prefer Greek and Latin derivatives to Anglo-Saxon. Thus instead of saying: "I like short clear words," many a professor would think it more seemly to say: "I prefer an abbreviated phraseology, distinguished for its lucidity." Your professor is sometimes right, the longer word may carry the meaning better—but not because it is long. Allen Upward in his book *The New Word* warmly advocates Anglo-Saxon English as against what he calls "Mediterranean" English, with its polysyllables built up like a skyscraper.

Professional pedagogy, still alternating between the Middle Ages and modern science, can produce what Henshaw Ward once called the most repellent prose known to man. It takes an iron will to read as much as a page of it. Here is a sample of what is known in some quarters as "pedageese":

> Realization has grown that the curriculum or the experiences of learners change and improve only as those who are most directly involved examine their goals, improve their understandings and increase their skill in performing the tasks necessary to reach newly defined goals. This places the focus upon teacher, lay citizen and learner as partners in curricular improvement and as the individuals who must change, if there is to be curriculum change.

I think there is an idea concealed here somewhere. I think it means: "If we are going to change the curriculum, teacher, parent, and student must all help." The reader is invited to get out his semantic decoder and check on my translation. Observe there is no technical language in this gem of pedageese, beyond possibly the word "curriculum." It is just a simple idea heavily ververbalized.

In another kind of academic talk the author may display his learning to conceal a lack of ideas. A bright instructor, for instance, in need of prestige may select a common sense proposition for the subject of a learned monograph—say, "Modern cities are hard to live in"—and adorn it with imposing polysyllables: "Urban existence in the perpendicular declivities of megalopolis . . ." et cetera. He coins some new terms to transfix the reader—"mega-decibel" or "stratocosmopolis"—and works them vigorously. He is careful to add a page or two of differential equations to show the "scatter." And then he publishes, with 147 footnotes and a bibliography to knock your eye out. If the authorities are dozing, it can be worth an associate professorship.

While we are on the campus, however, we must not forget that the technical language of the natural sciences and some terms in the social sciences, forbidding as they may sound to the layman, are quite necessary. Without them, specialists could not communicate what they find. Trouble arises when experts expect the uninitiated to understand the words; when they tell the jury, for instance, that the defendant is suffering from "circumorbital haematoma."

Here are two authentic quotations. Which was written by a distinguished modern author, and which by a patient in a mental hospital? You will find the answer at the end of [this selection].

1. Have just been to supper. Did not knowing what the woodchuck sent me here. How when the blue blue blue on the said anyone can do it that tries. Such is the presidential candidate.
2. No history of a family to close with those and close. Never shall he be alone to be alone to be alone to be alone to be alone to lend a hand and leave it left and wasted.

REDUCING THE GOBBLE

As government and business offices grow larger, the need for doing something about gobbledygook increases. Fortunately the biggest office in the world is working hard to reduce it. The Federal Security Agency in Washington,[4] with nearly 100 million clients on its books, began analyzing its communication lines some years ago, with gratifying results. Surveys find trouble in three main areas: correspondence with clients about their social security problems, office memos, official reports.

Clarity and brevity, as well as common humanity, are urgently needed in this vast establishment which deals with disability, old age, and unemployment. The surveys found instead many cases of long-windedness, foggy meanings, clichés, and singsong phrases, and gross neglect of the reader's point of view. Rather than talking to a real person, the writer was talking to himself. "We often write like a man walking on stilts."

Here is a typical case of long-windedness:

> *Gobbledygook as found:* "We are wondering if sufficient time has passed so that you are in a position to indicate whether favorable action may now be taken on our recommendation for the reclassification of Mrs. Blank, junior clerk-stenographer, CAF 2, to assistant clerk-stenographer, CAF 3?"
>
> *Suggested improvement:* "Have you yet been able to act on our recommendation to reclassify Mrs. Blank?"

Another case:

> Although the Central Efficiency Rating Committee recognizes that there are many desirable changes that could be made in the present efficiency rating system in order to make it more realistic and more workable than it now is, this committee is of the opinion that no further change should be made in the present system during the current year. Because of conditions prevailing throughout the country and the resultant turnover in personnel, and difficulty in administering the Federal programs, further mechanical improvement in the present rating system would require staff retraining and other administrative expense which would seem best withheld until the official termination of hostilities, and until restoration of regular operations.

The FSA invites us to squeeze the gobbledygook out of this statement. Here is my attempt:

---

[4]Now the Department of Health and Human Services.

> The Central Efficiency Rating Committee recognizes that desirable changes could be made in the present system. We believe, however, that no change should be attempted until the war is over.

This cuts the statement from 111 to 30 words, about one-quarter of the original, but perhaps the reader can do still better. What of importance have I left out?

Sometimes in a book which I am reading for information—not for literary pleasure—I run a pencil through the surplus words. Often I can cut a section to half its length with an improvement in clarity. Magazines like *The Reader's Digest* have reduced this process to an art. Are long-windedness and obscurity a cultural lag from the days when writing was reserved for priests and cloistered scholars? The more words and the deeper the mystery, the greater their prestige and the firmer the hold on their jobs. And the better the candidates's chance today to have his doctoral thesis accepted.

The FSA surveys found that a great deal of writing was obscure although not necessarily prolix. Here is a letter sent to more than 100,000 inquirers, a classic example of murky prose. To clarify it, one needs to *add* words, not cut them:

> In order to be fully insured, an individual must have earned $50 or more in covered employment for as many quarters of coverage as half the calendar quarters elapsing between 1936 and the quarter in which he reaches age 65 or dies, whichever first occurs.

Probably no one without the technical jargon of the office could translate this: nevertheless, it was sent out to drive clients mad for seven years. One poor fellow wrote back: "I am no longer in covered employment. I have an outside job now."

Many words and phrases in officialese seem to come out automatically, as if from lower centers of the brain. In this standardized prose people never *get* jobs, they "secure employment"; *before* and *after* become "prior to" and "subsequent to"; one does not *do*, one "performs"; nobody *knows* a thing, he is "fully cognizant"; one never *says*, he "indicates." A great favorite at present is "implement."

Some charming boners occur in this talking-in-one's-sleep. For instance:

> The problem of extending coverage to all employees, regardless of size, is not as simple as surface appearances indicate.
> Though the proportions of all males and females in ages 16–45 are essentially the same . . .
> Dairy cattle, usually and commonly embraced in dairying . . .

In its manual to employees, the FSA suggests the following:

Instead of	Use
give consideration to	consider
make inquiry regarding	inquire
is of the opinion	believes
comes into conflict with	conflicts
information which is of a confidential nature	confidential information

Professional or office gobbledygook often arises from using the passive rather than the active voice. Instead of looking you in the eye, as it were, and writing "This act requires . . . ," the office worker looks out of the window and writes: "It is required by this statute that . . ." When the bureau chief says, "We expect Congress to cut your budget," the message is only too clear; but usually he says, "It is expected that the departmental budget estimates will be reduced by Congress."

> GOBBLED: "All letters prepared for the signature of the Administrator will be single spaced."
> UNGOBBLED: "Single space all letters for the Administrator." (Thus cutting 13 words to 7.)

Only People Can Read

The FSA surveys pick up the point . . . that human communication involves a listener as well as a speaker. Only people can read, though a lot of writing seems to be addressed to beings in outer space. To whom are you talking? The sender of the officialese message often forgets the chap on the other end of the line.

A woman with two small children wrote the FSA asking what she should do about payments, as her husband had lost his memory. "If he never gets able to work," she said, "and stays in an institution would I be able to draw any benefits? . . . I don't know how I am going to live and raise my children since he is disable to work. Please give me some information. . . ."

To this human appeal, she received a shattering blast of gobbledygook, beginning, "State unemployment compensation laws do not provide any benefits for sick or disabled individuals . . . in order to qualify an individual must have a certain number of quarters of coverage . . ." et cetera, et cetera. Certainly if the writer had been thinking about the poor woman he would not have dragged in unessential material about old-age insurance. If he had pictured a mother without means to care for her children, he would have told her where she might get help—from the local office which handles aid to dependent children, for instance.

Gobbledygook of this kind would largely evaporate if we thought of our messages as two way—in the above case, if we pictured ourselves talking on the doorstep of a shabby house to a woman with two children tugging at her skirts, who in her distress does not know which way to turn.

Results of the Survey

The FSA survey showed that office documents could be cut 20 to 50 percent, with an improvement in clarity and a great saving to taxpayers in paper and payrolls.

A handbook was prepared and distributed to key officials.[5] They read it, thought about it, and presently began calling section meetings to discuss gob-

---

[5]By Milton Hall.

bledygook. More booklets were ordered, and the local output of documents began to improve. A Correspondence Review Section was established as a kind of laboratory to test murky messages. A supervisor could send up samples for analysis and suggestions. The handbook is now used for training new members; and many employees keep it on their desks along with the dictionary. Outside the Bureau some 25,000 copies have been sold (at 20 cents each) to individuals, governments, business firms, all over the world. It is now used officially in the Veterans Administration and in the Department of Agriculture.

The handbook makes clear the enormous amount of gobbledygook which automatically spreads in any large office, together with ways and means to keep it under control. I would guess that at least half of all the words circulating around the bureaus of the world are "irrelevant, incompetent, and immaterial"—to use a favorite legalism; or are just plain "unnecessary"—to ungobble it.

My favorite story of removing the gobble from gobbledygook concerns the Bureau of Standards at Washington. I have told it before but perhaps the reader will forgive the repetition. A New York plumber wrote the Bureau that he had found hydrochloric acid fine for cleaning drains, and was it harmless? Washington replied: "The efficacy of hydrochloric acid is indisputable, but the chlorine residue is incompatible with metallic permanence."

The plumber wrote back that he was mighty glad the Bureau agreed with him. The Bureau replied with a note of alarm: "We cannot assume responsibility for the production of toxic and noxious residues with hydrochloric acid, and suggest that you use an alternate procedure." The plumber was happy to learn that the Bureau still agreed with him.

Whereupon Washington exploded: "Don't use hydrochloric acid; it eats hell out of the pipes!"[6]

---

[6] *Note:* The second quotation on page 56 comes from Gertrude Stein's *Lucy Church Amiably.*

William Zinsser

Writing in Your Job

William Zinsser is a noted author, columnist, editor, teacher of writing, and critic.

The memo, the business letter, the administrative report, the financial analysis, the marketing proposal, the note to the boss, the fax, the Post-it—all the pieces of paper that circulate through your office every day are forms of writing. Take them seriously. Countless careers rise or fall on the ability or inability of employees to state a set of facts, summarize a meeting or present an idea coherently.

Most people work for institutions: businesses, banks, insurance firms, law firms, government agencies, school systems, nonprofit organizations and other entities. Many of them are managers whose writing goes out to the public: the president addressing the stockholders, the banker explaining a change in procedure, the school principal writing a newsletter to parents. Whoever they are, they are often so afraid of writing that their sentences lack all humanity—and so do their institutions. It's hard to imagine that these are real places where real men and women come to work every morning.

But just because people work for an institution they don't have to write like one. Institutions can be warmed up. Administrators can be turned into human beings. Information can be imparted clearly and without pomposity. It's a question of remembering that readers identify with people, not with abstractions like "profitability," or with Latinate nouns like "utilization" and "implementation," or with inert constructions in which nobody can be visualized doing something ("pre-feasibility studies are in the paperwork stage").

Nobody has made the point better than George Orwell in his translation into modern bureaucratic fuzz of this famous verse from Ecclesiastes:

> I returned and saw under the sun that the race is not to the swift, nor the battle to the strong, neither yet bread to the wise, nor yet riches to men of understanding, nor yet favor to men of skill; but time and chance happeneth to them all.

Orwell's version goes:

> Objective consideration of contemporary phenomena compels the conclusion that success or failure in competitive activities exhibits no tendency to be commensurate with innate capacity, but that a considerable element of the unpredictable must invariably be taken into account.

First notice how the two passages look. The one at the top invites us to read it. The words are short and have air around them; they convey the rhythms of human speech. The second one is clotted with long words. It tells us instantly that a ponderous mind is at work. We don't want to go anywhere with a mind that expresses itself in such suffocating language. We don't even start to read.

Also notice what the two passages say. Gone from the second one are the short words and vivid images from everyday life—the race and the battle, the bread and the riches—and in their place have waddled the long and flabby nouns of generalized meaning. Gone is any sense of what one person did ("I returned") or what he realized ("saw") about one of life's central mysteries: the capriciousness of fate.

Let me illustrate how this disease infects the writing that most people do in their jobs. I'll use school principals as my first example, not because they are the worst offenders (they aren't) but because I happen to have such an example. My points are intended for all the men and women who work in all the organizations where language has lost its humanity and nobody knows what the people in charge are talking about.

My encounter with the principals began when I got a call from Ernest B. Fleishman, superintendent of schools in Greenwich, Connecticut. "We'd like you to come and 'dejargonize' us," he said. "We don't think we can teach students to write unless all of us at the top of the school system clean up our own writing." He said he would send me some typical materials that had originated within the system. His idea was for me to analyze the writing and then conduct a workshop.

What appealed to me was the willingness of Dr. Fleishman and his colleagues to make themselves vulnerable. Vulnerability has a strength of its own. We decided on a date, and soon a fat envelope arrived. It contained various internal memos and mimeographed newsletters that had been mailed to parents from the 16 elementary, junior, and senior high schools.

The newsletters had a cheery and informal look. Obviously the system was making an effort to communicate warmly with its families. But even at first glance certain chilly phrases caught my eye—"prioritized evaluative procedures," "modified departmentalized schedule"—and one principal promised that his school would provide "enhanced positive learning environments." Just as obviously the system wasn't communicating as warmly as it thought it was.

I studied the principals' material and divided it into good and bad examples. On the appointed morning in Greenwich I found 40 principals and curriculum coordinators assembled and eager to learn. I told them I could only applaud them for submitting to a process that so threatened their identity. In the national clamor over why Johnny can't write, Dr. Fleishman was the first adult in my experience who admitted that youth has no monopoly on verbal sludge and that the problem must also be attacked at the top.

I told the principals that we want to think of the men and women who run our children's schools as people not unlike ourselves. We are suspicious of pretentiousness, of all the fad words that the social scientists have coined to avoid making themselves clear to ordinary mortals. I urged them to be natural. How we write and how we talk is how we define ourselves.

I asked them to listen to how they were defining themselves to the community. I had made copies of certain bad examples, changing the names of the schools and the principals. I explained that I would read some of the examples aloud. Later we would see if they could turn what they had written into plain English. This was my first example:

Dear Parent:

    We have established a special phone communication system to provide additional opportunities for parent input. During this year we will give added emphasis to the goal of communication and utilize a variety of means to accomplish this goal. Your inputs, from the unique position as a parent, will help us to plan and implement an educational plan that meets the needs of your child. An open dialogue, feedback and sharing of information between parents and teachers will enable us to work with your child in the most effective manner.

                                                Dr. George B. Jones
                                                Principal

That's the kind of communication I don't want to receive, unique though my parent inputs might be. I'd like to be told that the school is going to make it easier for me to telephone the teachers and that they hope I'll call often to discuss how my children are getting along. Instead the parent gets junk: "special phone communication system," "added emphasis to the goal of communication," "plan and implement an educational plan." As for "open dialogue, feedback and sharing of information," they are three ways of saying the same thing.

Dr. Jones is clearly a man who means well, and his plan is one we all want: a chance to pick up the phone and tell the principal what a great kid Johnny is despite that unfortunate incident in the playground last Tuesday. But Dr. Jones doesn't sound like a person I want to call. In fact, he doesn't sound like a person. His message could have been tapped out by a computer. He is squandering a rich resource: himself.

Another example that I chose was a "Principal's Greeting" sent to parents at the start of the school year. It consisted of two paragraphs that were very different:

    Fundamentally, Foster is a good school. Pupils who require help in certain subjects or study skills areas are receiving special attention. In the school year ahead we seek to provide enhanced positive learning environments. Children, and staff, must work in an atmosphere that is conducive to learning. Wide varieties of instructional materials are needed. Careful attention to individual abilities and learning styles is required. Cooperation between school and home is extremely important to the learning process. All of us should be aware of desired educational objectives for every child.

Keep informed about what is planned for our children this year and let us know about your own questions and about any special needs your child may have. I have met many of you in the first few weeks. Please continue to stop in to introduce yourself or to talk about Foster. I look forward to a very productive year for all of us.

> Dr. Ray B. Dawson
> Principal

In the second paragraph I'm being greeted by a person; in the first I'm hearing from an educator. I like the real Dr. Dawson of Paragraph 2. He talks in warm and comfortable phrases: "Keep informed," "let us know," "I have met," "Please continue," "I look forward."

By contrast, Educator Dawson of Paragraph 1 never uses "I" or even suggests a sense of "I." He falls back on the jargon of his profession, where he feels safe, not stopping to notice that he really isn't telling the parent anything. What are "study skills areas" and how do they differ from "subjects"? What are "enhanced positive learning environments" and how do they differ from "an atmosphere that is conducive to learning"? What are "wide varieties of instructional materials": pencils, textbooks, filmstrips? What exactly are "learning styles"? What "educational objectives" are "desired," and who desires them?

The second paragraph, in short, is warm and personal; the other is pedantic and vague. This was a pattern I found repeatedly. Whenever the principals wrote to notify the parents of some human detail, they wrote with humanity:

> It seems that traffic is beginning to pile up again in front of the school. If you can possibly do so, please come to the rear of the school for your child at the end of the day.

> I would appreciate it if you would speak with your children about their behavior in the cafeteria. Many of you would be totally dismayed if you could observe the manners of your children while they are eating. Check occasionally to see if they owe money for lunch. Sometimes children are very slow in repaying.

But when the educators wrote to explain how they proposed to do their educating, they vanished without a trace:

> In this document you will find the program goals and objectives that have been identified and prioritized. Evaluative procedures for the objectives were also established based on acceptable criteria.

> Prior to the implementation of the above practice, students were given very little exposure to multiple choice questions. It is felt that the use of practice questions correlated to the unit that a student is presently studying has had an extremely positive effect as the test scores confirm.

After I had read various good and bad examples, the principals began to hear the difference between their true selves and their educator selves. The problem was how to close the gap. I recited my four articles of faith: clarity, simplicity, brevity and humanity. I explained about using active verbs and avoiding windy "concept nouns." I told them not to use the private vocabulary of education as a crutch; almost any subject can be made accessible in good English.

These were all basic tenets, but the principals wrote them down as if they had never heard them before—and maybe they hadn't, or at least not for many years. Perhaps that's why bureaucratic prose becomes so turgid, whatever the bureaucracy. Once an administrator rises to a certain level, nobody ever points out to him again the beauty of a simple declarative sentence or shows him how his writing has become swollen with pompous generalizations.

Finally our workshop got down to work. I distributed my copies and asked the principals to rewrite the more knotty sentences. It was a grim moment. They had met the enemy for the first time. They scribbled on their pads and scratched out what they had scribbled. Some didn't write anything. Some crumpled their paper. They began to look like writers. An awful silence hung over the room, broken only by the crossing out of sentences and the crumpling of paper. They began to sound like writers.

As the day went on, they slowly relaxed. They began to write in the first person and to use active verbs. For a while they still couldn't loosen their grip on long words and vague nouns ("parent communication response"). But gradually their sentences became human. When I asked them to tackle "Evaluative procedures for the objectives were also established based on acceptable criteria," one of them wrote: "At the end of the year we will evaluate our progress." Another wrote: "We will see how well we have succeeded."

That's the kind of plain talk a parent wants. It's also what stockholders want from their corporation, what customers want from their bank, what the widow wants from the agency that's handling her social security. There is a deep yearning for human contact and a resentment of bombast. Recently I got a "Dear Customer" letter from the company that supplies my computer needs. It began: "Effective March 30 we will be migrating our end user order entry and supplies referral processing to a new telemarketing center." I finally figured out that they had a new 800 number and that the end user was me. Any institution that won't take the trouble in its writing to be both clear and personal will lose friends, customers, and money. Let me put it another way for business executives: a shortfall will be experienced in anticipated profitability.

Here's an example of how organizations throw away their humanity with pretentious language. It's a "customer bulletin" distributed by a major corporation. The sole purpose of a customer bulletin is to give helpful information to a customer. This one begins: "Companies are increasingly turning to capacity planning techniques to determine when future processing loads will exceed processing capabilities." That sentence is no favor to the customer; it's congealed with Orwellian nouns like "capacity" and "capabilities" that convey no procedures that a customer can picture. What *are* capacity planning techniques? Whose capacity is being planned? By whom? The second sentence says: "Capacity planning adds objectivity to the decision-making process." More dead nouns. The third sentence says: "Management is given enhanced decision participation in key areas of information system resources."

The customer has to stop after every sentence and translate it. The bulletin might as well be in French. He starts with the first sentence—the one about capacity planning techniques. Translated, that means "It helps to know when you're giving your computer more than it can handle." The second sentence—"Capacity planning adds objectivity to the decision-making process"—means you should know the facts before you decide. The third sentence—the one about enhanced decision participation—means "The more you know about your system, the better it will work." It could also mean several other things.

But the customer isn't going to keep translating. Soon he's going to start looking for another company. He thinks, "If these guys are so smart, why can't they tell me what they do? Maybe they're *not* so smart." The bulletin goes on to say that "for future cost avoidance, productivity has been enhanced." That seems to mean that the product will be free: all costs have been avoided. Next the bulletin assures the customer that "the system is delivered with functionality." That means it works. I would hope so.

Finally, at the end, we get a glimmer of humanity. The writer of the bulletin asks a satisfied customer why he chose this system. The man says he chose it because of the company's reputation for service. He says: "A computer is like a sophisticated pencil. You don't care how it works, but if it breaks you want someone there to fix it." Notice how refreshing that sentence is after all the garbage that preceded it: in its language (comfortable words), in its details that we can visualize (the pencil), and above all in its humanity. The writer has taken the coldness out of a technical process by relating it to an experience we're all familiar with—waiting for the repairman when something breaks. I'm reminded of a sign I saw in the New York subway which proves that even a huge municipal bureaucracy can talk to its constituents humanely: "If you ride the subway regularly you may have seen signs directing you to trains you've never heard of before. These are only new names for very familiar trains."

Still, plain talk will not be easily achieved in corporate America. Too much vanity is on the line. Managers at every level are prisoners of the notion that a simple style reflects a simple mind. Actually a simple style is the result of hard work and hard thinking; a muddled style reflects a muddled thinker or a person too arrogant or too dumb or too lazy to organize his thoughts. Remember that what you write is often the only chance you'll get to present yourself to someone whose business or money or goodwill you want. If what you write is ornate or pompous or fuzzy, that's how you'll be perceived. The reader has no other choice.

I learned about corporate America by venturing out into it, after Greenwich, to conduct workshops for some major corporations, which also asked to be dejargonized. "We don't even understand our own memos anymore," they told me. I worked with the men and women who write the vast amounts of material these companies generate for internal and external consumption. The internal material consists of house organs and newsletters whose purpose is to tell employees what's happening at their "facility" and to give them a sense of belonging. The external material includes the glossy magazines and annual reports that go to stockholders, the speeches delivered by high executives, the releases sent to the press, and the consumer manuals that explain how the product works. I found almost all of it lacking in human juices and much of it impenetrable.

Typical of the sentences in the newsletters was this one:

Announced concurrently with the above enhancements were changes to the System Support Program, a program product which operates in conjunction with the NCP. Among the additional functional enhancements are dynamic reconfiguration and inter-systems communications.

There's no joy for the writer in such work, and certainly none for the reader. It's language out of *Star Trek*, and if I were an employee I wouldn't be cheered—or informed—by these efforts to raise my morale. I would soon stop reading them. I told the corporate writers that they had to find the people behind the fine achievements being described. "Go to the engineer who conceived the new system," I said, "or to the designer who designed it, or to the technician who assembled it, and get them to tell you in their own words how the idea came to them, or how they put it together, or how it will be used by real people out in the real world." The way to warm up any institution is to locate the missing "I." Remember: "I" is the most interesting element in any story.

The writers explained that they often did interview the engineer but couldn't get him to talk English. They showed me some typical quotes. The engineers spoke in an arcane language studded with acronyms ("Sub-system support is available only with VSAG or TNA"). I said that the writers had to keep going back to the engineer until he finally made himself intelligible. They said the engineer didn't *want* to be made intelligible: if he spoke too simply he would look like a jerk to his peers. I said that their responsibility was to the facts and to the reader, not to the vanity of the engineer. I urged them to believe in themselves as writers and not to relinquish control. They replied that this was easier said than done in hierarchical corporations, where approval of written reports is required at a succession of higher levels. I sensed an undercurrent of fear: Do things the company way and don't risk your job trying to make the company human.

High executives were equally victimized by wanting to sound important. One corporation had a monthly newsletter to enable "management" to share its concerns with middle managers and lower employees. Prominent in every issue was a message of exhortation from the division vice-president, whom I'll call Thomas Bell. Judging by his monthly message, he was a pompous ass, saying nothing and saying it in inflated verbiage.

When I mentioned this, the writers said that Thomas Bell was actually a diffident man and a good executive. They pointed out that he doesn't write the message himself; it's written for him. I said that Mr. Bell was being done a disservice—that the writers should go to him every month (with a tape recorder, if necessary) and stay there until he talked about his concerns in the same language he would use when he got home and talked to Mrs. Bell.

What I realized was that most executives in America don't write what appears over their signature or what they say in their speeches. They have surrendered the qualities that make them unique. If they and their institutions seem cold, it's because they acquiesce in the process of being pumped up and dried out. Preoccupied with their high technology, they forget that some of the most powerful tools they possess—for good and for bad—are words.

If you work for an institution, whatever your job, whatever your level, be yourself when you write. You will stand out as a real person among the robots, and your example might even persuade Thomas Bell to write his own stuff.

Alan Siegel

The Plain English Revolution

Alan Siegel, a leading consultant in the Plain English movement, is the Chairman and Chief Executive Officer of Siegel & Gale, a New York firm specializing in language simplification.

> *LOUIS XVI: C'est une grande révolte?*
>
> *DUC DE LA ROCHEFOUCAULD—LIANCOURT: Non, Sire, c'est une grande révolution. (When the news arrived at Versailles of the Fall of the Bastille, 1789)*

Three-and-a-half years ago I wrote an article . . . on the movement away from legalese toward forms and documents that can be understood by ordinary citizens.* At that time, the movement consisted of isolated revolts against legalistic gobbledygook and bureaucratese. A handful of business documents—mainly insurance policies and loan notes—had been simplified by forward-looking companies. In government, the scattered advocates of clear communication shared the lament of Alfred Kahn, then head of the Civil Aeronautics Board: "There's nothing I can do but cry. I feel so lonely and futile."

Now the plain language movement is becoming a revolution. . . . But the Bastille has not fallen, yet. Part of the legal community remains resistant to change. Some supposed practitioners of Plain English have confused simplicity with simplemindedness. And the greatest obstacle has yet to be fully overcome—a misunderstanding of what Plain English is all about.

That misunderstanding was dramatized when New Jersey's Plain English Bill, passed by the legislature, expired in February 1980, through a pocket veto by

---

*"To lift the curse of legalese—Simplify, Simplify," *Across the Board*, June 1977.

Gov. Brendan Byrne, who said: ". . . although regulating the protection of consumers is a worthy goal, legislating the style of a society's prose is another thing."*

But the movement has nothing to do with "legislating the style of a society's prose." We are not trying to turn English into one-syllable words, or to translate Saul Bellow into baby-talk. We are not even trying to do away with professional jargon, though that is a tempting target. Let lawyers talk to lawyers, or accountants to accountants, as they please. So long as they understand one another, that is just fine.

Well, what is the Plain English movement all about? *Its aim is to make functional documents function, whether they are put out by business or by government. If a consumer is expected to abide by a formal document—an insurance policy, a mortgage, a lease, a warranty, a tax form—then the consumer should be able to understand the document.* Carl Felsenfeld, vice president and counsel for Citicorp N.A., expresses the concept in legal terms: "There is growing dissatisfaction with contracts where consumers merely 'sign here' and can't, under any reputable system of contract law, be deemed to have agreed to all the printed verbiage." In social terms, I would say that people are learning a new right: the right to understand.

It's been argued that the fault lies with consumers, or with the educational system that does not train them to cope with legalese, or accounting, or insurance terminology. But that argument is neither reasonable nor realistic. In fact, documents meant for consumers have been made much more difficult than they really need to be. Redressing the balance—making sure that the documents consumers are expected to understand are made understandable—is a matter of simple fairness, and simple efficiency. In the long run it's in the interests of business as well as the consumer.

Five years of experience have proved that consumer contracts and forms of all kinds can be made much more understandable without sacrificing legal effectiveness. The very first plain language loan note, which I helped write for Citibank of New York, was introduced in 1975. Since then, Citicorp counsel Felsenfeld notes, "We've lost no money and there has been no litigation as a result of simplification."

That loan note remains a good example of just what plain language means. Among other things:

- A personal tone was used throughout—"I" and "me" rather than "the undersigned" or "Borrower," "you" and "yours" rather than "the Bank."
- Language was radically simplified. For example, "To repay my loan, I promise to pay you . . ." instead of "For value received, the undersigned (jointly and severally) hereby promise(s) to pay . . ."
- When unfamiliar terms couldn't be eliminated, we added explanatory phrases. For instance, ". . . if this loan is refinanced—that is, replaced by a new note—you [the bank] will refund the unearned finance charge, figured by the rule of 78—a commonly used formula for figuring rebates or installment loans."

---

*Rewritten and reintroduced in June, the bill was passed again. This time Gov. Byrne signed it (on October 10th), out of deference to its "unusually large" number of sponsors—65 of the Assembly's 80 members.

- We shortened the sentences wherever possible and even used contractions ("I'll pay this sum . . ."). To enhance clarity, we chose active instead of passive verb forms where possible.
- Improvements in design included the use of larger (12-point) type printed in green on light brown stock. Compared with the previously intimidating format, this visually appealing approach suggests immediately that the document is supposed to be read.

The Citibank note taught us a fundamental lesson about simplifying language. Consumer contracts have traditionally been adapted from mercantile contracts, which feature verbose protective clauses accumulated over the years in an effort to cover all possible contingencies. Many of these provisions have no practical value in the consumer marketplace. So the first task in simplifying a document is not rewriting it in Plain English, but identifying clauses taken from commercial contracts that can be eliminated from the consumer contract without jeopardizing its validity. The secret to doing this lies in analyzing actual business experience to see which provisions are really being used.

In the case of the Citibank note, the provisions describing the lender's protections in case of default proved to be the biggest challenge. The traditional note listed a string of contingencies more than 180 words long:

Before

In the event of default in the payment of this or any other Obligation or the performance or observance of any term or covenant contained herein or in any note or other contract or agreement evidencing or relating to any Obligation or any Collateral on the Borrower's part to be performed or observed; or the undesigned Borrower shall die; or any of the undersigned become insolvent or make an assignment for the benefit of creditors; or a petition shall be filed by or against any of the undersigned under any provision of the Bankruptcy Act; or any money, securities or property of the undersigned now or hereafter on deposit with or in the possession or under the control of the Bank shall be attached or become subject to distraint proceedings or any order of process of any court; or the Bank shall deem itself to be insecure, then and in any such event, the Bank shall have the right (at its option), without demand or notice of any kind, to declare all or any part of the Obligations to be immediately due and payable. . . .

But analysis of Citibank's business experience disclosed that the typical consumer loan transaction needs only one event of default—failure to pay. With one additional protection added, the following replaced all the fine print above:

After

Default   I'll be in default—

1.   If I don't pay an installment on time, or
2.   If any other creditor tries by legal process to take any money of mine in your possession.

A number of insurance companies, whose product, after all, is words on paper, continue to get good marks for simplification. The Massachusetts Savings Bank Life Insurance (SBLI) Whole Life Policy used to begin like this:

IN CONSIDERATION OF THE APPLICATION for this policy (copy attached hereto) which is the basis of and a part of this contract and of the payment of an annual premium as hereinafter specified for the basic policy as of the Date of Issue as specified herein and on the anniversary of such date in each year during the continuance of this contract until premiums have been paid for the number of years indicated in the POLICY SPECIFICATIONS or until the prior death of the Insured . . .

By contrast, the cover of their simplified policy begins with this message:

Please take the time to read your SBLI policy carefully. Your SBLI representative will be glad to answer any questions. You may return this policy within ten days after receiving it. Deliver it to any SBLI agency. We'll promptly refund all premiums paid for it.

The insurer's promise to pay the consumer used to be phrased in these forbidding terms:

The Bank Hereby Agrees upon Surrender of this Policy to Pay the Face Amount Specified Above, less any indebtedness on or secured hereunder . . . upon receipt of due proof of the Insured's death to the beneficiary named in the Application herefor or to such other beneficiary as may be entitled thereto under the provisions hereof, or if no such beneficiary survives the Insured, then to the Owner or to the estate of the Owner.

In Plain English, policyholders can understand what they are buying:

We will pay the face amount when we receive proof of the Insured's death. We will pay the named Beneficiary. If no Beneficiary survives the Insured, we'll pay the Owner of this policy, or the Owner's estate.

Any amount owed to us under this policy will be deducted. We'll refund any premiums paid beyond the month of death.

Perhaps the most ambitious simplification program so far has been that undertaken by the St. Paul Fire and Marine Insurance Company, of St. Paul, Minnesota. In 1975 they became one of the first large insurers to begin simplifying their policies. As a pilot project, the company produced an easy-to-read "personal liability catastrophe policy" that used personalized examples in colloquial English:

Before

a.   Automobile and Watercraft Liability:

1.   any Relative with respect to (i) an Automobile owned by the Named Insured or a Relative, or (ii) a Non-owned Automobile, provided his actual operation or (if he is not operating) the other actual use thereof is with the permission of the owner and is within the scope of such permission, or

2.   any person while using an Automobile or Watercraft, owned by, loaned or hired for use in behalf of the Named Insured and any person or organization legally responsible for the use thereof is within the scope of such permission.

After

We'll also cover any person or organization legally responsible for the use of a car, if it's used by you or with your permission. But again, the use has to be for the intended purpose.

You loan your station wagon to a teacher to drive a group of children to the zoo. She and the school are covered by this policy if she actually drives to the zoo, but not if she lets the children off at the zoo and drives to her parents' farm 30 miles away.

By 1978, St. Paul had reduced the number of policy forms in its commercial business package from 366 to 150. Plans now call for most of its commercial insurance policies to be rewritten and reprinted in a more readable format by 1982. The company has appointed a Manager of Forms Simplification, with his own staff and a detailed style manual, to oversee the effort.

The financial field is a particularly promising new area for Plain English efforts. A case in point: Sanford C. Bernstein & Co. is a New York broker-dealer advising clients on investments as well as trading for them. It is subject to regulation under the Federal Securities Exchange Act of 1934 and the Investment Advisers Act of 1940, as well as regulation by the New York Stock Exchange, the National Association of Securities Dealers, the Board of Governors of the Federal Reserve System, and state securities laws in various jurisdictions in which it does business or has clients. These various regulations extend to every aspect of the firm's business, from the content of advertising and soliciting materials to client reports, statements, and the consents that must be obtained before the firm can effect certain types of transaction on clients' behalf. Nonetheless, the firm was able to replace jargon with Plain English, while satisfying legal and technical requirements.

Before

It is the express intention of the undersigned to create an estate or account as joint tenants with rights of survivorship and not as tenants in common. In the event of the death of either of the undersigned, the entire interest in the joint account shall be vested in the survivor or survivors on the same terms and conditions as theretofore held, without in any manner releasing the decedent's estate from the liability provided for in the next preceding paragraph.

After

Other signers share your interest equally. If one of you dies, the account will continue and the other people who've signed the agreement will own the entire interest in it.

Where obscure terms could not be eliminated, the new forms explain them. The meaning of the term "clearing agents," for instance, is described in this way:

Clearing agents. We can use other broker/dealers as clearing agents to execute transactions, to hold securities in custody on your behalf and to perform other routine procedures in connection with your account. These clearing agents will act at our direction, and they won't have any part in investment decisions.

Especially complex ideas are not only explained but accompanied by illustrative examples:

Buying on margin. In managing your account, we may buy securities on margin. This means that we'll loan you part of the cost of the securities and charge you interest on the loan. By buying on margin, we can purchase more securities on your behalf than we could

Plain Language Scorecard

- Thirty-four states have laws or regulations setting standards for clear language in insurance policies. In 1978, New York became the first state to require that business contracts "primarily for personal, family, or household purposes" be plainly written in everyday language. Since then, New Jersey, Maine, Connecticut, and Hawaii have passed similar laws requiring consumer contracts to be understandable to the public, and a score of other states are considering such laws.

- Hundreds of corporations across the country are simplifying their documents—employee benefit manuals, brokerage account agreements, trusts, the notes to financial statements, customer correspondence, billing statements, and internal communications, ranging from corporate policy manuals to the humble memo.

- President Carter, following up a promise for "regulations in plain English for a change," issued two unprecedented Executive Orders to Federal agencies telling them to simplify paperwork and eliminate gobbledy-gook from regulations. Agencies that have started to do so include the Environmental Protection Agency and the Department of Health and Human Services (formerly HEW). The Civil Aeronautics Board, where Alfred Kahn started a push for Plain English, is rewriting the notices it requires to be posted at airline counters and printed on the backs of tickets. The Federal Acquisition Regulation Project (FARP), in the Office of Federal Procurement Policy (Department of Management and Budget), offers great potential; FARP is rewriting and recodifying the massive Federal regulations on procurement of good and services.

- Major law firms, such as Shearman & Sterling in New York, have launched programs to train their young lawyers in clear legal drafting.

- The Internal Revenue Service committed over one million dollars in an all-out effort to simplify the Federal income tax forms. Simplified prototypes were presented to the Congress in November.

- The 1978 Amendments to the Constitution of the State of Hawaii include a provision that, "Insofar as practicable, all governmental writing meant for the public . . . should be plainly worded, avoiding the use of technical terms."

- On April 15, 1980, Gov. Hugh Carey of New York issued an Executive Order directing all state agencies to write their forms and regulations in plain language.

- Last fall, New York City's Department of Consumer Affairs proposed simplified language for more than 40 of its regulations covering deceptive and unconscionable business practices.

- The prestigious Practising Law Institute held a program and published a course book on "Drafting Documents in Plain Language." Another such program is planned for 1981.

- A lobbying group for the movement, Plain Talk, Inc., has been established in Washington, D.C.

- The National Institute for Education, a Federal research agency, has funded the Document Design Project, which is being run by the American Institutes of Research in conjunction with Siegel & Gale and Carnegie-Mellon University. The project's purpose is to study and encourage the use of simplified public documents.

—A.S.

using only the cash and securities actually available in your account at the time. Because it increases the amount we can invest on your behalf, buying on margin can increase the return on your investment. However, it can also increase the amount of loss on your account.

> Example: Assume that the value of A Manufacturing Corp. stock in your portfolio has increased from $5,000 to $10,000. Because of changed market conditions, we decide to sell it and buy Z Industries common stock for your portfolio. If, for instance, the margin requirement is 50%, in theory we could purchase $20,000 worth of Z Industries stock for your account by using the $10,000 proceeds from the sale of your A Manufacturing stock and loaning you the additional $10,000.

In government, the promise—and difficulty—of the Plain English movement is illustrated by the case of the Food Stamp program. That program, begun in 1961, was substantially revised by the Food Stamp Act of 1977. As a result, the forms that had been used by the states were made obsolete. There was a bewildering variety of these forms to begin with. Wisconsin's, for instance, was 37 pages long. To help the states design simple, client-oriented forms reflecting the changes in the law, the Food and Nutrition Service (FNS) of the Department of Agriculture commissioned my firm to come up with new forms to offer to the states as models.

Our approach is illustrated by the new titles we gave our forms:

Old	**New**
Notice of Expiration	Continuing Your Food Stamps
Notice of Eligibility, Denial or Pending Status	Action Taken on Your Food Stamp Case
Tax Dependency Form	Student Tax Report

Significantly, the new forms ran into a special kind of opposition. First, *we* discovered that not all state governments agreed that the forms should *help* people to use the program, as the law intended. Food Stamp recipients are seen as one measure of a state's poverty, so some officials were reluctant to "encourage" their use by simplifying the paperwork. Some states had actually sought to discourage applicants by using forms that needlessly demanded embarrassing information, such as whether the applicant was a drug user or had a criminal record. Some state officials also felt that the language of the new forms, which included words such as "please," "thank you," and "sincerely," was too nice to people wanting money from the government.

Second, we found that state caseworkers themselves often appeared threatened by new, simplified forms. Some felt that their authority was being undermined. For example, people whose applications for food stamps are turned down or who are cut off are allowed by law to request a fair hearing. We embodied this feature of the law in the forms themselves: letters bearing bad news to Food Stamp clients included a perforated tear-off portion that could be sent back to appeal the decision. Some caseworkers objected heatedly: they felt the tear-offs would cast doubts on their competence and would generate "unnecessary" hearings, increasing their already heavy workload.

Plain Research Is Needed

The writing of Plain English documents is still a developing field. Researchers have discovered that the reader's comprehension has as much to do with what is inside his or her head as it has to do with what is on the printed page. Diagnostic research can help clarify what information must be explained at greater length.

A study conducted by law professor Jeffrey Davis, described in *Virginia Law Review* 63: No. 6, 1977, provides an excellent example. Davis prepared a simplified installment purchase contract for a refrigerator, showed it to shoppers from various backgrounds, and then asked them questions about its contents. The contract included this definition of default: "I will be in default if I fail to pay an installment on time or if I sell or fail to take proper care of the collateral." Yet most of the readers—even the well educated— failed to understand that they would be in default if they carelessly damaged the refrigerator. They persisted in the common but mistaken belief that they could not be in default if they made all their payments on time.

—A.S.

After more than two years, the objections seem to have been overcome. Our model forms are used now by about half the states, and one FNS official commented earlier this year: "I've never heard a critical comment about the forms themselves." In fact, as in other simplifications projects, researchers got a strong impression that staff objections had as much to do with the fact the forms were new and unfamiliar as with their style or content.

Earlier in this piece I mentioned that some Plain English efforts have confused simplicity with simplemindedness. A major source of this confusion is the quick-and-dirty "readability formula" that equates clarity with short words and sentences. Here is a rider to a health insurance policy obviously written to such a formula. It is both uncommunicative and insulting to the reader's intelligence.

THE DRUG SPECIAL ENDORSEMENT

The DRUG SPECIAL is a small Endorsement. It goes with your [name of company] Contract. It depends on that other Contract to say who's covered. It is added coverage, with big extra benefits.

The DRUG SPECIAL gives you special help in paying for DRUGS . . .

Some attempts at simplification produce a kind of black humor. One life insurance company offers a "simplified" policy requiring that, "The Insured must die while this policy is in force." In a grisly parody that might be called "Dick and Jane Become Underwriters," the policy continues:

The Insured's death may be caused by accidental Bodily Injury. If so the Beneficiary may be paid an additional amount. It will be equal to the Face Amount. Three things must all happen. (1) Death must have been caused by Accidental Bodily Injury. Not by sickness. Not by anything else. . . .

Another misuse of the movement is the way some lawyers are cynically using "simplified" language, notably in apartment leases, to misinform consumers and to mislead them by playing on their ignorance. The nation's first plain language law went into effect in New York State in November 1978. One year later, the State Consumer Protection Board found that most revised lease forms "force tenants to surrender nearly every right they have under law." Rosemary S. Pooler, the board's executive director, referring to the lease prepared by the New York City Bar Association, commented: "It is ironic that tenants were in some respects better off with the 'legalese' of the 1965 lease since the . . . 'plain language' lease seems to have taken almost every opportunity to resolve legal issues in favor of landlords, sometimes at the expense of existing statutory and decisional law."

To illustrate her point, Mrs. Pooler gave, among others, the following examples:

RENT PAYMENT PROVISION

Pre-plain language version (1965): "The tenant will pay the rent as herein provided."

Plain language version (1978): "Tenant will pay the rent without any deductions, even if permitted by law."

The new document flies in the face of New York State's Multiple Dwelling Law (Sec. 302-a); Real Property Law (Sec. 235-2); Real Property Actions and Proceedings (Sec. 756 and Article 7-A); and several recent court decisions, all of which give tenants the right to reduce or withhold rent under certain circumstances—for instance, when these are serious building code violations.

ASSIGNMENT AND SUBLETTING

Pre-plain language version (1965): "(Tenant) will not assign this lease or underlet the leased premises or any part thereof without the landlord's written consent, which landlord agrees not to withhold unreasonably."

Plain language version (1978): "Tenant shall not assign this lease or enter into a sublease unless it is allowed by a law of the State of New York."

Since 1975, New York's Real Property Law (Sec. 226-b) has given tenants the right to request their landlord's permission to sublet or assign their lease. If the landlord withholds permission unreasonably, he must let the tenant break the lease. The 1965 version is a better statement of existing law than the plain language version, which implies that subletting is allowed by state law only in very special situations.

INSTALLATION OF LOCKS, CHAINS, ETC.

Pre-plain language version (1965): None.

Plain language version (1978): "Tenant will not, without landlord's written approval: . . . 3. Put in any locks or chain guards or change any lock-cylinders on the doors of the Apartment."

But New York State's Multiple Dwelling Law (Sec. 51-C) gives tenants of multiple dwellings the right to install additional locks without the landlord's permission, provided that the landlord who requests a key is given one.

Though many lawyers have seen the light, some of them—and their corporate clients—misguidedly try to protect themselves by insisting on too-precise standards for compliance. Such traditionalists do not like the Plain English laws that have been passed in New York, Maine, and Hawaii. Those laws simply require that each document for a residential lease or for a loan, property or services of less than $50,000 for personal, family or household purposes be:

1. Written in a clear and coherent manner using words with common everyday meanings;
2. Appropriately divided and captioned by its various sections.

Ironically, some lawyers and executives who are quick to decry regulatory minutiae in other areas object to this general approach to plain language legislation. They yearn for the false security of the simplistic "readability formula" approach, which ignores less easily quantified elements such as grammar, logic, and organization. They fear that without an exact definition of "clear, coherent, everyday language," compliance will elude them. Lawyers such as Wilbur H. Friedman, chairman of the New York County Lawyers' Association Special Committee on Consumer Agreements, direfully predicted that New York's law would create "upheaval" among businesses and an "absolutely staggering" burden on the courts.

Safe English

The *New York Times* (Aug. 31, 1980) reported the first case under the Plain English law on contracts:

"The New York State Attorney General was puzzled by this sentence.

'The liability of the bank is expressly limited to the exercise of ordinary diligence and care to prevent the opening of the within-mentioned safe deposit box during the within-mentioned term, or any extension or renewal thereof, by any person other than the lessee or his duly authorized representative and failure to exercise such diligence or care shall not be inferable from any alleged loss, absence or disappearance of any of its contents, nor shall the bank be liable for permitting a colessee or an attorney in fact of the lessee to have access to and remove the contents of said safe deposit box after the lessee's death or disability and before the bank has written knowledge of such death or disability.'

"Saying, 'I defy anyone, lawyer or lay person, to understand or explain what that means,' Attorney General Robert Abrams sued the Lincoln Savings Bank in New York City . . . demanding that it simplify a customer agreement on safe-deposit boxes.

"The case is settled . . . The former 121-word sentence now says: 'Our liability with respect to property deposited in the box is limited to ordinary care by our employees in the performance of their duties in preventing the opening of the box during the term of the lease by anyone other than you, persons authorized by you or persons authorized by the law.'"

But these fears were unfounded: there has been no flood of lawsuits. The arguments of the technically minded still have their appeal, however, as demonstrated by Connecticut's plain language law. It lists two alternate sets of criteria for plain language, with nine and eleven different tests respectively, including these:

1. The average number of words per sentence is less than twenty-two; and
2. No sentence in the contract exceeds fifty words; and
3. The average number of words per paragraph is less than seventy-five; and
4. No paragraph in the contract exceeds one hundred fifty words; and
5. The average number of syllables per word is less than 1.55; and . . .

Cookbook detail like this only guarantees that the spirit of plain language legislation will be lost in attempts to follow it to the letter: the clarity of a sentence becomes less important than seeing that it has "less than 1.55" syllables per word. And what would happen to interstate commerce if one of Connecticut's neighbors were to require less than 20 words per sentence instead of 22, or 160 words per paragraph instead of 150?

Keeping plain language laws simple, like the law in New York State, is essential. The courts should be left free to judge particular cases according to a general standard, as they do now with the concept of the "reasonable man." Lawyers who hang back, in favor of the traditional approach to legal language, should remember the traditional result: as Harold Laski put it, in every revolution the lawyers lead the way to the guillotine.

Mark Mathewson

A Critic of Plain Language Misses the Mark

Mark Mathewson is Director of Publications for the Illinois State Bar Association.

Open-minded soul that I am, I'm ever on the lookout for articles and essays that purport to defend legal writing, eager to learn the good news about *hereby, wheretofore,* doublets, embedded clauses, and the like. But even though articles occasionally appear under titles like "In Defense of Legalese," they are rarely what they claim to be. Most are not so much defenses of legal writing as attacks on the aspirations and expectations of plain-language advocates—critiques of the critique of legal writing.

One such contribution to the genre, titled *A Defense of Legal Writing,* was written by Richard Hyland and appeared in the *University of Pennsylvania Law Review,* 134 U. Pa. L. Rev. 599 (1986). It's a lengthy and subtle essay and deserves a more ambitious response than the one I'll offer here, but I do feel impelled to answer this broadside against plain-language advocates.

First, let me offer a crude summary of the attack. According to Hyland, critics of legal writing say that lawyers should write like novelists, particularly novelists with a no-frills prose style (Hemingway being the obvious example), and should write so that laypeople can understand them. But, these critics charge, lawyers will not and cannot write in an engaging, "novelistic" prose style for two reasons. First, lawyers have an economic interest in confounding clients with convoluted prose. (It makes the law appear mysterious, remote, indecipherable, and hence makes lawyers appear to be worth $150 an hour.) Second, lawyers deal in abstractions rather than flesh and blood, grist and grit, and so cannot employ clear, compelling, down-to-earth language.

The problem, Hyland writes, is that the critics of legal writing miss the point. At one level, they are right: lawyers *do* deal chiefly in abstractions, not flesh and blood. Indeed, the law is all about applying abstractions—rules—to facts. But the critics are wrong in implying that there's something sinister in the fact that lawyers don't use plain language. It isn't sinister. Legal writing cannot be plain, down-to-earth, and compelling, because plain, down-to-earth, compelling language cannot convey abstract thought. "Because legal concepts are elements of legal theory," Hyland writes, "lawyers do not—and may not—use language as it is used in literature." What's more, that's why ordinary people cannot understand legal writing; they are not trained in abstract, conceptual thought.

The *real* problem with legal writing, Hyland suggests, is that most lawyers don't know how to think conceptually and thus cannot write well. Sloppy writing results from sloppy thinking. If lawyers could only think clearly, if they were masters of conceptual thought, they would write effectively (though not plainly, of course). So instead of teaching "tips," which "are either wholly arbitrary—such as the suggestion that sentences should average no more than a certain number of words in length—or meaningless platitudes, like the reminder that sentences should be no longer than necessary," those who would improve legal writing should strive to teach conceptual thought, say by requiring law students to read good books or to learn a classical language.

That's Mr. Hyland's argument in a nutshell. So where has he gone wrong? How has he unfairly assailed the assailers of legalese?

For one thing, he has misrepresented the collective wisdom of the critics of legal writing and thus has set up a pair of false premises. First, few, if any, critics of legalese insist that laypeople be able to understand all legal writing. Second, few, if any, still think that lawyers should write like novelists.

Lawyers should write like novelists? If that were truly what plain-language advocates thought, Hyland could rightly dismiss them, but that's not what they think, at least none I've read. I do think I know where Hyland got this curious notion. Several prominent critics of the current state of legal prose have cited Hemingway's work approvingly, and because Hemingway was a novelist, Hyland apparently inferred that critics think lawyers should write like novelists, or at least like novelists who write like Hemingway.

In fact, I think most critics of legalese would have lawyers write like good technical writers, or perhaps good newswriters. Indeed, Hemingway's prose style is a popular model for plain-language advocates precisely because it is spare, plain, simple, transparent, designed to transmit information with a minimum of noise. The fact that he made his name as a novelist is not the point. After all, some novelists write ornately, densely, in a style that is anything but lean and transparent. Plain language, not literary language, is the goal of plain-language advocates.

Beyond that, novelists must entertain as they inform and enlighten, and no plain-language advocate that I know of demands that lawyers write entertainingly—*clearly*, yes, but not entertainingly. No one could reasonably insist that contracts be gripping narratives, that legal memos fire the imagination. Lawyers should be able to write persuasively when the occasion demands, but it's usually enough that their prose not be a source of confusion, that it not draw attention to itself, that it not get in the way of content.

As for Hyland's other false premise, no one could seriously demand that Joe Ordinary Citizen, completely untutored in the law, be able to read and understand every piece of legal writing. Even with my populist leanings, I wouldn't hold legal writers to such a standard. Most plain-language enthusiasts would insist that consumer contracts, jury instructions, and other legal writing designed for public consumption be comprehensible to readers of ordinary intelligence. But when lawyers are writing for other lawyers, it stands to reason that legal writing, like any technical writing, will contain information (including some professional jargon) that lay readers can't understand. The point is that the substance, not the syntax, should present the challenge. Even other lawyers deserve relatively simple, clear, plain prose.

But the crux of Hyland's argument is that legal prose *cannot* be clear and plain, because plain language cannot convey conceptual thought. Hyland asserts (without offering supporting evidence) that plain or "concrete" language is fine for relating everyday occurrences but that a simple prose style is incapable of conveying abstract thought. Well, I just don't buy it. I'm no linguist, but I can write a legal brief in simple declarative sentences if I choose to. It won't read like a novel. You won't want to curl up beside the fireplace with it. It probably won't be fully comprehensible to people without legal training; that is to say, it may refer to legal principles unfamiliar to lay readers. But it will take "abstract, conceptual thought"—rules of law—and apply them to the facts. What's more, even lay readers will be able to navigate my sentence structure if I've done my job properly; the content might defeat them, but the form will not.

In short, I don't believe that the prose style that expresses concepts must be more difficult than the prose style that expresses facts. Moreover, I don't think the common failings of legal writing—*hereby, wheretofore,* doublets, lengthy embedded sentences, and the rest—have anything to do with conveying abstract thought. If you don't believe me, consider that the fact summaries in judicial opinions are often as obtuse as the rules of law, and there's nothing abstract about a fact summary. If you still don't believe me, dip into a set of plain-language jury instructions sometime, and compare the simplified version to the original. I'll bet you'll agree that the plain-language version conveys the same meaning as the original, but does it more clearly.

As for Hyland's suggestion that good legal thinking will lead to good legal writing—that all one must do to become a better legal writer is to become a better legal analyst—it just won't wash. While sloppy writing is usually a by-product of sloppy thinking, good legal analysts are not necessarily good writers; some are, some aren't. Open any law review and you'll find some excellent legal analysis expressed in impossibly convoluted prose.

Bad writing is not simply the result of sloppy thinking. There's nothing "sloppy," really, about an embedded sentence, in which intervening clause after intervening clause piles up between the main subject and verb. Such a sentence can be grammatically correct and is comprehensible after you break it down and chart it out. But why write such a sentence? Why force your reader to parse and chart your prose? The writer should be doing that work, not the reader. Think of it in economic terms: there will almost always be fewer writers of a document than readers, and the interests of efficiency will surely dictate that the writers, not the readers, translate the prose into simple form.

The point is, critics of legalese aren't asking for what Hyland says they are. They aren't asking that legal prose sing or entertain, and they aren't demanding that the public be able to understand every legal document. They're only asking that legal writing be clear, or at least more clear than most of it is now. How can anyone feel impelled to defend against so modest a proposal?

University of Wisconsin–Extension Equal Opportunities Program Office and Department of Agricultural Journalism

---

A Guide to Nonsexist Language

With a little thought, you can use accurate, lively, figurative language in your classrooms, publications, columns, newsletters, workshops, broadcasts and telecommunications—and still represent people fairly. Breaking away from sexist language and traditional patterns can refresh your style.

You can follow two abbreviated rules to check material for bias: Would you say the same thing about a person of the opposite sex? Would you like it said about you? That's the bottom line. Use your own good sense on whether a joke, comment or image is funny—or whether it unfairly exploits people and perpetuates stereotypes.

Most fairness rules improve communication. Use explicit, active words; give concrete examples, specifics and anecdotes to demonstrate facts; present your message in context of the "big picture"; draw your reader in; use parallel forms; present a balanced view; and avoid clichés and generalizations that limit communication. Balanced language rules also guide you to choose or create balanced visual images.

Good communication respects individual worth, dignity, integrity and capacity. It treats people equally despite their sex, race, age, disability, socioeconomic background or creed. And it expresses fairness and balance. To communicate effectively, use real people, describe their unique characteristics and offer specific information. Using stereotypes or composites stifles communication and neglects human potential.

Routinely using male nouns and pronouns to refer to all people excludes more than half the population. There have been many studies that show that when the generic "he" is used, people in fact think it refers to men, rather than men *and* women. Making nouns plural to ensure plural pronouns can help you avoid using the singular "generic" male pronoun.

Many professional titles and workplace terms exclude women and unfairly link men with their earning capacity, while others patronize and subordinate women. Such nongeneric titles reinforce assumptions restricting women and men to stereotypical roles, inaccurately identify people, and give false images of people and how they live and work.

As a communicator, teacher or illustrator, you can help correct and eliminate irrelevant and inaccurate concepts about what it means to be male or female, black or white, young or old, rich or poor, healthy or disabled, or to hold a particular belief.

Editing, publishing and style manuals recognize the need for creating accurate, quality messages without slighting anyone and are beginning to prescribe standards for writing and evaluating manuscripts that represent people without stereotyping them.

Colleges and universities may want to develop their own booklets on nonsexist communication. For example, Franklin and Marshall College (PA) and Michigan State University have developed their own materials on bias-free language.

PRONOUNS: EACH PERSON, TO THE BEST OF HER OR HIS ABILITY

1. *Address Your Reader*
   No. If he studies hard, a student can make the honor roll.

   Yes. If you study hard, you can make the honor roll.

2. *Eliminate the Pronoun*
   No. Each nurse determines the best way she can treat a patient.

   Yes. Each nurse determines the best way to treat a patient.

3. *Replace Pronouns with Articles*
   No. A careful secretary consults her dictionary often.

   Yes. A careful secretary consults a dictionary often.

4. *Use Plural Nouns and Pronouns*
   No. Teach the child to walk by himself.

   Yes. Teach children to walk by themselves.

   He is expanding his operation.

   They are expanding their operation.

   Everyone needs his own space.

   All people need their own space.

5. *Alternate Male and Female Pronouns Throughout Text*
   No. The baby tries to put everything he finds in his mouth.

   Yes. The baby tries to put everything she finds in her mouth.

6. *Use Both Pronouns and Vary Their Order*
   No. A worker with minor children should make sure his will is up to date.

   Yes. A worker with minor children should make sure her or his will is up to date.

7. *Use Specific, Genderless Nouns*
   No. The average man on the street speaks his mind on the issues.

   Yes. The average voter speaks out on political issues.

8. *Substitute Job Titles or Descriptions*
   No. He gave a test on Monday.

   Yes. The professor gave a test on Monday.

9. *Repeat the Noun or Use a Synonym*
   No. The professor who gets published frequently will have a better chance when he goes before the tenure board.

   Yes. The professor who gets published frequently will have a better chance when faculty tenure is granted.

(*Note:* We don't recommend using "their" to refer to a singular noun.)

(*Note:* Nations, battleships, gas tanks and other objects have no gender.)

TITLES: PEOPLE WORKING

Replace Language Stereotyping Men

No	Yes
Businessman/men	Business person/people, people in business, executive, merchant, industrialist, entrepreneur, manager
Cameraman	Camera operator, photographer
Chairman	Chairperson, chair, moderator, group leader, department head, presiding officer
Congressmen	Members of Congress, Representatives, congressmen and congresswomen
Craftsman	Craftsperson, artisan
Deliveryman/boy	Delivery driver/clerk, porter, deliverer, courier, messenger
Draftsman	Drafter
Fireman	Firefighter
Foreman	Supervisor
Guys	Men, people
Headmaster	Principal
Kingpin	Key person, leader
Lumberman	Wood chopper, tree/lumber cutter
Male nurse	Nurse

Manhole/cover	Sewer hole, utility access/cover
Man-hours	Labor, staff/work hours, time
Man-made	Manufactured, handbuilt, hand made, synthetic, simulated, machine-made
Night watchman	Night guard, night watch
Policeman	Police officer, detective
Pressman	Press operator
Repairman, handyman	Repairer (Better: plumber, electrician, carpenter, steam fitter's apprentice)
Salesman/men	Salespeople, salesperson(s), sales agent(s), sales associate(s), sales representative(s), sales force
Spokesman	Representative, spokesperson, advocate, proponent
Sportsman	Sports/outdoor enthusiast (Better: hunter, fisher, canoer)
Sportsmanship	Fair play
Statesman	Political leader, public servant, diplomat
Statesmanship	Diplomacy
Steward/stewardess	Flight attendant
Weatherman	Weather reporter, meteorologist
Workmen	Workers

Replace Titles Stereotyping Women

No	Yes
Authoress	Author
Aviatrix	Pilot, aviator
Career girl/woman	Professor, engineer, mathematician, administrative assistant
Coed	Student
Gal, Girl, Girl Friday	Woman, secretary, assistant, aide (Better: full name)

Housewife, lady of the house	Homemaker, consumer, customer, shopper, parent
Lady/female doctor, lawyer	Doctor, lawyer
Little lady, better half	Spouse, partner
Maid, cleaning lady	Houseworker, housekeeper, custodian
Poetess	Poet
Sculptress	Sculptor
Usherette	Usher
Waitress	Wait person, waiter
Working wife/mother	Worker

Replace Stereotypical Adjectives and Expressions

No	Yes
Act like a lady and think like a man	Act and think sensitively and clearly
Act like a gentleman/man	Be polite, brave, keep your chin up
Dear Sir	Dear Madam or Sir, Dear Personnel Officer/Director, Dear Executive/Manager (Better: name)
Fatherland	Homeland, native land
Founding fathers	Pioneers, colonists, patriots, forebears, founders
Gentleman's agreement	Informal agreement, your word, oral contract, handshake
Lady luck	Luck
Ladylike, girlish, sissy, effeminate	Tender, cooperative, polite, neat, fearful, weak, illogical, inactive (Both male and female characteristics)
Layman, layman's terms	Lay, common, ordinary, informal, nontechnical
Maiden name	Birth name
Maiden voyage	First/premiere voyage
Male chauvinist	Chauvinist

Male ego	Ego
Man-sized	Husky, sizable, big, large, voracious
Man-to-man defense/talk	Player-to-player, person-to-person, face-to-face, one-to-one
Manly, tomboy	Courageous, strong, vigorous, adventurous, spirited, direct, competitive, physical, mechanical, logical, rude, active, messy, self-confident (Both female and male characteristics)
Mother doing dishes, father reading the paper	Men and women doing dishes, women and men reading the paper (Note: Also applies to visual images)
Mother Nature, Father Time	Nature, time
Mothering, fathering	Parenting, child-rearing
Motherly	Protective, supportive, kind
Unwed mother	Mother
Woman did well for a woman/ as well as a man	Woman did well, woman performed competently
Woman's/man's work	Avoid (Too broad; use specifics)
Women's page	Lifestyle, living section

Gwyneth Olofsson

International Communication and Language

Gwyneth Olofsson owns Communico, an international training and consulting firm based in Sweden.

English has become the *lingua franca* of the business world, and people from Amsterdam to Zanzibar use it every day as a "tool of the trade." They also spend a lot of time and money trying to eliminate their language mistakes, not realizing that the fewer they make the more dangerous the errors are likely to become, because people aren't expecting them. Furthermore, just because someone has mastered the grammar and vocabulary of a language and pronounces it better than some native speakers does not mean he or she *uses* it in the same way.

Communication is not only about what the words mean in the dictionary, it's also about how you string them together. There is, after all, a certain difference between "Do that job tomorrow," "I'd appreciate it if you did that job tomorrow," and "Do that job tomorrow or I'll have your guts for garters," even if all three phrases are designed to achieve the same end. Those of us who are native English speakers have a responsibility not to use expressions that are likely to confuse non-native speakers (e.g., "Have you cottoned on, or do I have to spell that out to you?"). We also have to ensure that when "born" English speakers encounter a communication style that seems brusque, unfriendly, or arrogant in someone whose native language is not English, they will not assume that this is a true reflection of this person's personality or intention. It may well be that the speaker hasn't mastered the many nuances of words and body language that a native speaker interprets without even thinking about it. So in an unfamiliar culture, newcomers may find themselves wondering if the downcast eyes that accompany a statement are a sign of modesty or dishonesty.

Recently I ran an intercultural simulation, one part of which involved a group of ten British participants "learning" to be members of a fictitious culture. This made-up culture valued touch, and as part of the exercise participants were encouraged to touch each other at every opportunity, especially when communicating with each other. The simulation was a nightmare for everyone involved. The older male members of the group in particular found it extremely difficult to touch their colleagues at all. It wasn't surprising. Their physical contact with non-family members over the last forty years had been limited to a handshake with customers and a quick elbow in the ribs from strangers on a crowded subway, so to learn to communicate with colleagues in a tactile way that is the norm for millions of people in Latin America or Africa was just too much of a challenge.

Communication is about your facial expression, gestures, and actions. This was brought home to me a few years ago when a young family moved in to the next farm. My Swedish husband was born and brought up on a farm located on an island off the Swedish coast, and the new family had moved there from an outlying island and had two young children, as we did.

The four kids started to play together one day and were having a wonderful time when it started to rain. I went out and asked them, in Swedish, if they wanted to come into the house to play. The two new children looked at me and said nothing, then suddenly turned tail and ran as fast as they could in the direction of their home.

I couldn't make any sense of this, but when I went in and told my husband what had happened he showed no surprise. Without looking up from his newspaper he said, "They've gone home to ask their mother if they can come in." I was amazed. How did he know? He'd never even met them. But sure enough, in a couple of minutes there was a knock at the door and there they stood. Thinking about it, there were two things that surprised me. The first was that the two children hadn't said a word when I'd asked them a question, and the second was that my husband had understood the whole situation without even having seen what had happened.

The explanation was, of course, that he and the two children shared the same cultural roots. He had grown up, as they had, in a community where everyone knew everyone else; a homogenous community where people understood what their neighbors would do before they did it. If you grow up in a society like this you don't need to spell things out. Communication takes place without words because the situation is familiar and is governed by a set of unwritten rules that everyone understands.

If, on the other hand, you look at a country with an entirely different profile, like the U.S., for example, a relatively new country where enormous numbers of people immigrated from other cultures, communication patterns developed quite differently. With high levels of mobility as thousands of people headed west across the continent, individuals were forced to get to know one another quickly and establish their own rules as they went along. It's clear that in such a situation good communication skills were vital, because you couldn't expect the people you met to share your background or assumptions, so your communications with your peers had to be clear, unambiguous, and explicit. This explains why today many people in the U.S. have a very different communication style than the natives of the small island off the west coast of Sweden—and many other places where people have known each other all their lives.

MORAL

The way we communicate, and what we do or do not say, may be entirely mystifying to people from other cultures, even though we believe we have made ourselves perfectly clear.

WHAT TO SAY AND HOW TO SAY IT

Even those of us who pride ourselves on being direct don't always say what we mean. If English speakers were to phone a colleague's secretary and ask "Is David in?" we would be surprised if she answered, "Yes" and put the phone down. We assume she would answer the question we *didn't* ask, "May I speak to David?"

Different cultures have different attitudes to directness. I remember a time several years ago when I was in England and having problems with my car. I drove to a garage, parked the car in front, and went inside to report the problem. There was a long line, and as I waited a truck driver came in and addressed the woman waiting behind me in a broad Newcastle accent. "Thanks for moving your car, pet. The other wife just walked away and blocked me in."

In fact, "the other wife" was me. I hadn't seen the truck arrive behind me, and by leaving my car where I did had managed to block his exit. We're talking here about a Newcastle-upon-Tyne truck driver, with tattoos, beer belly, and shaven head, wearing a T-shirt with a picture of a man, not unlike himself, strangling a big snake. But because of the way he had been brought up, this poor guy could not bring himself to speak to me directly and tell me I was blocking his exit, but had to speak to the woman behind me to give him a pretext to tell the world of my stupidity. I mean, it wasn't as if he looked like he was afraid of conflict or had spent his formative years at Eton with Prince William learning how to conduct himself correctly in court circles. But somewhere in his cultural soft-wiring he'd learned that in certain situations, and addressing a certain type of person (e.g., a middle-aged woman, as opposed to a young man), he should use an indirect communication style.

Your own personal communication style will be affected by many factors. Obviously, the culture you come from plays a large part, as does your own native language. Even climate may have a role to play in how we express ourselves. One interesting (although not entirely serious) observation on this theme was made by the English writer Ford Madox Ford who wrote, "You cannot be dumb [silent] when you live with a person, unless you are an inhabitant of the North of England or the State of Maine." As someone with roots in the North of England I don't know if I can agree wholeheartedly with his conclusion that the colder the climate, the more taciturn the people. However, he's not alone in his conclusion: in both Italy and France the people of the south regard those in the cooler north as reserved and antisocial.

Other considerations affect both what we say and how we say it. For example, the CEO of a large corporation might mutter to a few friends over a drink at the club. "Well, guys, we really made a balls up of the last year's sales, didn't we?" However, he probably wouldn't make the same comment at the annual general meeting (although it might wake up the shareholders). He is more likely to say, "Due to circumstances beyond our control, our sales performance in the last year was disappointing." No matter where we come from, we all know that how we speak depends on the audience we are speaking to.

And speaking of audiences, if you gave a presentation and asked for questions, would you be pleased or worried if there weren't any? Would you take the silence to mean that you had made your point so clearly that everyone understood everything or as a warning sign that trouble was brewing? Would you assume that the audience had found your talk so boring they'd all dropped off to sleep? Or would [you] expect questions to emerge later during the informality of the coffee break? It depends, among other things, on whether the audience was comfortable with silence and whether they came from a culture where asking questions in public is about losing face. Or perhaps they all came from the State of Maine or the North of England. . . .

Letters 1–2

Many of us ask questions if we don't understand something. However, in some cultures this is not a step to be taken lightly.

Asking Questions   Letter 1

From the U.S. about **MEXICO**

> The company is introducing a complicated new process in one of its workshops in Mexico. We know it's difficult, and we have a training and support package we can offer if needed. I strongly suspect that they're having problems down there, but we haven't received a single request for advice or support. Why not?

As you know the process is a complicated one, why don't you provide the support package automatically instead of waiting for a request? Admitting you need help can be a difficult thing to do no matter what culture you come from. Questions of prestige and fear of losing face can mean that people are unwilling to expose themselves to possible criticism. Also, if in your culture you have learned that good employees know all the answers, you may well hesitate to tell your bosses that you don't! This problem can be compounded if headquarters is located abroad, especially in a country that is bigger or richer than your own; this can make national sensitivities even worse.

He Asked What?   Letter 2

From CANADA about CHINA

> I enjoyed my trip to China, but I was very surprised by some questions business acquaintances I hardly knew asked me. Two questions they asked me during a meal were how much my watch cost and how old my wife was. (I'm just glad she wasn't there to hear it!)

It's odd what different cultures regard as acceptable questions. In France and many other European countries, they regard the North American exchange of personal information (Do you have any children? What do you do in your free time?) as rather intrusive, though the French will quite happily discuss matters of religion, which are regarded as taboo by, for example, many people from the Middle East. Canadians and North Americans, of course, simply see such inquiries as a friendly way of building a relationship, and they expect to answer the same questions themselves. At the same time, North Americans usually find questions about money and age too personal to ask business acquaintances. However, for many Chinese, whether in China or elsewhere in Asia, and for people in the Middle East these questions form part of ordinary conversation and are just one way of getting to know you better. Indeed, such questions are seen as a natural way to show you're interested in your new acquaintance. People in countries as far apart as China, India, and Mexico might even think it rather unfriendly if people they met did *not* show any interest in their personal concerns.

Letters 3–4

The way people communicate with each other at work is affected by the structure of the organization they work for and by the expectations of fellow employees.

Communication Stop   Letter 3

From SWEDEN about GERMANY

> I work for a multinational company and am involved in a project that requires a lot of technical input. I contacted a German colleague I'd met at a conference for a little help. When I spoke to him on the phone he was quite pleased to help us, but the next day my manager got an e-mail from the German guy's boss saying that my colleague was too busy to help me.

I think the problem here is that you didn't use the "correct" channels of communication, according to the German company, anyway. In Germany, and indeed in the majority of European and American companies, the manager wants to be informed of what his or her department members are doing, as it's an important part of his or her role to co-ordinate their efforts. What you should have done first was to contact the manager and ask if you could approach your German colleague for some assistance. Not doing so might be interpreted by his or her manager as very rude, and even a bit underhanded.

I understand that you come from a country, Sweden, where it's the norm to delegate an enormous amount of power to non-managerial staff and give them a high degree of independence, especially if they are technical specialists. However, this is certainly not the case in most countries, which tend to be much more hierarchical. Indeed, most managers from the U.K. to the United Arab Emirates, by way of the U.S., would want to be informed of such an approach to a subordinate.

I suggest your manager make a formal request to his German counterpart asking if you may contact the specialist. You should include a description of the kind of questions to be tackled, and a description of the benefits your project will make to the company. And be *very* polite. After all, you are asking the manager for a favor—to be allowed to use the valuable time of one of the department's members.

Communication Breakdown   Letter 4

From NEW ZEALAND about **FRANCE**

> We're having real problems with our French subsidiary. We want a couple of departments in the French head office to collaborate in preparing a program for some visiting customers who want to see production operations. Naturally, this will involve consultation with the factory staff to see what is practicable. However, arrangements seem to be at a standstill. We can't understand what the problem can be.

What you have asked your French managers to do is to communicate in ways they may not be used to. First, you are asking your managers to operate across departmental boundaries; hence, it's not clear who is responsible for what. Second, they are being asked to communicate across hierarchical boundaries, because the managers will not be able to arrange a trip to see production facilities without some collaboration and discussion with the factory personnel.

The French, as well as Latin American and Southern European business cultures, tend to have very clear hierarchies where each person's responsibilities are spelled out. The same applies to cultures with a Confucian heritage like Japan, China, and South Korea, where respect is awarded to age, education, and rank in the company. The French also have rather compartmentalized communication patterns, and information is not freely shared as a matter of course, but tends to remain the property of those higher up the ladder. "Knowledge is power" is the name of the game, and one likely to hinder interdepartmental collaboration. Your culture (which is more tolerant of uncertainty) is more like that of the Scandinavians, the British, and Irish in your belief in a free flow of information, but many other cultures find this difficult to deal with. You are more likely to get a positive result if you give *one* of the managers responsibility for arranging the visit, and instruct him or her to involve the factory in the plans.

Letters 5–6

You may like to have things out in the open, or prefer to leave them unsaid.

A Major Error

Letter 5

From MEXICO about **GERMANY**

> We have a new German manager who is making himself extremely unpopular here. He has introduced a new quality control system that is complicated and takes time to learn. Inevitably mistakes are made. However, when he finds an error, he seems to delight in pointing this out to the person involved in front of everyone. Several people are already thinking of handing in their notices.

Your new manager is certainly not trying to offend people intentionally. In his own direct way, a way shared by U.S. Americans who also believe that it is better to "tell it like it is," he might even be trying to help by identifying the problem. He obviously does not understand that Mexicans regard this very direct approach as fault-finding, confrontational, and aggressive. Mexicans, like most Central and South Americans and East Asians, are skilled at avoiding confrontations and situations that involve a loss of face, but this is still something your new manager has to learn. Until he does, try not to take his criticism personally.

No No

Letter 6

From the U.S. about **INDONESIA**

> I found it very difficult working in Indonesia because I couldn't get a straight answer to a straight question, and this often led to misunderstandings. As far as I could see, they often said yes when they meant no. Why?

Most Indonesians find it hard to give a straightforward *no* to a request. If you ask for something to be done that is difficult or even impossible your Indonesian colleague, instead of saying *no* or *sorry,* may say instead that he will try. Also, a promise to do something that keeps getting postponed can be another indirect way of refusing a request. There is no intention to deceive, but simply a wish to avoid situations leading to open disagreement or disappointment that would cause you to lose face. And bear in mind that people from cultures with this indirect communication style are perfectly well understood by each other. They are simply tuned in to "reading between the lines" in a way you are not.

This communication pattern is not confined to Indonesia. In countries as far away from Indonesia as Pakistan, India, and Japan the word *no* is regarded as impolite and is rarely heard in a business context. In Mexico and South America, too, politeness and diplomacy are valued as useful ways of avoiding conflict.

But bear in mind that speakers of English can be indirect sometimes too. If invited to a party they don't want to attend, the vast majority of English speakers will say they have a cold rather than admit that they're planning to spend the evening in front of the TV. This is just another variation on the "white lie" theme, and as such is remarkably similar to the indirect response you mentioned in your question.

Letters 7–9

It's easy to create the wrong impression if you choose an inappropriate communication style—and what is inappropriate is in the ear of the listener.

Aggressive   Letter 7

From SWEDEN about **FRANCE**

I find it extremely difficult to discuss business with the French. It is impossible to talk about things with them calmly and sensibly. They are very critical of any ideas that they have not originated themselves, but take any criticism of their own plans personally and get angry.

If you come from a country like Sweden, where open conflict is frowned on, you may find the French debating style very aggressive. For the French, a love of words is combined with a liking for verbal combat, and they are used to organizing their case logically and presenting their arguments with force and conviction, not necessarily because they believe in them, but because they consider that it is through argument and counter-argument that you will eventually arrive at the truth or the best solution to a problem. And if you don't, the debate has been an enjoyable chance to flex your intellectual muscles anyway!

However, the bad feelings that may result from such spectacular clashes will usually quickly be forgotten, which is also hard for people from more low-key cultures to understand. Of course, the French are not alone in their love of discussion. Greeks, Israelis, Argentineans, and Poles all enjoy a good debate too, and North Americans and Australians are no shrinking violets when it comes to putting their points forward. For the French and Australians in particular, debate is a way of taking the measure of a new acquaintance.

In your particular case, at a meeting with the French you should emphasize the most important points of your argument and repeat them patiently. Don't get tied up with details or try to score debating points. Instead, focus on the most important points you want to achieve and keep the meeting focused on them. Be very well prepared, and if in a corner, be ready to use a weapon to which the French have no defense—silence.

Patronizing Pommie   Letter 8

From AUSTRALIA about the **U.K.**

We have a new boss from the U.K. with one of the most affected upper-class English accents I have ever heard. Every time he opens his mouth I can just see him at the Queen's garden party in a tuxedo and top hat. I just can't take him seriously, and I wonder how he expects to communicate with the other guys in the company.

For historical reasons an upper-class English accent in Australia is associated with money and power, and the use and misuse of both. Australia is a proud new multiethnic country and many Aussies find reminders of their colonial past, that includes the accent of the former ruling class, embarrassing and even painful.

But it's true that this particular type of British accent (RP, which is short for Received Pronunciation) is linked to a certain powerful social group in a way that different U.S. regional accents are not. It also continues to be an accent that dominates the boardrooms of many companies. Even in England itself people with strong regional accents may associate RP with snobbery and privilege, which is why younger members of the upper classes try to tone it down a bit. But give your boss a chance. It would be unfair to judge how well he's likely to do his job on the basis of his vowel sounds!

Just Making Conversation   Letter 9

From BRITAIN about JAPAN

> I met several Japanese businesspeople who visited Britain recently, and I tried to be pleasant and help them relax. I told a few jokes that seemed to go down well, but I later heard that they hadn't been appreciated. Yet at the time everyone laughed!

Your mistake was to treat your visitors as if they were from your own country. I'm sure this was done from the best of motives, but it is a mistake to assume that every culture shares the same kind of humor. Just because your Japanese visitors laughed didn't necessarily mean that they found your joke funny—people from different cultures tend to laugh at different things. Research about what people of different nationalities find funny concluded that the Irish, British, Australians, and New Zealanders thought that jokes involving word play were funniest. Canadians and U.S. Americans preferred jokes where there was a sense of superiority—either because a person looked stupid or was made to look stupid by another person. Many European countries, like France, Denmark, and Belgium, liked rather surreal jokes and jokes about serious topics like death and illness.

You don't say whether you told your jokes during a business meeting or after work in the pub. However, in many countries humor is confined to non-work situations, and joking in an important meeting, for example, is seen as a sign that you are not treating the subject (or the individual) with respect. This would certainly apply to Germany and Finland as well as Japan, where humor when business matters were being discussed would be regarded as inappropriate. And of course it might well be that your visitors didn't understand your English but did not want to lose face by showing it, because even if you are fluent in a foreign language, jokes are always the last things you understand.

Finally, you need to know that people from East Asian countries as widely apart as Japan, South Korea, and Thailand may laugh if embarrassed or nervous as well as when they're happy.

Letters 10–11

Rudeness may be what the listener hears, rather than what the speaker intends.

Rude, or Just Informal?   Letter 10

*From DENMARK about **DENMARK***

> In Denmark we tend to communicate in an informal way and consequently leave out titles like "Mr." or "Dr." We also like to communicate directly rather than "beating about the bush." But I know this isn't the case in other cultures and wondered just how rude we are perceived to be.

It depends where you're going and who you're meeting. In Northern Europe, Australia, and the U.S., communication styles are quite relaxed and informal, and people take pride in talking to both manual workers and top managers in more or less the same way. They also tend to be rather pragmatic in their understanding of what language is for—generally it's to get things done. So they say clearly what they mean so the message comes over loud and clear. This group won't regard your informal and direct style as at all rude.

In other cultures, however, what you say may be secondary to how you say it, and the British, along with the Arabs and people from many Asian cultures, put a lot of weight on how the message is delivered. Words are regarded as an important way of establishing and building relationships, not simply a tool for getting things done. If your "tone" is wrong and you are perceived as rude, people from these cultures can take offense, and, for example, not using the right titles for an individual can be regarded as a sign of disrespect.

As a general rule, it's better to err on the side of formality when communicating with people of other nationalities, even if you've worked together for quite some time. Words define your relationship with an individual, and if you want to ensure that the relationship is one of mutual respect, your communication style must reflect that.

Let Me Finish!   Letter 11

*From SOUTH AFRICA about **ITALY***

> I travel often in Italy and in other Mediterranean countries, and I find it very irritating to be constantly interrupted. What can I do to stop this?

The short answer is—not a lot. What you as a South African would call a rude interruption, nationals from Southern European countries may regard as perfectly acceptable. They may instead see an interruption as an expression of interest and involvement in what the speaker is saying and in his or her ideas. In short, in countries such as Italy, if you wait for a pause in the conversation in order to present your own point of view, you'll never open your mouth! You'll find that the nationals of these countries interrupt each other too, so don't take it personally. This is because silence does not have an important role in the communication patterns of most Latin countries. Indeed, the tempo of conversation may simply be too fast to allow for a pause between speakers.

If you are interrupted in the middle of a presentation, don't show annoyance but say that you'll deal with the points raised at the end of your talk; don't let yourself be thrown off track. If the interruption occurs in the middle of an informal meeting, accept that this is regarded as a legitimate way of raising relevant points and practice your debating skills.

Letters 12–13

When to remain silent is a decision we make almost unconsciously when operating in our own culture. But in another culture this decision may be interpreted in a way we don't expect.

Stuck Dumb   Letter 12

*From POLAND about **SOUTH KOREA***

During my recent trips to South Korea I have built up a good relationship with an engineer of about my own age who works in my own area of expertise. He speaks good English, and we have had a number of informal meetings where we've made tentative decisions about some technical developments. However, when his boss is present he hardly ever opens his mouth, even though this manager has to use an interpreter and does not have a technical background.

It is quite usual in South Korea, and neighboring Japan, that a younger employee will be quiet in front of older managers as a sign of respect. It would be regarded as immodest to display his superior knowledge of English or the technical matter at hand in front of his boss. This manager will not be directly involved in the technical side of things, but will want to know a little about you personally and see you "in action" so he can come to some conclusion about whether you and the company you represent are likely to make good working partners.

Small Talk Versus Silence   Letter 13

*From FINLAND about the **U.K.***

We hear a lot about the importance of "small talk" when doing business with the British. But if you don't have anything particular to say, why should you keep on talking? Surely it makes more sense to keep your mouth shut.

In cultures where conversation is an art form, as in France and Italy, a firmly shut mouth may be equated with a firmly shut mind. You may be regarded as rude if you are not prepared to make an effort to get to know your counterparts on a personal rather than simply on a business level. However, you are not the only one to find this need for "small talk" difficult. In addition to Finns, Swedes and Norwegians also have a problem with it. In your cultures silence is accepted as a part of conversation in a way it is not in many others (although the Japanese are more like you in their acceptance of silence). To many Europeans and Americans, general social conversation is a prelude to more serious discussions and is regarded as a way of getting to know your colleague before you get down to brass tacks.

If you are stuck about what to talk about, non-controversial topics are best to start with. In 1758, Samuel Johnson wrote, "It is commonly observed that when two Englishmen meet, their first talk is of the weather." Some things just don't change, and not only the English find this subject a useful "icebreaker" with strangers. Other useful subjects are the journey to the meeting, sports, and questions about your visitor's hometown or area, but the real secret is to relax and allow yourself to show you are interested in your partner and what he or she has to say. Feel free to ask questions, as long as they don't get *too* personal. People usually enjoy talking about themselves. Neither should you be afraid to talk about yourself and your own interests. Conversation is like dancing the tango (surprisingly, perhaps, this is very popular in Finland) in that it needs practice. It also requires sensitivity to what your "partner" is feeling and anticipation of the next move.

Letter 14

Giving presentations at home can be bad enough, but speaking to people of other cultures can be even harder.

Political Correctness   Letter 14

From AUSTRALIA about the U.S.

> I've just returned from the U.S. where I gave a number of lectures on a technical matter. During one of my talks I used the expression "to call a spade a spade." One of my listeners raised his hand and said that he found the expression offensive—he had taken it as a racist comment! Is this political correctness run wild?

To put it bluntly, yes it is. The expression "to call a spade a spade" simply means to describe something truthfully and honestly. However, in the U.S. *spade* is a derogatory term for a black person; it comes from the expression "as black as the ace of spades." Your listener obviously confused the two.

When you speak in public on any subject, it is simple good manners to ensure that what you say does not unintentionally offend any particular group, hurt their feelings, or show them disrespect, especially if this group has been given a hard time by society at large over the years; women, black people, homosexuals, and handicapped people are some groups that spring to mind. It's obvious that people belonging to these groups are just as deserving of consideration and courtesy as the traditional top dogs—white heterosexual able-bodied males.

However, this respect for the dignity of others should not stop you from getting your own message across. The term *political correctness* has unfortunately come to be associated with a "holier than thou" attitude, and some North Americans use it to beat less politically correct fellow citizens over the head. Luckily, it is primarily a North American phenomenon, but one that the rest of us should be aware of when we have contact with Canadians or U.S. Americans.

IN A NUTSHELL: *What to say and how to say it*

Global Business Standards

Good small talk topics:

Weather is always safe, although boring, especially in countries that don't have a lot!

Sports are usually safe too, unless the city or country has suffered a spectacular defeat in the national sport recently.

The art and cultural history of the country is usually safe (but watch out for any historical discussion that can lead to a political debate.)

Global Warnings

No swearing in your own or any other language.

Keep humor to a minimum until you are sure your partners/guests laugh at the same things as you.

Don't comment negatively about another culture—especially on religion, politics, or sexual matters. (Occasionally requests for information on the first two may be interpreted favorably, but be careful.)

- **Argentina:** People like to express opinions and love to debate. Voices may be louder than elsewhere in South America. (See Letters 4, 5, 6, 7, 11, and 13.)
- **Australia:** People enjoy talking and debating. There is an informal style of communication that is not based on hierarchy. (See Letters 7, 8, 9, 10, and 14.)
- **Austria:** Communication within companies is inhibited by departmental and hierarchical boundaries. There is a direct yet formal communication style. May be an adversarial approach to debate among peers. (See Letter 3.)
- **Belgium:** Communication within companies is inhibited by departmental and hierarchical boundaries. French speakers' adversarial style in discussions may appear very negative or aggressive. Flemish speakers are more low-key. (See Letters 3 and 9.)
- **Brazil:** Relatively personal questions (in more reserved cultures) about income, age, and so on are acceptable. Emotions are expressed openly. (See Letters 4, 5, 6, 11, and 13.)
- **Canada:** There are different communication styles depending on whether you are in English- or French-speaking Canada. (See U.K. and France.) (See Letters 2, 7, 9, and 14.)
- **China:** Personal questions about income, age, and so on are acceptable. Ordinary conversations can be loud and may sound unintentionally rude or angry. (See Letters 2, 4, and 5.)
- **Denmark:** Informal communication style is the norm. (See Letters 4, 9, and 10.)

- **Finland:** Small talk is not usual. Silence is accepted. The verbal style is very quiet and restrained. (See Letters 4, 9, and 10.)
- **France:** Communication within companies is inhibited by departmental and hierarchical boundaries. Adversarial style in discussions may appear to outsiders to be very negative or aggressive. (See Letters 2, 3, 4, 7, and 9.)
- **Germany:** Communication within companies may be inhibited by departmental and hierarchical boundaries. There is a direct yet formal communication style. Adversarial style in discussions may appear very negative or aggressive. Negative messages are given directly; tact is not a priority. (See Letters 3, 5, and 9.)
- **Hong Kong:** Personal questions about income, age, and so on are acceptable. Ordinary conversations can be loud, and may sound unintentionally rude or angry. (See Letters 2, 4, and 5.)
- **India:** Personal questions about income, age, and so on are acceptable. In these "high context" cultures a straight *no* is regarded as rude. Explanations and communication styles may be indirect. (See Letters 2 and 6.)
- **Indonesia:** Quiet, calm polite conversation style is the norm. This is also appreciated in others. (See Letters 2, 5, and 6.)
- **Italy:** Overlapping conversational style is the norm. Interruptions are not regarded negatively. Emotions are expressed openly. (See Letters 3, 4, 11, and 13.)
- **Japan:** Deference to senior and older colleagues (when present) may inhibit Japanese from communicating. Self-consciousness about their English may be another inhibiting factor. There is an oblique and indirect communication style and modesty is important. A straight *no* is regarded as rude. (See Letters 4, 5, 6, 9, 12, and 13.)
- **Mexico:** There is an indirect communication style. Direct confrontation is avoided. It's important to "save face." (See Letters 1, 2, 4, 5, 6, 11, and 13.)
- **Netherlands:** People have a rather blunt and straightforward speaking style and are quite informal.
- **Norway:** There is an informal and direct communication style. Silence is an accepted part of communication. (See Letters 4, 10, and 13.)
- **Poland:** People enjoy debate and discussions. Politeness and formality are quite important. (See Letter 7.)
- **Russia:** The first response to any question is usually *no*, but persistence is often rewarded. It is important for Russians not to lose face in discussions. They may show disagreement or anger quite openly.
- **Saudi Arabia:** Ordinary conversations can be loud and may sound unintentionally rude or angry to outsiders. Emotions are expressed openly. (See Letters 2 and 10.)
- **South Africa:** Lots of sports analogies (from rugby, cricket, etc.) used. Different ethnic groups use different communication styles. (See Letter 11.)
- **South Korea:** When getting to know you, people may ask personal questions, but they are not intending to be rude. (See Letters 5, 9, and 12.)
- **Spain:** A straight *no* is regarded as rude. Explanations and communication styles may be indirect. (See Letters 3, 4, 11, and 13.)
- **Sweden:** Communication across hierarchical boundaries is common. Written communication in English may sound brusque, even rude, because of first language

interference. Silence is an accepted part of communication. (See Letters 3, 4, 7, 10, and 13.)

- **Switzerland:** Humor has little place in business. German speakers will not make small talk, but French and Italian speakers will.
- **Taiwan:** See China.
- **Thailand:** There is a very tactful communication style, and heated debates are not popular. (See Letters 5 and 9.)
- **Turkey:** People may be reluctant to say *no*. It is more important to be polite than to be accurate or clear. (See Letters 3 and 4.)
- **UK:** Small talk is an important social skill. Humor is used widely to defuse tension and to create positive social contacts. People are judged according to how they use language. An oblique style, including understatement or irony, may be used. (See Letters 3, 4, 8, 9, 10, and 13.)
- **US:** Political correctness (and good manners) means that you should be very careful how you express yourself. This applies to all references to gender, age, race, religion, or sexual orientation. Communication is generally direct and explicit. (See Letters 1, 2, 3, 5, 6, 7, 9, 10, and 14.)
- **Venezuela:** People like to debate but rarely admit they are wrong or do not know something. (See Letters 4, 5, 6, 11, and 13.)

A GLOBAL LANGUAGE?

There are over 400 million speakers of English as a first language in the world, with about the same number of people using it as a second language. However, over 700 million people speak one of the many dialects of Chinese. The world also contains almost 300 million Spanish speakers, and about 180 million speakers of Hindi and Arabic, respectively. (And undoubtedly included in these figures are a good few thousand gifted people who speak *all* these languages.)

However, English speakers can take comfort from statistics that say 75 percent of the world's mail, telexes, and cables are in English, that it is the medium for 80 percent of the information stored on the world's computers, and that it is the language of over half the world's technical and scientific periodicals. In fact, it can be said with justice that English is on the way to becoming the first truly global language.

The need for a language in which people from Siberia to Santiago can communicate directly with each other has long been acknowledged, and the establishment of artificial languages such as Esperanto has tried unsuccessfully to fulfill this need. Now, due to a series of accidents of history, it looks as if English is likely to step into the breach. But if a language is "global," it is no longer the exclusive property of its native speakers. Indeed, it is claimed that there is a European variety of English, sometimes called *Euro-English*, which is already evolving, and some people believe that it will eventually become the European language of business. It even has an official name: English as a lingua franca in Europe (ELFE). This version of English regards as acceptable some "mistakes" that most teachers of the language spend their careers trying to eradicate. For example, "He goes to work every day at 8:00 o'clock" would be accepted as correct, as the meaning of the sentence remains clear.

Some academics believe that this modified version of English, which would turn increasingly to continental Europe rather than to the U.S. or the U.K. for its standards of correctness and appropriateness, is the future. Whether that is true remains to be seen, but whatever happens, the message is clear: English is a useful tool for international communication, but it is no longer the exclusive property of people who speak it as a first language.

And what about this privileged group: Those of us who by an accident of birth have learned to speak the global language of business and industry without effort? Can we just rest on our laurels in the knowledge that our customers, suppliers, and even our employers will communicate with us in *our* native language, rather than in *theirs*?

That might be a mistake. I know of at least one international company of management consultants that will not employ anyone who does not speak at least one foreign language fluently. The reason given is that each language gives you a new perspective on the world, and if you are going to work with people not from your own culture you need to be able to shift away from your "native" perceptions from time to time, because language affects how you think.

Letters 15–16

It's inevitable that when speaking English as a foreign language you will make mistakes, and these mistakes can take many forms.

Rude Writers   Letter 15

From SPAIN about **SWEDEN**

In the office where I work we have often had visitors from Sweden, and we've been very impressed both by their English and by their pleasant and friendly manners. However, we have received some letters from these very same people lately and have been amazed by the poor standard of their English and by the tone of the letters, which we find rather arrogant.

You'll be wiser to trust your first impressions. There's a major difference between how we speak and how we write, and whether we're using our native language or someone else's. For example, Swedish children learn English from about the age of eight and quickly become fluent and accurate speakers, but there isn't the same emphasis on written skills (the reverse is true in Japan and South Korea, where writing is prioritized).

When they write in their own language, Swedes are often very informal and rather blunt; this reflects their egalitarian approach to their fellow citizens. When they transpose this style into written English they can unintentionally sound very rude, especially as there isn't a Swedish equivalent for *please* as there is, for example, in Spanish (*por favor*).

It is often difficult to establish the right "tone" in written communication when body language and tone of voice are missing from the communication equation. I have noticed when people from French-, Arabic-, and Spanish-speaking countries write to me, although the grammar and vocabulary may be less than perfect, the tone is extremely polite and rather more formal than letters and e-mails from the U.S. or the U.K. This is because the writers are imitating the more formal and courteous written styles of their own languages and transposing them to English.

Misunderstandings such as you describe, which arise from the tone of a letter or written material, are often the result of "first language interference" and can be hard to identify and correct. It's easier if you make the wrong impression during a face-to-face encounter, because then you get immediate feedback from your listener's body language or facial expression.

The moral is that when writing in any language you should be more formal than when you're speaking, and most importantly, ensure that the tone of the letter is polite and friendly. This is hard to do in a foreign language, but it is even more important than getting the grammar or vocabulary right. If you feel that you cannot judge the tone of your letter yourself, try to get a native speaker to read it before sending it off to ensure that you're not going to offend anyone by appearing less charming than you actually are!

Thin Skin   Letter 16

From the NETHERLANDS about **FRANCE**

> I made a mistake the other day when a French visitor used a wrong word when he was speaking English. He told a group of us when we arrived at this office to "Please sit down, and I'll enjoy you in a minute." We Dutch laughed a little about this, and thought he would too, for we know him well and have always worked well together. However, he was extremely offended. We are sorry for our tactlessness but also surprised at his sensitivity.

His reaction is not hard to account for. There is a lot of prestige involved in how well you speak a foreign language, and if the corporate language is English but it isn't your native language, you can feel threatened if you are concerned that your English isn't up to standard. And when people feel threatened, they can become both defensive and aggressive. Speaking a foreign language means that, like it or not, you have to give a public display of how well you command one of the most important tools of your profession, and that can be a nerve-wracking experience.

The standard of English in the Netherlands and in Northern Europe is extremely high, and this fact may have made your French colleague's reaction worse. Until relatively recently the French have not taken English-language learning seriously (although they have not been as bad as the British and Americans about learning foreign languages). He may have been able to accept a native speaker's superiority, but to have another non-native speaker laughing at his errors was humiliating.

Letters 17–18

There are many countries with more than one national language and most nations have linguistic minorities. To forget these facts is to show an unacceptable degree of ignorance of the culture you are dealing with.

One Country—Two Languages   Letter 17

From BRAZIL about CANADA

> I'll be going to Quebec soon but speak only English. How important is it to be able to speak French as well?

I'd take at least a few lessons in French if you intend to do a lot of business in Canada, for this is one country where English is not regarded simply as an efficient tool for international business communication. Instead, it's regarded by some of its French-speaking citizens as a symbol of the oppression by the English-speaking majority of the French-speaking minority.

Canada is divided into ten different provinces, and they have both French and English as their official languages. Today you will find both languages on maps, tourist brochures, and product labels. Historically there has been friction between the French-speaking Québécois and the English-speaking people who have surrounded them for centuries. The Québécois have seen French speakers in other provinces become assimilated into the English-speaking culture, and they take great pains to preserve their language and culture so the same thing doesn't happen to them. So if Quebec is your destination I suggest learning as much French as possible before departure, both as a goodwill gesture and as a survival measure in case you meet some of the Québécois who can't or won't speak English. But be warned: The French they speak in Canada is not the same as that spoken in France, and even some of the English you hear in Québec may be unfamiliar, as many French words have been incorporated into the English they speak there.

One Country—Several Languages   Letter 18

From AUSTRALIA about BELGIUM

> I will probably be traveling to Belgium in the near future. I speak elementary French and my native language is English. Will that be enough?

A lot depends on where in Belgium you are going, for despite its small size and population of around 10 million, there are two completely different languages spoken. In Flanders, the northern part of the country, the people speak Flemish, which is a variation of Dutch, and all employers in Flanders are required by law to use Flemish in the workplace.

In Wallonia, the southern part of the country, they speak French, as do many of the inhabitants of Brussels. For Belgians, which language they speak is very much a part of their national identity. The situation in the country is made even more complicated because many Walloons cannot speak Flemish and some Flemish people are reluctant to speak French! However, in the capital about a quarter of the residents are non-Belgian, so there English is increasingly accepted. Be grateful that English is your native language, because it can be regarded as a sort of "neutral territory" outside the political and historical issues that otherwise make the language question in Belgium such a hot potato.

Letters 19–20

There are many different "Englishes," two of which are described here.

British versus U.S. English   Letter 19

From FRANCE about the U.S.

I've recently come back from the U.S. where I attended a conference. One lecture dealt with different human resources issues, and I was surprised to hear the term *attrition* used in this context. The only time I've heard it before is in war of *attrition,* meaning a war involving total destruction of the enemy. When I got home I checked in my English dictionary and found *attrition* means "the state of wearing away." I'm none the wiser!

I'm not surprised. This is an excellent example of what George Bernard Shaw meant when he wrote "England and America are two countries separated by the same language." I imagine you learned British English rather than American, and there is a little area where the two don't correspond. Don't be alarmed: *Attrition* doesn't refer to a particularly drastic (and permanent) way of getting rid of unwanted staff! It's a human resources term describing the process by which people leave their jobs at a company when they move to another position, retire, decide to study, and so on and are not replaced. The term for the same phenomenon in England is *natural wastage* (which most Americans think sounds like some sort of sewerage system).

Don't blame your dictionary. Apart from the British-English and American-English differences, the English language is in a constant state of change and dictionaries cannot possibly keep up with all developments.

"International English" for Presentations   Letter 20

From the U.S. about the Rest of the World

I'm used to giving presentations in the U.S., but I will soon be going abroad for the first time. I'll be presenting information in a number of different countries where I guess most people do not speak English as their first language. Are there any changes I should make to my presentations to adapt them?

Speaking to non-native English speakers certainly requires extra thought, although in certain parts of Asia, for example, Singapore and Hong Kong, which are former British colonies, people may speak English as a first language.

To give a clear message speak slowly and clearly and pause often. In addition, use a tape recorder or ask someone not from your own hometown to establish whether you have a strong accent and if you do, try to tone it down. It's important to be confident and believe in what you are presenting, but make sure you don't come over as too loud (aggressive) or too relaxed (casual). In the more restrained cultures of Eastern Asia or Northern Europe you could appear to be trying to dominate your audience.

To give non-English native speakers a chance to absorb the key facts, repeat your main points in different ways. Try not to use sports metaphors. Violent metaphors are also inappropriate, especially in cultures that value gentle and controlled behavior, so don't use phrases like "bite the bullet," "twist your arm," or "ride roughshod over someone."

If you want your listeners to understand you, avoid the latest buzzwords, idioms, and slang. The use of initials and abbreviations can also be confusing, so use the full form instead. Two more things: don't use even the mildest swear words, and be careful in your use of humor.

It would also be wise to avoid using hand gestures to illustrate a point as they may not be interpreted the same way internationally. One example would be the way a Mexican speaker brought a presentation to a speedy halt in the U.K. by indicating the number two by two raised fingers with the back of his hand facing the audience. He had inadvertently told his British audience to f*** off.

What you *should* do is to make sure that you take plenty of visual material, as this can remove the need for words, and clarify points for people whose native language is not English. Another idea is to distribute written information (in English or the home language) before the meeting so participants have time to read it and translate it if necessary. Remember that it is hard work listening to a foreign language, so keep your presentation shorter than you would at home and make sure you have lots of breaks. This also gives people the chance to ask you questions, something they may not wish to do in front of a large audience if their English is shaky, or if they feel such questions would entail a loss of face by revealing they haven't followed everything you have said.

And a final word of advice: If you don't already speak a foreign language, start to learn one. It will give you an insight into what your Asian colleagues are up against.

Letter 21

Native speakers of English have an enormous business advantage, but they should not misuse it, or they will cause resentment.

Sensitive Speakers Sought   Letter 21

*From MEXICO about the **U.K.**, the **U.S.**, **AUSTRALIA**, etc.*

> Why can't native English-speakers show a little more sensitivity in their dealings with non-English speakers? They often use their superiority in the language to dominate meetings, and if there are two or more present they speak far too fast and use words and expressions we are not familiar with.

Your question is a useful reminder to everyone who has English as his or her first language. People who speak no foreign languages themselves, and this includes many British and American people, often forget what a strain it is listening to a foreign tongue, and when speaking to foreigners they make no concessions when it comes to their choice of words. Not only that, they forget that their listeners may have learned to speak British RP (Received Pronunciation) or Network Standard American English at school and are not used to strong regional accents. Ironically, it's when non-native speakers speak really good English that the worst problems arise, for it's then that Aussies, Kiwis, or Brits forget they're talking to a foreigner and speak in exactly the same way they would to someone from back home, while their poor listeners struggle to keep up.

One of the most important things for native speakers to remember is to listen. Don't treat a person's silence as a sign for you to continue to speak, but wait. Your colleague has to formulate his or her ideas in a foreign language, and that takes time.

Letters 22–23

As long as there are different languages there will inevitably be problems with translation.

Language Mistake   Letter 22

*From SOUTH KOREA about **BRAZIL***

> My company employed an agency to translate our material for the Brazilian market. We'd already sent away the material when we discovered that it had been written in Spanish and not Portuguese. Our Brazilian agents have told us that it's useless and they require new material. Are the languages really so different?

As well as being the language of Brazil, Portuguese is widely spoken in Venezuelan cities, and elsewhere in South America that Spanish isn't the primary language. It was lucky that your agents spotted the mistake before the material was printed, for national language forms a vital part of national identity, and not respecting this is asking for trouble. Spanish and Portuguese are closely related languages but they are far from being identical, and Brazilians dislike foreigners who do not appreciate this fact. I can imagine that a similar assumption about the inter-changeability of Swedish, Norwegian, and Danish or the different Chinese languages would cause the same sort of resentment. You really have no choice but to recall the Spanish version and provide a Portuguese version as quickly as possible. If you are interested in doing business in Brazil it would be wise to show an interest in, and a certain background knowledge of, the country so you avoid "putting your foot in it" again. You can consult appropriate books, and the Internet is a great source of useful information.

Interpreters   Letter 23

From MEXICO about JAPAN

> I'm going to be traveling to Japan with a small group of other managers. We don't speak Japanese and were wondering if we should take an interpreter with us, which would be very expensive, or if we can ask the Japanese firm if they can arrange one for us.

It depends on how much money is at stake. If you're hoping to build a solid long-term relationship that is going to earn your company a fat profit, then it's worth thinking about developing a working relationship with a fluent Japanese speaker (preferably a native speaker) who is bicultural as well as bilingual and knows what your company does.

You can hire an interpreter from an agency in Japan, but then you'd have to make sure you allowed sufficient time in Japan to get to know each other before you met your potential partners. She (most Japanese translators are female) needs to know in advance what ground the talks are going to cover so she can prepare herself. She also needs to become familiar with the communication style of the person or people she's translating for. One more thing: if you do decide to hire an interpreter in Japan, book her well in advance as there are not many Japanese-Spanish translators, and you may have to accept a Japanese-English substitute.

Asking the Japanese company to provide an interpreter may not be a good idea, because even though you can be quite sure she will translate the Japanese side's message correctly (she will probably know their business very well), there's no guarantee your message is going to be expressed as you intended. For example, she may not want to take on the responsibility of delivering a message from you that will not please her fellow citizens. They may not have heard the expression "Don't shoot the messenger," but many interpreters are only too familiar with the meaning behind it.

To minimize the possibilities of misunderstandings, have a written summary of the points you are going to make at the meeting translated and distributed *before* the meeting, and get a written summary of the proceedings translated into Japanese shortly *after* the meeting.

IN A NUTSHELL: *A Global Language?*

Global Business Standards

For native English speakers: learn at least one foreign language as well as you can.
For non-native English speakers: learn English as well as you can.
For everyone: learn a few words of the language of any country you visit and of any foreign visitor you are going to meet.

- **Argentina:** The official language is Spanish, but it is influenced somewhat by Italian. (See Letters 15 and 22.)
- **Australia:** The language is influenced by both British and American English, but it has a distinctive accent and a special Aussie vocabulary. (See Letter 21.)
- **Austria:** German is spoken with a distinctive accent.
- **Belgium:** Official languages are Flemish (similar to Dutch), French, and German. The language spoken is closely tied to a person's ethnicity, and group loyalty is strong. (See Letters 15 and 18.)
- **Brazil:** Portuguese is spoken here—not Spanish like most of the rest of South America. (See Letters 15 and 22.)
- **Canada:** There are two official languages: English and French. The language spoken is closely tied to a person's ethnicity, and group loyalty is strong. (See Letters 15, 17, and 21.)
- **China:** The official spoken language is Mandarin, a language based on tones. It is also the only form of written language. In some provinces people speak one of four major dialects, but these aren't understood by speakers of the other dialects. (See Letter 20.)
- **Denmark:** Danish is almost indistinguishable to Norwegian in written form. Norwegians, Danes, and Swedes can often understand each other.
- **Finland:** The language is similar to Hungarian (!). In some areas Finns also speak Swedish.
- **France:** You are judged according to how well you speak French, and your command of the language is seen as an indicator of your education and intelligence. There is a big difference between using the familiar *tu* (informal) and the more formal *vous*. (See Letters 15 and 16.)
- **Germany:** There is a big difference between using the familiar *Du* and the formal *Sie*.
- **Hong Kong:** English, Cantonese, and Mandarin are widely spoken. (See Letter 20.)
- **India:** There are eighteen official languages and about as many dialects distributed geographically (e.g., Hindi, Punjabi, and Gujarati, and Urdu, which is spoken mostly by Muslim minority). English is widely spoken by educated people. Many people are bilingual or multilingual.

- **Indonesia:** There are more than 300 ethnic languages. Bahasa Indonesia, the major unifying language, is adapted from Bhasa Melayu (Malay). (See Letter 20.)
- **Italy:** About 60 percent of Italians speak a dialect, which may be impossible for other Italians to understand. The vast majority also speaks standard Italian.
- **Japan:** Spoken Japanese and Chinese are quite different. Basic literacy requires mastery of three alphabets, one of which is derived from Chinese and contains about two thousand characters. (See Letters 20 and 23.)
- **Mexico:** Spanish is spoken by 98 percent of the population. (See Letters 15 and 21.)
- **Netherlands:** Dutch is spoken. It is almost identical to Flemish, which is spoken in Belgium. It is also the ancestor of South Africa's Afrikaans. The Dutch are some of the best speakers of English as a foreign language in the world. (See Letter 16.)
- **Norway:** There are two distinct and rival versions of Norwegian. Norwegian is almost indistinguishable to Danish in written form. Norwegians, Danes, and Swedes can often understand each other.
- **Poland:** Polish is a Slavic language, but unlike Russian, it uses the Latin script.
- **Russia:** Russian uses the Cyrillic alphabet. Words are pronounced as they are spelled. Russian is spoken by most people, but Russia is made up of about a hundred ethnic groups, many with their own languages.
- **Saudi Arabia:** Arabic is the official language of the country and is widely spoken in the whole region. (See Letter 15.)
- **South Africa:** There are eleven official languages. English, Afrikaans (related to Dutch), and Zulu are the main ones.
- **South Korea:** Compared to Chinese and Japanese, the alphabet is easy to learn. Foreign (English) words are readily integrated into Korean. There is much pressure on young Koreans to learn English. (See Letters 20 and 22.)
- **Spain:** The Castilian dialect is the accepted standard. There are also three regional languages. Catalan (as well as Castilian) is spoken widely in Barcelona, Spain's second-largest city. There are some differences from the Spanish of Latin America. (See Letter 15.)
- **Sweden:** A sharp intake of breath can mean *yes*. Norwegians, Danes, and Swedes can often understand each other. (See Letter 15.)
- **Switzerland:** There are four official languages and most Swiss speak at least two fluently. The result of the most recent census shows the breakdown of first language speakers as follows: (Swiss) German 63.9%, French 19.5%, Italian 6.6%, Romansh 0.5%, others 9.5%.
- **Taiwan:** Mandarin is the official language, but 70 percent of the population speaks Southern Fujianese, often called Taiwanese. They do not use the modernized Chinese script currently used in China.
- **Thailand:** Like Chinese, Thai is a tonal language. The written script is based on ancient Indian languages. Fellow Thais usually understand regional and ethnic dialects. (See Letter 20.)

- **Turkey:** Turkey is an oral culture. What is said and heard is taken more seriously than what is written.
- **UK:** Differences between British and American English may lead to misunderstandings. (See Letters 15, 19, 20, and 21.)
- **US:** Differences between British and American English may lead to misunderstandings. Spanish is widely spoken by Latin American immigrants in southern states and California. (See Letters 15, 19, 20, and 21.)
- **Venezuela:** Spanish is spoken. There is a distinctive Venezuelan accent, and some specifically Venezuelan vocabulary exists. In major cities Portuguese is quite common. (See Letters 15 and 22.)

Part 3

Business and Technical Correspondence

Traditionally, business and technical correspondence has taken two forms: the letter and the memo. More recently, it has assumed a third electronic form, e-mail. This electronic innovation has brought with it a myriad of issues, most of which unfortunately remain unresolved. There has been little discussion of, let alone agreement about, the etiquette of e-mail.

In the past, business and technical correspondence *seemed* easier on some levels: letters went to people outside of the company; memos, to people inside the company. Letters were advertisements for writers and their companies. Memos provided a record of decisions made and actions taken within a department of a company.

Letters and memos announce or reaffirm policies, confirm decisions and conversations, and send or request information. Some letters and memos are routine; others concern pressing issues. In either case, letters and memos require careful writing.

Given the volume of correspondence many companies produce, it is easy for business and technical writers to be tempted to use shortcuts when writing. Looking for a shortcut can, however, court danger if the shortcut involves relying on form responses or copies of previous—though not necessarily good examples of—similar correspondence.

There is nothing wrong with form letters or memos, as long as such correspondence is appropriate to the given writing situation. A letter reminding customers of bills past due or a memo transmitting attached data is often an effective and time-efficient way to communicate—provided such letters and memos are appropriate to the situations and audiences with which they are used.

The danger with form letters and memos is that they can become a crutch that writers depend on even in situations in which they are inappropriate. Such misuse of form letters and memos can create new problems for writers. Writers may have to send a follow-up letter or memo to fix what they tried to do in the first piece of correspondence. Or worse, they might have to send *several* follow-up messages.

A process approach to business and technical correspondence will help determine when form letters and memos are appropriate and when something more original is required. The careful use of the writing process can also make follow-ups unnecessary.

The advent of electronic correspondence would seem to offer a boon to business and technical writers, but every silver lining has its cloud. The ease with which electronic correspondence can be transmitted has led some writers to become more casual—if not careless and sloppy—in their electronic correspondence than they would be in producing correspondence in the more traditional forms of letters and memos.

Electronic correspondence can be sent efficiently and effectively within and outside companies and businesses to multiple audiences, even those across the globe. At the same time, a misstatement, a mistake, or an unintended slight can suddenly be transmitted to a very large global audience—an audience at times even much larger than the original, intended audience—making the writer look foolish, or worse, look foolish to a large body of readers. Just ask an internationally known electronics retailer that decided to lay off a number of employees using the following e-mail message:

> The workforce reduction notification is currently in progress. Unfortunately your position is one that has been eliminated.

Forgetting for a moment that such a message is no way to handle employee dismissals, this e-mail was subsequently posted on the worldwide web by someone, was then picked up by the newswires, and soon turned into a major international public relations fiasco for the company.

As indicated earlier, as of yet there is no established or generally adopted etiquette for electronic correspondence. In the absence of such etiquette, writers would be wise to be conservative in such correspondence, adhering closely to the rules and principles that work so well for letters and memos. Whether the product is a letter, a memo, or an electronic message, a process approach is the safest approach for a writer to take.

The selections in this part of *Strategies* follow a decidedly process-oriented approach to business and technical correspondence in all forms. Readers in the world of work are impatient with people who waste their time, and David V. Lewis offers some suggestions on how to write to people effectively by showing business writers how to save their readers' time while also selling themselves and the organizations and companies that they represent through their correspondence.

Saying "no" is, of course, always difficult and presents special problems for writers. As Allan A. Glatthorn suggests, writers can say "no" in their correspondence and still keep clients, make friends, or even develop further business contacts. Harold K. Mintz next offers a formula for writing better memos, and John S. Fielden and Ronald E. Dulek suggest a model for efficient professional writing, no matter what form the writing may take. Janis Fisher Chan ends this section of *Strategies* with an essay that details ways in which electronic communication can be used to deliver a variety of messages—while at the same time allowing writers to present a professional image.

David V. Lewis

Making Your Correspondence Get Results

When he wrote this essay, David V. Lewis was in-house consultant in sales and management train-ing with Western Company of North America in Fort Worth, Texas.

If you turn out an average of five letters a day, you'll produce nearly twice as many words during the year as the typical professional writer.

These letters reflect *your* attitude—and obviously your organization's—toward customers. Realizing this, many progressive organizations train key people in the art of writing readable, results-getting letters. For example, the New York Life Insurance Company produces more than a million letters a year and uses its correspondence as a public relations tool.

"Anyone who writes a letter for New York Life holds a key position in our orga-nization," says Nathan Kelne, vice president of the company. "By the letters our people write, they help determine how the public feels toward our company, and toward the life insurance business as a whole. Since they are instrumental in shaping the personality of the company, they are in a very real sense public relations writers."

For example, before New York Life launched its companywide letter-writing courses, here's how one of its executives tried to explain to a beneficiary the way the death claim was to be paid on a $3,000 policy:

> The monthly income per $1,000 under option 10 years certain is $7.93 per $1,000 total face amount of insurance, which total is $3,000.

Fortunately, the beneficiary was able to cut through the jargon. He replied:

> As I understand your letter, you seem to be saying that I should receive a monthly check for the amount of $23.79 for at least 10 years.

But another policyholder's response to a similarly bewildering letter is more typical:

Please tell me what you want me to do, and I'll be glad to do it.

Some universally proven principles that can help you sell yourself and your company to the public will now be examined.

WRITE FOR HIM, NOT TO HIM

General Foods is an organization that believes strongly in creating a favorable image through its correspondence. The public relations people came across this letter signed by a marketing executive and ready for mailing:

```
Dear Sir:
    Enclosed please find a questionnaire which we are sending to
all our retail contacts in this state.
    Will you please answer this as soon as possible? It's very
important that we have an immediate reply. We're delaying final
plans for our retail sales program in this area until we get
answers to the questionnaire.
    With thanks in advance, we are,

                                    Gratefully yours, . . .
```

The letter is clear and to the point; it *does* communicate readily. But there's a major flaw. It points out benefits the company will receive instead of suggesting how the program will help the recipient. The letter is writer- rather than reader-oriented.

Psychologists say that each of us is basically interested in himself or herself. We want to know "what's in it for me?" Once you routinely approach letter writing from this point of view, you'll find yourself telling your readers in specific terms how the letter will benefit them.

The questionnaire letter was rewritten this way:

```
Dear Sir:
    Enclosed is a questionnaire on our proposed retail sales pro-
gram. Your advance opinion on how this new plan would help you
and others will be very useful in our evaluation of the program.
    If you will fill out and return this questionnaire as soon as
possible, we can let you and other retail contacts in this area
know promptly of changes in the present program. Thank you.

                                    Very truly yours, . . .
```

The revised version asks for your advance opinion, suggests how quick action might help you and *others,* and promises to let you know promptly of changes. There's also another improvement. The tired and outdated "With thanks in advance" has been replaced by a modern "Thank you."

The best way to persuade your readers to your point of view is to show them that it will be worth their while to do so. This rule holds true for just about every successful letter, sermon, sales presentation, or advertisement. A good sales letter, for example, is much like a good ad in that it attempts to dramatize benefits to the reader.

What does it take to get—and hold—a reader's attention? The most power-ful letters appeal to basic needs and emotions rather than to purely logical rea-sons. Mainly, management people want to know how to save time, money, and effort. Show them how your product or service can help them do one or more of these things, and you're likely to get their attention.

With effort, almost any letter can be oriented to the reader, even the usually hard-to-write collection letter. Here's such a specimen, written almost entirely from the writer's point of view:

```
Dear Sir:
    Our records show that you are three months delinquent in pay-
ment of your bill for $37.50.
    Perhaps this is an oversight on your part. Otherwise we cannot
understand why you have not taken care of this obligation.


                                    Very truly yours, . . .
```

Keeping the reader's self-esteem in mind, the letter could have been rewritten this way:

```
Dear Sir:
    We know you'll want to take care of your small past-due
account for $37.50. This will help you to maintain the fine
credit you have built up with us over the years.


                                    Sincerely, . . .
```

To help develop the "you" attitude, put yourself in your reader's shoes, then write from his or her point of view. Once you've developed this attitude, you'll auto-matically start telling your readers what's in it for them. Instead of saying, "I wish to thank you," you'll write, "Thank you." Instead of writing, "We'd like to have your business," you'll write, "You'll find our service can help your business in many ways."

PERSONALIZE YOUR LETTERS (THERE'S POWER IN PRONOUNS)

Some years ago, many would have considered this letter to have been perfectly proper and very effective:

```
Gentlemen:
    Enclosed herewith are the subject documents which were
requested in yours of the 10th. The documents will be duly
reviewed and an opinion rendered as to their relevancy in the
involved litigation.


                                    Very truly yours, . . .
```

The letter *was* all right—back in the horse-and-buggy days! Phrases like "enclosed herewith," "duly received," and "involved litigation" would have marked the writer as learned. But the executive who makes a habit of writing like that today is generally regarded as an anachronism.

Current usage calls for clear, to-the-point letters, written mostly in conversational language. Like good conversation, your letters should generally be friendly, filled with personal references, and almost always informal in tone and language. Here's how the horse-and-buggy letter might have been rewritten by the modern executive:

```
Dear Sam:
    Here are the documents you asked for. I'll look them over and
let you know if we can use them in our lawsuit.

                                             Cordially, . . .
```

When experts write a letter, even a form letter, they generally try to make it sound as personal as possible. Almost always, it contains a sprinkling of personal pronouns. In orienting your letter to the reader, fill it with "you," "your," and "yours." Use "I" and "me" sparingly.

Here's a case in point, a letter sent out by a mortgage company (emphasis added by the author):

```
Dear Mr. Jones:
    We want to thank you for your query about our new mortgage
insurance plan.
    We are enclosing a pamphlet which outlines benefits of our new
policy and gives testimonials from some of our policyholders.
    We would like very much to enroll you within the next 30 days,
since we are offering a special low premium rate as our intro-
ductory offer.

                                             Very truly yours, . . .
```

The repeated use of *we* and *our* clearly shows the letter is written with the mortgage company's interests at heart. But notice how substituting *you* and *yours* orients the letter to the reader's interest. (Emphasis has again been added by the author.)

```
Dear Mr. Jones:
    Thank you for your recent inquiry about our new mortgage
insurance plan.
    You'll find that the enclosed pamphlet outlines benefits of
the policy in detail. You'll also probably be interested in the
testimonials furnished by some policyholders.
    If you sign up within the next 30 days, you'll be able to take
advantage of special premiums we're offering on an introductory
basis. Thank you.

                                             Cordially, . . .
```

The word is out now in enlightened business circles. Companies are telling their executives to regard every letter as a personal contact: to write your own way, to develop your own style (within bounds), to make your letter distinctively *you*.

MASTERING TONE (YOUR PERSONALITY IN PRINT)

Writing a "rejection letter" that leaves the recipient's self-esteem intact and preserves his or her goodwill is a difficult task. It requires tact, diplomacy, and empathy—all of which must be effected through appropriate *tone*. For example, here's the way one banker turned down a builder, a long-time customer:

```
Dear Mr. Jones:
    We regret to inform you that your request for an additional
loan in the amount of $250,000 must be rejected. It was the
judgment of the loan committee that, with your present commit-
ment, such a loan would present too much of a risk for us.
    Naturally, we look forward to doing business with you on your
existing commitment.


                                    Sincerely, . . .
```

What's wrong: Tone, mainly—the *way* the rejection is phrased. It deflates the reader's ego and makes it virtually certain the builder will do his banking elsewhere in the future.

True, the rejection was "justified." But why not let the reader down more gently and leave the door open for future business? For example:

```
Dear Mr. Jones:
    One of the most distasteful tasks we have is to turn down a
loan application, particularly when it is from a regular and
respected customer like you.
    We know you'll be disappointed, and so are we. We share your
excitement about your plans to expand the Roseland Project, and
hope that circumstances will later warrant our working with you
on this and other projects.
    However, the current economic outlook, combined with your
delinquent status on your existing loan, makes your new loan
application a questionable venture for us at the present time.
    We will be glad to work with you in any way we can to resolve
your current financial problems, and we look forward to helping
you meet your future financial commitments.


                                    Sincerely, . . .
```

The two letters say the same thing, but in vastly different ways. Suffice it to say that tone is as important to a letter as good muscle tone is to an athlete. A negatively phrased letter can have the same effect as an abrasive personality. As one New York Life executive put it:

> By its very nature, the "no" letter is a turndown and leaves the reader unsatisfied. Yet there are ways to soften the blow, and one of the best is simple candor. So, if you must say

"no," state the reason first: "Currently there are no vacancies in that department. There-fore, I cannot offer you a position with the company at this time."

A "no" response often requires positive alternatives if they are appropriate (for exam-ple, "Although, for the reasons mentioned, I cannot do as you ask, may I suggest that . . .". You have to be more diplomatic and more sensitive if you're to have any chance of salvaging your reader's goodwill or ego.

And that really is your goal in an effective "no" letter—to tell your readers something they don't want to hear in a way that compels understanding and, ideally, acceptance.

Unfortunately, tone requirements vary not only from person to person, but from one situation to another. For example, you would ordinarily use fast-paced, persuasive prose for a sales-promotion letter. In writing to a highly regarded law firm, you might use a slightly more "dignified" approach. If you're writing for a service-type organization, you might use a middle-of-the-road approach, as the government does.

Government letter writers are told to strive for a tone of "simple dignity," to make letters brief and to the point, and to avoid gobbledygook. "Don't act as if you're the only game in town," they're told. "On the other hand, don't bow and scrape just because you're a service organization."

Poor letters start with poor thinking on the part of the writer. Negative thoughts lead to negative words—and before you know it, there goes the old ball game. Here's a letter from a manager of a department store to a customer who complained that an appliance she had recently bought didn't work. (Negative words have been italicized by the author.)

```
Dear Sir:
    I am in receipt of your letter in which you state that the hair
dryer you purchased from us recently failed to meet the warranty
requirements.
    You claim that the dryer failed to do the things you say our
salesman promised it would do.
    Possibly you misunderstood the salesman's presentation. Or per-
haps you failed to follow instructions properly. We positively
know of no other customer who has made a similar complaint about
the dryer. The feeling is that it will do all that is stated if
properly used.
    However, we are willing to make some concessions for the
alleged faulty part. We will allow you to return it; however, we
cannot do so until you sign the enclosed card and return it to us.

                            Very truly yours, . . .
```

The tone is unmistakably negative. "State" and "failed to meet" in the open-ing paragraph imply there's some doubt that the claim is valid. In the second paragraph, "you say" suggests the salesman didn't make the statement at all.

Such phrases as "you misunderstood," "you failed to follow instructions," and "if properly used" tell you in so many words that you're not too bright. And to wrap it up, the writer uses the negative "we cannot . . . until" instead of the positive "we will . . . as soon as."

Most letters aren't this negative, of course. But it doesn't take much to offend. Any of these negative words or phrases, in themselves, could have spoiled the tone of an otherwise effective letter.

If you've been guilty of taking a negative approach, study the following examples. In each case, the negative thought (emphasis added by the author) has been converted into a positive one.

Negative	Since you *failed* to say what size you wanted, we cannot send you the shirts.
Positive	You'll receive the shirts within two or three days after you send us your size on the enclosed form.
Negative	We *cannot* pay this bill in one lump sum as you requested.
Positive	We can clear up the balance in six months by paying you in monthly installments of $20.
Negative	We're sorry we *cannot* offer you billboard space for $200.
Positive	We can offer you excellent billboard space for $300.
Negative	We are *not* open on Saturday.
Positive	We are open from 8 A.M. to 8 P.M. daily, except Saturday and Sunday.

Negative words aren't the only cause of poor letter tone. Many letters are made more or less "neutral" by mechanical, impersonal, or discouraging language. Here's an example of each fault with a preferred alternative:

Interpersonal	Many new names are being added to our list of customers. It is always a pleasure to welcome our new friends.
Personal	It's a pleasure to welcome you as our customer, Mr. Jones. We will make every effort to serve you well.
Mechanical	This will acknowledge yours of the 10th requesting a copy of our company's annual report. A copy is enclosed herewith.
Friendly	Thanks for requesting a copy of our annual report, which is enclosed. We hope you will find it helpful.
Discouraging	Since we have a shortage of personnel at this time, we won't be able to process your order until the end of the month.
Encouraging	We should have more help shortly, which will enable us to get to your order by the end of the month.

Ideally, the tone of your letter should reflect the same ease in conversation that you enjoy when talking about your favorite hobby, business, or pastime.

HOW TO WRITE (MORE) THE WAY YOU TALK

This letter was sent out by a large department store. Imagine yourself as the recipient.

```
Dear Sir:
    We are in receipt of yours of July 10 and contents have been
duly noted.
```

```
As per your request, we are forwarding herewith copies of our
new fall brochure. Thanking you in advance for any business you
will be so gracious as to do with us,

                                            Yours truly, . . .
```

Conversational? Of course not. Who uses such language as "We are in receipt of . . . ," "duly noted . . . ," and "as per your request"? Practically no one. People just don't talk that way. Most of these phrases went out with the Model-T—or should have.

Face to face, the writer probably would have said, "Here's the fall brochure you requested. Let me know if we can serve you." It says the same thing, in less space, and without all the fuss.

Some business and professional people still feel it isn't quite proper to write conversationally, mainly because they have seen so much stilted business writing. But many progressive companies are telling their people to communicate in plain English, using only those technical terms that are absolutely necessary. As one executive of a major company said, "The best letters are more than just stand-ins for personal contact. They bridge any distance by the friendly way they have of talking things over person to person." And this from a U.S. Navy bulletin: "At best, writing is a poor substitute for talking. But the closer our writing comes to conversation, the better our exchange of ideas will be."

The consensus clearly is that informal, natural business writing is *in;* stilted business writing is *out.*

One word of caution about writing (more) the way you talk. Since World War II, readability experts have urged business people to "write more the way they talk" or "write as they talk."

Detractors have soft-pedaled the idea, claiming that most conversation is rambling, often incoherent, and frequently a bit too earthy. These are valid objections, but they miss the point. You're not being asked to write *exactly* the way you talk. Rather, you're being asked to bridge the gap between the spoken and the written word—to narrow the difference in the way you would *give* an order verbally and the way you would *write* that same order in a memo. Naturally, writing requires more restraint than speech, and the writer must normally use fewer words. But you can do these things and still capture the tone and cadence of spoken English.

The first step in making your correspondence more conversational is to rid your vocabulary of worn-out business phrases. Here are some of the more flagrant offenders. They were stylish once, but they've done their duty and need to be honorably discharged.

Old Hat	Conversational
At a later date	Later
If this should prove to be the case	If this is the case
This will acknowledge receipt of	Thank you for
Attached herewith please find	Here is; Enclosed is
We shall advise you accordingly	We'll let you know

Old Hat	Conversational
Due to the fact that	Because
With regard to	About
Please notify the writer as to	Please let me know
Enclosed please find a stamped envelope	I've enclosed a stamped envelope
We are submitting herewith a duplicate copy	Here is a copy
In compliance with your request	Here is
We are submitting herewith our check in the amount of $75	Here is our check for $75
We beg to advise (acknowledge) that	[Begging is unnecessary]
The information will be duly recorded	We'll record the information
The subject typewriter	This typewriter
In compliance with your request	As you requested
We will ascertain the facts and advise accordingly	We'll let you know
The writer wishes to state	[Just say it]

This list is far from complete but you get the idea. Once you're mentally geared to writing more conversationally, you'll detect many other clichés. Try to eliminate them.

Contractions will also make your writing more conversational. They play a part—a very large part—in almost everyone's everyday conversation. Even your most learned associate doesn't say, "We shall endeavor to be there at eight o'clock"; he's more likely to say, "We'll try to be there at eight o'clock." Instead of saying, "I am going to the ball game," he'll probably say, "I'm going to the ball game." It takes less effort to say, "You needn't bother to call," than to say, "You need not bother to call." Contractions tend to make the spoken words flow more smoothly. That's why they make writing appear more natural.

Probably the most common contractions are here's, there's, where's, what's, let's, haven't, hasn't, hadn't, won't, wouldn't, can't, couldn't, mustn't, don't, doesn't, didn't, aren't, isn't, and weren't. Then there are the pronoun contractions: I'll, I'm, I'd, I've, he's, he'll, he'd, and so forth.

Using these and other contractions when they facilitate the flow of words will do much to give your writing a quality of spontaneity and warmth. But they must be used with discretion. Using too many contractions can sometimes make your writing too informal. And used in the "wrong" place, they might not "sound right." Indeed, the key is whether the contraction sounds right when the sentence is read.

Take this section from the Gettysburg Address: "But in a larger sense, we cannot dedicate, we cannot consecrate, we cannot hallow this ground." The passage would undoubtedly have lost its historic tone if *can't* had been used instead of cannot.

On the other hand, the advertising slogan "We'd rather fight than switch" would lose some of its punch if phrased, "We would rather fight than switch." Appropriate usage depends on how the contraction makes the sentence sound when it's read aloud.

Next time you get ready to write a letter, ask yourself, "How would I say this if I were talking to the person?" Then go ahead and write in that vein.

Allan A. Glatthorn

"I Have Some Bad News for You"

A consultant in management communications and the author of more than thirty books on writing, Allan A. Glatthorn is retired senior faculty member in the Teaching of Writing Program at the University of Pennsylvania.

One of the most difficult letters or memos to write is the one containing bad news. A subordinate requests a salary increase which you cannot grant. A community organization asks for a contribution which you do not wish to make. An unqualified applicant asks for an appointment to discuss a position with your company, and you do not wish to take the time. You decide to have to dismiss a good worker because declining business mandates cutbacks. In each case you have some unpleasant news to deliver—but you wish to deliver it in a way that does not offend or alienate. How do you accomplish this difficult task? Let me offer some general guidelines and then explain the specific techniques.

The first guideline is to remind you that the successful manager is people-sensitive, able to empathize with others. When you receive a request, you stand in the shoes of the petitioner. You realize, to begin with, that from the standpoint of the asker there are no foolish requests. People ask for things that they need, that they believe they are entitled to, that they hope to receive. And you also realize that there is really no good way to break bad news. No matter how empathic and tactful you may be, your bad news will inevitably cause disappointment and will often cause distress.

The second guideline derives from the first: remember that bad news is best delivered face to face. People getting bad news have some strong needs. They want an opportunity to express their negative feelings. They want the chance to press the request or appeal the decision. They want to explore the reasons more fully to be sure that there is no hidden message. These needs can best be met in a face-to-face meeting. So if you have bad news to give to an employee or a valued client, the best method is to confer with the individual, deliver the bad news in person, and then, if desirable, follow up with a written statement for the record. Don't avoid the unpleasant confrontation by writing a memo, hoping that the petitioner will go away quietly.

Finally, remember that everyone values honesty and forthrightness, especially when being disappointed. Typically, the receiver of bad news is inclined to be cynical, often resenting your attempts to be tactful and politic. You therefore should be sure that your desire to soften the blow does not beguile you into distorting the truth. Explain as tactfully as possible the real reasons for the bad news, instead of offering lame excuses.

With these general guidelines in mind, you should be able to send two kinds of "bad news" messages: the indirect and the direct.

THE INDIRECT BAD NEWS MESSAGE

The indirect message of bad news uses the soft and gentle approach. It tries to cushion the blow by burying the bad news in the middle of the letter or memo, surrounded by positive expressions of appreciation. You send the indirect message under one or more of these circumstances:

- You want further contact with the petitioner.
- You want to project the image of a caring individual.
- You believe that the petitioner won't be able to handle a more direct statement.

The formula is a simple one: THANKS . . . BECAUSE . . . SORRY . . . THANKS.

Thanks

You begin with a positive statement. You express appreciation for the idea offered, the interest expressed, the petition received. Here's how an indirect bad news letter from an insurance company begins:

> Thank you very much for your recent inquiry about our automobile insurance coverage. We are pleased that you have considered transferring your account to our company.

Because

You then continue with the reasons for the bad news. The strategy here is that stating the reasons first cushions the shock, preparing the reader for the bad news to come. Remember to use reasons that will make sense to the petitioner and that will at least project an image of sincerity. If at all possible, state reasons that will depersonalize the rejection, as the next paragraph of that letter does:

> In order to keep our premiums as low as possible, we find it necessary to accept only a small number of new accounts. And in fairness to our present customers, we accept applications only from those whose claims records are comparable to those of drivers we presently insure.

Sorry

You next present the bad news itself, but you state it in a positive fashion. If you can suggest an alternative, do so. If you can find a way to leave the door open, make that clear—but do not give false hopes. Notice how the insurance letter continues:

> Even though we would like to include you among our insured drivers, we will not be able to do so at the present time. It seems to us that your best choice at this time is to remain with your present company. If your claims record improves during the coming year, we would be happy to have you reapply for coverage with us.

Thanks

The indirect message of bad news closes with another expression of appreciation—and ends on a positive note, as this example illustrates:

> We do appreciate your considering our company. And we hope you will be able to reapply under more positive circumstances.

Figure 1 [on the next page] shows another example of this formula at work.

THE DIRECT MESSAGE OF BAD NEWS

The direct message is a tough no-nonsense statement. While courteous, it gets right to the point and does not try to bury the bad news. It would be used under one or more of these circumstances:

- You want to slam the door shut, discouraging any other request from that petitioner.
- You want to project an image of toughness and directness.
- You are addressing an individual who prefers forthrightness and equates indirectness with softness or dishonesty.

The formula for the direct message is THANKS . . . SORRY . . . BECAUSE . . . THANKS. It uses the same ingredients as the indirect message, but it changes the order. And the language is less subtle and more direct.

Thanks

You begin with a courteous expression of appreciation, since courtesy is expected even in the most direct communication. So a firm and direct letter rejecting an applicant might begin like this:

> Thank you for your letter and resume. We appreciate your interest in joining Mutual Life.

Sorry

The direct message moves quickly to the bad news, to be sure that the message is heard as intended. The bad news is stated forthrightly—but without offending. So the rejection letter continues:

Unfortunately, we do not have a position available that would match your qualifications, and it seems unlikely that such a position will develop in the future.

Because

Now the reasons come, after the bad news has been delivered. The reasons are stated directly but not offensively, as in this example:

We have found that our most successful middle-level managers are those who have had the benefit of working with Mutual in a variety of nonmanagerial positions. We therefore tend to promote from within and do not encourage applications for managerial positions from those who have not previously worked with us.

Thanks

The message ends courteously—but the door is firmly closed. There must be no mistake about it: the news is bad. So the rejection letter ends like this:

We hope you will be able to find employment with a company that can make use of your excellent qualifications, and we do appreciate your interest in applying with us.

The successful manager projects both people-sensitivity and toughness. The indirect message emphasizes sensitivity; the direct message affirms the toughness.

To: Bill Harkins
From: Joanne Clemens
Subject: Flex-Time Proposal
Date: February 19,1985   *File:* A-342

Bill, thanks very much for forwarding your proposal for a flex-time schedule for your department. I appreciate the creative energy—I can always count on getting good ideas from you.

Our data suggest that flex time would result in an uneven distribution of worker hours at a time when uniformity seems desirable. As you are aware, the recent increase in customer demand is presenting us with the right kind of problem: our people and equipment are working at maximum capacity. My experience with other companies using flex time indicates that our productivity would suffer if we instituted it now.

And my concern for maintaining productivity makes me reluctant to implement your excellent suggestion now. However, I have asked our personnel department to review the data. Because of my respect for your leadership, I want to give all of your ideas the most careful consideration.

Thanks again for taking the time to share your ideas with me. I appreciate your developing a very sound proposal at a time when I know you are quite busy.

FIGURE I   The Indirect Bad News Message

Harold K. Mintz

How to Write Better Memos

Harold K. Mintz was Senior Technical Editor for the RCA Corporation when he wrote this article.

Memos—interoffice, intershop, interdepartmental—are the most important medium of in-house communication. This article suggests ways to help you sharpen your memos so that they will more effectively inform, instruct, and sometimes persuade your coworkers.

Memos are informal, versatile, free-wheeling. In-house they go up, down or sideways.* They can even go to customers, suppliers, and other interested outsiders. They can run to ten pages or more, but are mostly one to three pages. (Short memos are preferable. Typed single-space and with double-space between paragraphs, Lincoln's Gettysburg Address easily fits on one page, and the Declaration of Independence on two pages.) They can be issued on a one-shot basis or in a series, on a schedule or anytime at all. They can cover major or minor subjects.

Primary functions of memos encompass, but are not limited to:

- Informing people of a problem or situation.
- Nailing down responsibility for action, and a deadline for it.
- Establishing a file record of decisions, agreements and policies.

Secondary functions include:

- Serving as a basis for formal reports.
- Helping to bring new personnel up-to-date.
- Replacing personal contact with people you cannot get along with. For example, the Shubert brothers, tyrannical titans of the American theatre for 40 years, often refused to talk to each other. They communicated by memo.

---

*We will return to this sentence later.

- Handling people who ignore your oral directions. Concerning the State Department, historian Arthur Schlesinger quoted JFK as follows: "I have discovered finally that the best way to deal with State is to send over memos. They can forget phone conversations, but a memorandum is something which, by their system, has to be answered."

Memos can be used to squelch unjustified time-consuming requests. When someone makes what you consider to be an unwarranted demand or request, tell him to put it in a memo—just for the record. This tactic can save you much time.

ORGANIZATION OF THE MEMO

Memos and letters are almost identical twins. They differ in the following ways: Memos normally remain in-house, memos don't usually need to "hook" the reader's interest, and memos covering a current situation can skip a background treatment.

Overall organization of a memo should ensure that it answers three basic questions concerning its subject:

- What are the facts?
- What do they mean?
- What do we do now?

To supply the answers, a memo needs some or all of the following elements: summary, conclusions and recommendations, introduction, statement of problem, proposed solution, and discussion. Incidentally, these elements make excellent headings to break up the text and guide the readers.

In my opinion, every memo longer than a page should open with a summary, preferably a short paragraph. Thus, recipients can decide in seconds whether they want to read the entire memo.

Two reasons dictate placing the summary at the very beginning. There, of all places, you have the reader's undivided attention. Second, readers want to know, quickly, the meaning or significance of the memo.

Obviously, a summary cannot provide all the facts (see Question 1, above), but it should capsule their meaning, and highlight a course of action.

When conclusions and recommendations are not applicable, forget them. When they are, however, you can insert them either right after the abstract or at the end of the memo. Here's one way to decide: If you expect readers to be neutral or favorable toward your conclusions and recommendations, put them up front. If you expect a negative reaction, put them at the end. Then, conceivably, your statement of the problem and your discussion of it may swing readers around to your side by the time they reach the end.

The introduction should give just enough information for the readers to be able to understand the statement of the problem and its discussion.

LITERARY QUALITIES

A good memo need not be a Pulitzer Prize winner, but it does need to be clear, brief, relevant. LBJ got along poorly with his science adviser, Donald Hornig, because Hornig's memos, according to a White House staffer, "were terribly long and complicated. The President couldn't read through a page or two and understand what Don wanted him to do, so he'd send it out to us and ask us what it was all about. Then we'd put a short cover-memo on top of it and send it back in. The President got mad as hell at long memos that didn't make any sense."

Clarity is paramount. Returning to the asterisked sentence in the second paragraph of the introduction, I could have said: "Memoranda are endowed with the capability of internal perpendicular and lateral deployment." Sheer unadulterated claptrap.

To sum up, be understandable and brief, but not brusque, and get to the point.

Another vitally important trait is a personal, human approach. Remember that your memos reach members of your own organization; that's a common bond worth exploiting. Your memos should provide them with the pertinent information they need (no more and no less) and in the language they understand. Feel free to use people's names, and personal pronouns and adjectives: you-your, we-our, I-mine. Get people into the act; it's they who do the work.

Lastly, a well-written memo should reflect diplomacy or political savvy. More than once, Hornig's memos lighted the fuse of LBJ's temper. One memo, regarded as criticizing James E. Webb (then the head of NASA), LBJ's friend, infuriated the President.

Another example of a politically naive memo made headlines in England three years ago. A hospital superintendent wrote a memo to his staff, recommending that aged and chronically ill patients should not be resuscitated after heart failure. Public reaction exploded so overwhelmingly against the superintendent that shock waves even shook Prime Minister Wilson's cabinet. Result? The Health Ministry torpedoed the recommendation.

Two other courses of action would have been more tactful for the superintendent: make the recommendation orally to his staff or, if he insisted on a memo, stamp it "private" and distribute it accordingly.

Literary style is a nebulous subject, difficult to pin down. Yet if you develop a clear, taut way of writing, you may end up in the same happy predicament as Lawrence of Arabia. He wrote "a violent memorandum" on a British-Arab problem, a memo whose "acidity and force" so impressed the commanding general that he wired it to London. Lawrence noted in his *Seven Pillars of Wisdom* that, "My popularity with the military staff in Egypt, due to the sudden help I had lent . . . was novel and rather amusing. They began to be polite to me, and to say that I was observant, with a pungent style. . . ."

FORMAT OF THE MEMO

Except for minor variations, the format to be used is standard. The memo dispenses with the addresses, salutations, and complimentary closes used in letters. Although format is a minor matter, it does rate some remarks.

To and From Lines—Names and departments are enough.

Subject—Capture its essence in ten words or less. Any subject that drones on for three or four lines may confuse or irritate readers.

Distribution—Send the memo only to people involved or interested in the subject matter. If they number less than say, ten, list them alphabetically on page 1; if more than ten, put them at the end.

Text—Use applicable headings listed after the three questions under "Organization."

Paragraphs—If numbering or lettering them helps in any way, do it.

Line Spacing—Single space within paragraphs, and double space between.

Underlines and Capitals—Used sparingly, they emphasize important points.

Number of Pages—Some companies impose a one-page limit, but it's an impractical restriction because some subjects just won't fit on one page. As a result, the half-baked memo requires a second or third memo to beef it up.

Figures and Tables—Use them; they'll enhance the impact of your memos.

CONCLUSIONS

Two cautions are appropriate. First, avoid writing memos that baffle people, like the one that Henry Luce once sent to an editor of *Time*. "There are only 30,000,000 sheep in the U.S.A.—same as 100 years ago. What does this prove? Answer???"

Second, avoid "memo-itis," the tendency to dash off memos at the drop of a pen, especially to the boss. In his book, *With Kennedy*, Pierre Salinger observed that "a constant stream of memoranda" from Professor Arthur Schlesinger caused JFK to be "impatient with their length and frequency."

John S. Fielden and Ronald E. Dulek

How to Use Bottom-Line Writing in Corporate Communications

When they wrote this article, John S. Fielden and Ronald Dulek were Professors of Management Communications at the University of Alabama. Jointly, they also authored a series of books on effective business writing.

Every top executive complains about "wordy" memos and reports. From Eisenhower to Reagan, stories have circulated about their refusal to read any memo longer than one page. The CEO of one of the largest companies in the United States actually demands reports so short they can be typed on a three-by-five card. And J.P. Morgan is reputed to have refused audience to anyone who could not state his purpose on the back of a calling card.

"Don't be wordy!" "Be brief, brief, brief!" "Be succinct!" One writing expert after another exhorts business people with these slogans. And who will disagree?

We do.

As a result of an in-depth study of the writing done at the division headquarters of a very large and successful company, we are absolutely convinced that advice such as "Be brief!" is not only useless, it does not even address itself to the real writing problem.

What causes trouble in corporate writing is not the length of communications (for most business letters, memos, and reports are short), but a lack of efficiency in the organizational pattern used in these communications. And, as you will see, it is for the most part a lack of organizational efficiency on the part of writers that is often deliberate, or, if not consciously deliberate, so deeply ingrained in their behavioral programming that it causes an irresistible impulse to beat around the bush.

Put simply, people organize messages backwards, putting their real purpose last. But people read frontwards and need to know the writer's purpose immediately. That purpose is what we eventually came to call the message's bottom line.

At the beginning of the study we wondered, why do people write backwards? One possibility is that they are writing histories of their mental processes as they think their way through a problem. Since their conclusion could only be arrived at after analysis, the report therefore would state its conclusions last. But if that were the reason, it would only be analytical memos and reports that would be organized backwards.

Such was not the case in the study we did. Almost *all* memos and reports put their purpose last. Why? We determined to study the problem to see if we could design a cure for such blatantly inefficient writing.

COMPREHENSION IS THE KEY

In the division headquarters we worked with, 9,000,000 (internal and external) messages of all types are distributed annually. Our study began with an intense analysis of a sample of 2,000 letters, memos, and reports randomly drawn from company files. The typical communication was one page. Only in rare cases did any memo or report exceed three pages. These various documents were well-written in the sense of being above average in terms of mechanical correctness and aptness of word choice.

Yet only one in twenty of these communications was organized efficiently. Below is one of the actual memos we analyzed (disguised, of course). Look at your watch before you read it. Keep a record of how many times you have to read it before you really *comprehend* its message, and how long it takes to understand the report's purpose.

Memo A

The Facilities people have been working on consolidating HQ Marketing Functions into the new building at Pebble Brook. As presently envisioned, Marketing Research will remain in its current location but be provided with additional space for expansion. The following functions will be moved into the new facility—Business Analysis, Special Applications, and Market Planning. It is expected that Public Sector will be relocated in a satellite location. The above moves will consolidate all of Marketing into the Pebble Brook location with the exception noted above.

Attached is a preliminary outline of the new building by floor and whom it will house. I am interested in knowing if this approach is in agreement with your thoughts.

This memo seems brief on the surface. It contains only 115 words. But let's measure brevity not by words but by the length of time it takes a reader to comprehend a message.

We feel that if you were really the addressee, you would have had to read this memo twice. Why? Because you didn't know *why* you were reading it until the last paragraph. Once you discover the purpose—that you are being asked to approve a plan—you want to reread the memo to see if you do, in fact, agree with the moves. The memo suddenly (and, unfortunately, at its end) informed you that you were on the hook. Obviously, there is a big difference between the way you will read a memo containing information of general (and casual) interest and one which requires you to make a decision involving the physical moving and reshuffling of hundreds of powerful and sensitive people. Yet organizationally this memo as presented does not show even a foggy awareness of this difference.

Now read the revision. Notice how your comprehension time would have dropped significantly had you received this revision instead of the original.

Memo B

Attached is a preliminary outline—by floor—of the new building at Pebble Brook and a statement of whom it will house. I am interested in knowing if this approach is in agreement with your thinking.

Our suggestion is that we make the following changes:

1.   Business Analysis, Special Applications, and Market Planning will. . . .
2.   Public Sector will be. . . .
3.   Marketing Research will remain. . . .

In terms of comprehension, the revision lets you know right away that you are expected to make a decision. Whether or not you would mull over that decision, we cannot tell. But we do know this: in terms of actual time expended in comprehending what is being asked of the reader, Memo B can be comprehended in one-third the time required by Memo A. And, if we measure brevity in terms of comprehension time, rather than number of words, Memo B produces a 66 percent savings in comprehension time.

We learned immediately in our study that comprehension time drops dramatically when a memo states its purpose—why it is being written for the reader—at the very beginning.

Confirm this point by reading Memo C, another disguised memo drawn from the division headquarters. Time your comprehension as before.

Memo C

The first of a series of meetings of the Strategic Marketing planning group will be held on Thursday, September 7, from 1 to 4 P.M. in Conference Room C. These important meetings are for the purpose of monitoring and suggesting changes in overall market strategies and product support. Attached is a list of those managers who should attend on a regular basis. These managers should specifically be prepared to review alternative strategies for the new product line. The purpose of this reminder is to ask your help in encouraging attendance and direct participation by your representatives. Please have them contact Frank Persons for any further information and to confirm their attendance.

Now read Memo D and compare comprehension time once again.

Memo D

Please encourage those of your managers whose names are listed on the attachment to attend regularly and directly participate in the meetings of the Strategic Marketing Planning group.

The next meeting is to be held on Thursday, September 7, from 1 to 4 P.M. in Conference Room C.

Please have your representative(s):

1. Contact Frank Persons for any further information and to confirm attendance.
2. Be prepared specifically to review alternative strategies for the new F–62 line.
3. Be ready to discuss changes in overall market strategies and product support.

Memo D is obviously more efficient. Why? Not only because it begins by stating its purpose but also because it itemizes the actions requested of the reader, organizing them in an easy-to-digest checklist. The original gives extensive background about the meetings but buries the requested action in a fat paragraph. Most readers would have to read the original at least twice just to be able to ferret out exactly what is being asked of them.

The time being saved, of course, seems insignificant on communications as short as Memos A and C. But consider how significant the savings would be corporate-wide if every manager's comprehension time in reading all messages could be reduced by even a small percentage.

HIGH COST OF COMPREHENSION

A recent study done by International Data Corporation states that managers in information industries spend an astonishing 60 percent of their time reading and writing; professionals spend 50 percent.[1] If cost accounted, how much would, say, a 20 percent to 30 percent savings in reading and writing time amount to for companies in this industry alone? The possibilities are arresting.

We made some cost estimates for the communications undertaken by the division headquarters we studied. The 9,000,000 messages distributed annually by the division headquarters included, of course, all sorts of mailings and multiple copies of such things as new product announcements, price changes, and the like, often running into the thousands of copies. Therefore, it was not fair to assume all 9,000,000 messages mailed were individually composed. Instead, we determined through conservative estimates that 12 percent of the 9,000,000 mailings were individual communications. And for each of these we will assume, for the purpose of this article, the ridiculously low figure of $10 to be the cost of creating, typing, and distributing. Based on these estimates, the minimum total composition cost for this one divisional headquarters would be $10,800,000 a year (see Table 1).

TABLE I Estimated Division Headquarters Writing Costs

Number of messages sent annually	9,000,000
Percent individually composed	12%
Total individually composed	1,800,000
Composition cost (per message)	$10
Minimum total composition cost	$10,800,000

[1] *Automated Business Communications: The Management Workstation.* (Framingham, Mass.: International Data Corporation, 1981): 21.

But writing time is, of course, only part of the story. What about the cost of comprehending all 9,000,000 of these messages? For while we estimate that only 12 percent of these messages were individually composed (that is, not copies), all 9,000,000 messages were presumably intended to be read. Again, in an attempt to dramatize through understatement, we will use a low salary figure: $20,000 per year. You can, of course, substitute the actual salary and other figures for your own company and determine for yourself at least roughly the magnitude of your company's reading costs.

As Table 2 shows, we approximated the division's minimum reading costs to be over $4,500,000. Of course, this figure is bound to be far below actual costs. Not only are our salary estimates unrealistically low, but we haven't taken into account the fact that many of the documents were read by multiple readers. In fact, many memos urged recipients to pass information on to colleagues and subordinates.

TABLE 2 Estimated Division Headquarters Reading Costs

Low-median salary	$20,000
	$10/hour
	17¢/minute
Average* reading time	3 minutes
Reading cost per document	$0.51
Reading cost for 9,000,000 messages	$4,590,000

* Assumed that some messages read in depth; some not given more than a glance; some barely looked at.

USING DIRECT PATTERNS

While these dollar figures were somewhat astonishing and the possibilities of dollar savings enticing, the company under study evidenced the greater concern about the waste of productivity involved in such inefficient communications having become the norm. The specter of hard-working employees' time being wasted by an inundation of inefficiently organized memos and letters was distressing. The company asked us to teach people how to report in a "bottom-line" fashion. Therefore, we taught people to tell readers immediately what was their purpose in writing and what they expected of the reader, if anything. If people had no purpose in writing, they probably shouldn't write in the first place. If they didn't expect anything of the reader and were just offering possibly useful information, we told them to say so right away.

In short, we were teaching people to use a direct organizational pattern. We were urging them to eschew the circuitous pattern in which writers, because of some sensitivity (real or imagined), withhold their purpose and do not let their readers know why they are being written to and what is being asked of them until their minds have been conditioned to accept the points the writers are trying to get across.

Obviously, there is nothing wrong with a circuitous organizational pattern in certain circumstances. But in this company, and we suspect in many other companies across the country, the circuitous pattern has become the norm for all types of communication in all situations.

Just ask yourself: how sensible is it to always write backwards, in a way that is just the opposite of how people comprehend information? Look at one more illustrative letter from our study:

Memo E

This is in reference to the letter sent you by Joe Smith of ABC Materials, Inc.

Mr. Smith requested information available from Product Analysis Reports (PAR's). As soon as information was made available to me from this source, I orally relayed the response to Mr. Smith.

Making use of the Planning Application Model, I was able to respond to Mr. Smith's request for further information about potential new products of possible interest to ABC. It was not until I received the copy of Mr. Smith's letter that I was aware that the data provided for him was not sufficient.

I have used all the resources that I am aware of to resolve Mr. Smith's concerns. Mr. Smith has informed me he is more than satisfied with the work done and considers the project completed. He has also announced an intention of doing further business with us.

Attached are copies of all the requests that I've been asked to submit during the six months that I have been assigned to this account. Also attached is a copy of an ABC analysis, submitted by my predecessor, which related to one of the items referenced in his letter. Upon request, I will forward copies of all of the relevant analysis that are in my files.

Where's the bottom line? What's this writer trying to get across? Isn't it the following?

Memo F

I have reviewed and acted upon the letter sent to you by Joe Smith of ABC Materials. Mr. Smith has informed me he is more than satisfied with the work done and considers the project completed. He has also announced an intention of doing further business with us.

Here in some detail are the steps I have taken for Mr. Smith. . . .

Since what this writer is reporting is good news, the communication is not sensitive. There's no need to report this information as circuitously as we might well be tempted to do if we had bungled the situation with Mr. Smith and had lost his business.

What percentage of all communications would you estimate to fall into the sensitive category that may call for a circuitous organizational pattern? Ten percent is the outer limit of possibility, unless one has a specialized job such as handling complaints, writing sales letters, dealing with shareholders, or the like. Why, then, upon analyzing these 2,000 sample letters, memos, and reports from this corporate division, did we find that almost all documents were organized circuitously? Why, in this extremely well-managed and successful company, was it the exceedingly rare letter or memo that did not bury its purpose somewhere in the third or fourth paragraph, and most frequently in the last sentence?

At the time, we had no idea of the etiology of the disease, but we felt we had a simple cure. We would tell people how to organize their thoughts so that the bottom line of their message would be immediately highlighted and promptly presented in everything they wrote. To facilitate this goal, we invented a series of bottom-line reporting principles which, if followed, would enable writers to communicate in a direct, straightforward, no-nonsense fashion in all situations that were not fraught with sensitivity. This would, we thought, save writing time and expense (see Table 3).

TABLE 3   Principles of Bottom-Line Reporting

Principle 1:	State your purpose first unless there are overriding reasons for not doing so.
Principle 2:	State your purpose first, even if you believe your readers need a briefing before they can fully understand the purpose of your communication.
Principle 3:	Present information in order of its importance to the reader.
Principle 4:	Put information of dubious utility or questionable importance to the reader into an appendix or attachment.
Principle 5:	In persuasive situations, where you do not know how your reader will react to what you ask for, state your request at the start in all cases except: a. Those where you don't (or barely) know the reader, and to ask something immediately of a relative (or absolute) stranger would probably be perceived as being "pushy." b. Those where the relationship between you and your reader is not close or warm.
Principle 6:	Think twice before being direct in negative messages upward.

And readers, too, would benefit—instead of having to search through a memo to find out what purpose the writer had in writing to them and what the writer wanted or expected them to do, they could look at the first paragraph and see the answer to these vital questions. Moreover, if the purpose and topic seemed irrelevant to their interests or needs, they could reject reading it, or merely give it a glance. Significant reading time and dollar savings should certainly result.

The program was instituted throughout the division. Did it work? Yes, in terms of getting the principles across and in terms of getting intellectual acceptance of these principles.

But getting emotional commitment to these principles was quite another story. We sensed in discussions that writers' commitment to being circuitous was not merely a bad habit. It was something else, something that was so ingrained that forcing personnel to be direct actually caused disquiet in many people. What could have been the reason?

PROGRAMMING FOR INEFFICIENCY

Obviously, people who work in large organizations were not born there. They have come to those organizations programmed by their social upbringing and by their educational experiences. And both of these earlier programmings strongly contribute to resistance to bottom-line reporting.

Social Upbringing

People seldom are conscious of how their social upbringing programs them to be indirect. Yet almost every sensitive social situation reinforces the wisdom of being circuitous, of not being direct. Aren't most brief answers to sensitive questions regarded as brusqueness or curtness, as being short with someone?

It begins early. The children are asked, "Do you want to go to Aunt Alice's house?" The children answer, honestly and directly, "No!" Unacceptable! The children are scolded and soon learn to beat around the bush the next time, all the while searching for some plausible excuse to forestall the visit. It is not surprising that as adults, the same children, when asked by the boss, "What do you think of my new plan? Think it'll work?" think twice before responding, "No!"

Educational Programming

Having been thoroughly programmed by their families that being direct is being impolite, the children now go to school. Here they soon learn that a twenty-page term report gets a high grade; a two-page report gets a low grade. A five-page answer to a test question is good; a one-paragraph answer is bad. Regardless of what teachers may profess, they invariably give extra credit for "effort." And effort is most easily measured by numbers of words or by pounds of pages. A premium is placed on long-windedness, and long-windedness is achieved by being circuitous rather than direct.

Indoctrination into Anxiety

On their first jobs in a large organization, young people are naturally nervous. They are very concerned that whatever they write or say not make people upset. They are also very concerned about "getting good grades." Therefore, they fall back upon the same behavior that was rewarded in school. They are going to do everything possible not to look lazy. They are going to be thorough in everything they write. Every chance to write a report to a superior is a chance to write that blockbuster of a term paper that could not fail to impress the boss. They are going to get that "A."

The fact that young people enter organizations at the bottom provides a final step in their programming for being circuitous and indirect. Everybody knows that writing *up* in an organization is far different from writing *down*. When young people enter an organization, the only direction they *can* write is up. Therefore, all the early experiences received in corporations consist of writing situations where they have to write information to people who are in fact, or may someday be, their superiors. Naturally, they become very uneasy.

Now let's suppose the company institutes a program to encourage personnel to be more direct. Imagine yourself as that newly hired young person in the organization. Are you going to believe any program suggesting that you be blunt and direct in your upward or lateral communications, when your entire lifetime programming has proved to you over and over again that bluntness is all too frequently suicidal? No chance!

Young people may give lip service to such a program, but in any real-life situation in which they feel threatened (in actuality, almost all situations) they will avoid coming to the point with an almost religious passion. And in negative or sensitive situations, the last thing they are going to do is state their purposes and requests directly.

By contrast, their higher level superiors, having enjoyed years of power positions in the hierarchy (from whence they could write down to anyone in any fashion they pleased), take a far different view of writing. The superiors now pride themselves on directness and bewail long-windedness on the part of their subordinates. But the subordinates' desires for self-preservation (reinforced by all their preorganizational programming) force them to give lip service at best to corporate attempts to "get to the point."

And, let's face it, in the corporate pyramid almost everybody is somebody's subordinate and, perhaps because of files, one never knows who is going to read what has been written. Therefore, circuitous writing is partly the habit of a lifetime and partly CYA.

WHAT YOUR ORGANIZATION CAN DO

Is a cure then impossible? Is inefficient, circuitous writing simply to be endured and its costs in lost productivity and wasted dollars merely written off?

No; a cure for inefficient writing is possible. But a thorough organization-wide cure requires that:

- People recognize and reject their social and educational programming for being circuitous in all non-sensitive writing situations. This deprogramming is the responsibility of the individual.
- People learn to write efficiently; that is, learn to organize their messages in such a way as to make it easy (and fast) for readers to comprehend the message. Teaching the bottom-line principles will impart this skill. But implementing the bottom-line principles requires strong high-level management support.
- People must develop the self-confidence necessary to send bottom-line messages upward in nonsensitive (or slightly sensitive) messages. A long-range cure depends to a great extent on attitude, on reducing the tensions inherent in superior-subordinate communications. Most writing insecurities stem from real, not imagined, failures on the part of superiors to communicate clearly and unequivocally their willingness to accept bottom-lined messages from subordinates.

Higher level executives have to appreciate how threatening directness can be to subordinates. Superiors, therefore, need to be persuaded not only to have the following credo taped to the wall above their desks but also communicated to all subordinates with whom they relate:

The Superior's Credo

1. I will ask all subordinates to be direct in their messages to me and I will not become angry if subordinates do so politely—even when those thoughts run counter to mine.
2. I will recognize and appreciate subordinates' attempts to conserve my time (and other readers' time) in all memos and reports they write to me, or for my signature.
3. I will work out with subordinates some general understanding of how much detail I require in various circumstances.
4. I will make clear to subordinates that I judge their communications not by length and weight, but by directness and succinctness.
5. And when I myself report up in the organization, I will be as direct as I expect my subordinates to be.

Once senior executives have adopted and put this credo into practice, all subordinates should recognize that there is now no excuse for them not to live by the following credo:

The Subordinate's Credo

1. I will have the courage (in all but the most sensitive or negative situations) to state at the beginning of messages my purpose in writing, exactly what information I am trying to convey, and/or precisely what action(s) I want my reader to take (if possible, itemized in checklist form for easy comprehension).
2. My readers, especially if they are my superiors, are extremely busy. I must not waste their time by making them read unnecessary undigested detail any more than I would waste their time chattering on in a face-to-face interview.
3. I will make a judgment as to how much my readers need to know in order to take the action required by the communication.
4. If I am in doubt as to whether specific information is necessary to my readers, I will either put this information in summary form in attachments, or tell readers that I stand ready to offer more information if so requested.
5. I will avoid the arsenal of the con man. If I want something of a superior, I will ask for it forthrightly.

Janis Fisher Chan

E-Mail: Presenting a Professional Image

The author of multiple books, Janis Fisher Chan is a developmental editor for Jossey-Bass Pfieffer and a member of the faculty at the University of California Business Extension where she teaches business and professional writing.

I was invited to submit a proposal to develop a strategic plan for a local nonprofit. I was swamped with work, but consultants never say no, so I put together a proposal really quickly and attached it to an e-mail to the executive director. A few days later I got a reply telling me that I was not one of the finalists. Actually—and I am ashamed to say this—the director told me that there were multiple mistakes in my e-mail and the attachment and that lack of detail made her concerned that I would not be able to handle all the details in the strategic planning process. I was very embarrassed. I wish I had spent the few extra minutes to proofread what I wrote.

—Belinda Kim, Consultant

Suppose you received the two e-mail messages that follow from people you'd never met. What would be your image of the writers? Of their organizations?

FROM: designer43@aol.com

TO: c.reilly@brcm.com

SUBJECT: files

i am sosorry I didn't get these to yu mon or tues, retrieving txt files isnt much trble & I told paul I would do it, they want these files for their archives so, as

(continued)

(*continued*)

longas i dont run into any tech probs ican get them 2 u t-day. Cant make RTF files frm Quark finals, can make ASCII txt files, will they do u any good atthis point. ill plan on getting them to u ASAP if u can still use them.

Fredo

FROM: customerservice@@Argona.com

TO: Nellie Michaels

SUBJECT: Problem

Your message to Argona about your problem has been recieved and forwarded to me for reply. Please be advised that you are one of our most valuable customers and and we are axious to be of any assistance that we can be. With regard to your e-mail, I am able to offer you the following information. We will be pleased to repair your computer accessory that would involve shipping it to our repair center but prior to shipping the equipment it is necessary that you first speak with a technician in advance. A shipping address cannot be given for our repair centers, however, a case number needs to be created, details about the problem need to be documented in that case. In order to reach a techncian please, consult the instruction manual that was received with your accessory.

I hope you will find this information helpful and that answers all of your questions. Thank you for visiting our site and contacting Argona. Please feel free to contact the undersigned should you have any additional questions.

Sincerley,

Mina

Fredo's message-mail is just plain sloppy, as if he had written it while doing something else. Mina's is wordy, difficult to follow, and filled with errors. Both messages are not only hard to read, they give recipients a poor impression of the writers and, by extension, of their organizations.

The e-mail you write says a lot about you. It tells readers that you are thorough, accurate, and attentive—or not. It indicates that your message is to be taken seriously—or not. It implies that you know what you're talking about—or not.

In this . . . [essay], you'll find guidelines for writing e-mail messages that convey a positive, professional image, including how to:

- Use active, concise, specific language and plain English that communicate clearly and accurately
- Write grammatically correct sentences that convey complete thoughts and flow smoothly
- Use gender-neutral language when possible
- Avoid common errors of punctuation

FOR A SEAMLESS STYLE: USE LANGUAGE THAT COMMUNICATES CLEARLY, ACCURATELY, AND CONCISELY

Sometimes you come across corporate gibberish so tortured, so triumphantly incomprehensible, you can only shake your head in admiration. . . .

—David Lazarus, "Gobbledygook Boils Down to Loss of Grace"
San Francisco Chronicle, April 3, 2005

Use Active Language

Active language is energetic and clear, while passive language weakens your writing and can confuse readers.

Here's the difference: In a passive sentence, the subject is acted upon, while in an active sentence, the subject does the acting. Notice that in the passive sentence below, the subject, or actor (the "executive committee"), comes after the verb, or action ("was made").

Passive   The decision was made by the executive committee.

In an active sentence, the actor comes first, before the action:

Active   The executive committee made the decision.

Here's another example. In this case, the "actor" in the active sentence is the implied "you":

Passive   It would be appreciated if a summary of your speech could be received by Thursday.

Active   [You] Please send us a summary of your speech by Thursday.

You can see that active sentences get the message across quickly and clearly, while passive sentences are longer and flatter in tone. In fact, one of the worst things about passive writing is that it implies the message isn't important, as in this example:

Passive   A fire is being experienced in my building. It would be appreciated if this could be put out as soon as possible.

Clearly, that message would be more effective in active language:

Active   My building is on fire—put it out right away!

Active language is particularly important when you are giving instructions. Which of these instructions could you follow more easily?

Passive   The battery should be charged for a minimum of four hours before the equipment is put into operation. The charging of the battery can be initiated by the insertion of the dock connector into the cable that is then connected to the power source.

Active   Charge the battery for at least four hours before using the equipment. Insert the dock connector into the cable and then connect the cable to the power source.

Obviously, the instructions written in active language would be easier to follow. Some passive sentences omit the actor altogether. In fact, that omission is often deliberate.

Passive   Last quarter's budget figures were overstated.

Leaving out the actor in sentences like that one raises the question of who took, is taking, or will take the action.

It's not wrong to use passive language occasionally. Sometimes you want to focus on the action, and the actor isn't very important:

Passive   After a three-hour search, the missing laptop was found on a shelf in the storage cabinet.

You might occasionally use passive language to vary the rhythms of your sentences. Keep in mind, however, that too much passive language saps energy from your writing and makes it wordy. Passive language can also confuse readers.

Pointers for Using Active Language

To make sure that you use active language when you write, do the following:

Put the actor before the action.

Passive   The proposal was prepared by Martine.
Active   Martine prepared the proposal.
          (actor ➜ action)

Say who acted, not just what was done.

Passive   A new procedure has been designed for processing invoices.
Active   Accounting has designed a new procedure for processing invoices.

When giving instructions, talk directly to readers. In this example, the actor is the implied "you."

| *Passive* | Your comments on the draft report are to be sent to Alex, not to me. |
| *Active* | [You] Please send your comments on the draft report to Alex, not to me. |

. . . Use Plain English

Most of us read our e-mail very quickly. If a message is hard to understand, we're likely to skip it and go on to something that's easier to read. Even when we do take the time to read a message that's written in inflated language, jargon, overly technical language, or unfamiliar slang, we might miss or misunderstand important points.

Avoid Inflated Language

How easily can you read this e-mail excerpt?

> Per your request, attached hereto are documents describing the parameters of the proposed system modifications. Prior to implementing these modifications, we request your assistance in facilitating distribution of said documents to the appropriate personnel.

Perhaps what that writer is really trying to say is this:

> As you asked, I've attached a description of the proposed system changes. Before we begin making these changes, we would appreciate your help in distributing the description to all the project participants.

. . . [Strunk and] White said it clearly in *The Elements of Style:* "Do not be tempted by a twenty-dollar word when there is a ten-center handy, ready and able."* Language that is more complicated than it needs to be forces readers to translate the writing into ordinary words, and their translation might not be what you intended to say. When we write, we sometimes forget that ordinary words communicate clearly and directly and still convey a professional image. . . .

Use Jargon and Technical Language Cautiously

Webster's *New Collegiate Dictionary* defines jargon as a "hybrid language" that is "used for communication between specific people"—or "the technical terminology or . . . idiom of a special activity or group."

The following sentence is an example of the way in which jargon fails to communicate clearly to people outside a specific group. Can you understand it?

> Since we replaced the BRK used for Wildhorse, we have not had a single re-occurrence of report timeouts with Seabrook in the ALNIB environment.

Jargon can be a handy form of shorthand—assuming that your readers understand the words, phrases, and acronyms you use. But if they don't, you might as well write in Martian. They won't even be able to use a dictionary to find the meaning of the terms they don't understand.

---

*William Strunk, Jr., and E. B. White, *The Elements of Style* 4th ed. (1999). p. 69.

Similarly, for readers who understand the terminology, technical language is often the most concise, accurate way of conveying certain concepts or information. But if people don't understand the technical language you use, your writing won't make much sense. . . .

Use Abbreviations with Caution

> Don't send messages littered with "This is 4 U" and words with no vowels, like "pls snd cmmts." People at work don't speak "teenage text message."
>
> —Gabriel Kasper, Consultant, Kasper and Associates

Can you understand this writer's message?

> Attchd inclds nms & adds of all sbcntrctrs in this prjct & $$ of assessment, pr yr req.

You could probably decipher the message if you worked hard enough:

> Attached includes names and addresses of all subcontractors in this project and cost of assessment, per your request.

By abbreviating words instead of spelling them out, the writer saved a few seconds at the reader's expense. But that kind of shorthand forces readers to figure out your abbreviations, and there's a good chance they'll miss—or misunderstand—something important.

Curtail Colloquialisms

What image do you get of the person who wrote the following?

> I'll pass this on to the head honcho, but in the end gotta say that's life here in the Wild West. I'm guessing there won't be any more probs but with deadlines hanging fire I hear ya.

It's a safe bet that most people would not think of this writer as a professional. Slang and colloquial language might be fine for an e-mail to your friends, but it's not very businesslike. Also, people in other countries—and in other age groups—probably won't understand the terms. You can never be sure where your e-mail messages will end up, so it's a good idea to use standard language. You can still keep the tone friendly and casual. . . .

Cut Out the Clutter: Eliminate Unnecessary Words

Words that aren't necessary to convey meaning or tone slow readers down. Too many unnecessary words also make your writing boring and shift the focus away from the important message.

Cluttered   The attachment to this e-mail message is an example of the Member Preferences Questionnaire which is used by the Member Services

Department for the purpose of conducting an evaluation of the preferences of our members for services of various types. It would be appreciated if this example questionnaire could be forwarded to the customer service team in order for it to serve as a model for the customer services questionnaire that they are in the process of developing.

(By the way, did you notice that the cluttered example uses a lot of passive language?)

Concise   I've attached an example of the questionnaire that Member Services uses to evaluate our members' preferences for various services. Please send it to the Customer Service team so they can use it as a model for the form they are developing.

Pointers for Getting Rid of Clutter

For strong, focused writing, try to eliminate every word that serves no clear purpose—every word that is not needed to convey meaning or carry tone.

Use only one word for a one-word idea.

Cluttered   We are in agreement with you about the contract terms.
The client made an offer to host the meeting.
Human Resources has made a decision to redesign its Web site.

Concise   We agree with you about the contract terms.
The client offered to host the meeting.
Human Resources decided to redesign in Web site.

Avoid unnecessary repetition. Writers have a tendency to say the same thing more than once. In particular, we tend to use repetitive phrases—several words when one or two would say the same thing. For example:

past experience   regular weekly meetings
end result   equally as effective as
ten a.m. in the morning future plans

Here are some more examples:

Cluttered   The future site of the building site is located at the northwest corner of 7th and B streets.
Until last week, our team had the best record to date.
The sales figures during the period of January 1 and June 30 of this year compare favorably with the sales figures during the period of January 1 and June 30 of the previous year.

Concise   The building site is at the northwest corner of 7th and B streets.
Until last week, our team had the best record.
This year's January 1–June 30 sales figures compare favorably with those of the same period of last year.

. . . Use Specific Language

Which version of this sentence conveys the most useful information?

Version 1   The building was destroyed in a disaster some time ago.
Version 2   Fire destroyed the apartment house in March 2003.

Notice how much more information Version 2 conveys: the kind of disaster, the kind of building, and what "some time ago" means.

Here's another example:

Version 1   Please complete the paperwork in a timely manner.
Version 2   Please fill out the benefits application and return it by June 7.

Vague words such as "in a timely manner" are likely to achieve vague results. The more precise and specific your language, the easier it is for readers to understand your message—and the more likely you are to get the results you need. . . .

STRENGTHEN THOSE SENTENCES: USE SHORT, SIMPLE, FOCUSED SENTENCES AND GOOD GRAMMAR

E-mail is a party to which English teachers have not been invited.

—Former university professor R. Craig Hogan
in "What Corporate America Can't Build: A Sentence"
The New York Times, December 7, 2004

Using active, concise, specific language and plain English will go a long way toward making sure that your writing communicates clearly and forcefully. Communicating clearly and conveying a professional image also require paying attention to the way you structure your sentences.

We're not going to pretend to cover all the grammar rules . . . [here]. Lots of goods reference books and self-teaching guides are available (we've listed several on our Web site). But . . . [in what follows] you'll find descriptions of several problems that commonly result in unprofessional e-mail messages:

- Misused modifiers
- Incomplete sentences
- Awkward and overly long sentences
- Incorrect subject-verb agreement
- Incorrect and unclear use of pronouns

Read Your Organization's Style Guide

Does your organization have a style guide for e-mail and other types of written communication? If so, be sure you have a copy and are familiar with its contents. If not, consider working with your colleagues to develop one.

Watch Those Modifiers

A modifier is a word or group of words that refers to or changes the meaning of (modifies) another word or group of words. Confusing or awkward sentences result when modifiers "dangle" or are misplaced.

Dangling Modifier   After meeting with me over lunch to discuss the contract, the project was cancelled.

As written, that sentence implies that the *project* met with the writer over lunch. The modifier is "After meeting with me over lunch to discuss the contract." But it's not clear what those words are modifying—they don't relate logically to the rest of the sentence.

To repair the sentence, you'd need to say *who* met over lunch. Here's one possibility:

Revision   After meeting with me over lunch to discuss the contract, the client canceled the project.

Here's another example:

Misplaced Modifier   The water from the broken water pipe almost flooded the entire warehouse.

Because of the placement of "almost," that sentence seems to say that the warehouse narrowly escaped being flooded. But the writer actually meant to say that the water flooded most of the warehouse. Moving the word "almost" makes the meaning clear:

Revision   The water from the broken water pipe flooded almost the entire warehouse.

Now it's easy to see that by putting "almost" in the wrong place in the original sentence, the writer sent the wrong message.

Misplaced Modifier   The customer service representatives were informed that they would be assigned to rotating shifts for the next two months by their manager.

Can you tell whether the manager informed the representatives or will assign them to rotating shifts—or both? The sentence reads more clearly when the modifying words, "by their manager," are in the right place:

Revision   The customer service representatives were informed by their manager that they would be assigned to rotating shifts for the next two months.

But you probably noticed that both the original sentence and the revision in that example use passive language. Passive language often yields unclear, awkward sentences. Here's an even better revision:

Revision   The manager informed the customer service representatives that she would be assigning them to rotating shifts for the next two months.

. . . Use Complete Sentences

A complete sentence conveys a complete thought. A fragment is just that—part of a thought.

> Ran into John Posner the other day. Interested in joining the new venture. Would like copy of business plan ASAP. What a good opportunity!

The reader could probably figure out most of this message if he or she knew the writer well. But the last line would still be unclear—what's the "good opportunity"?

For e-mail, it's not always essential to write full sentences. For example, if you're engaging in an e-mail conversation with a colleague or a small group, or answering a quick question, a few words might convey your message just fine:

> To consider at tomorrow's meeting: deadlines, deadlines, deadlines.
> Monday, March 6—OK.
> Sure—when we get budget approval.

But even if readers can easily figure out the meaning of your abbreviated sentences, keep in mind that fragmented writing can convey a sense that your message isn't very important. If it were, you'd probably have taken the time to write complete sentences that express complete thoughts. Also, keep in mind that your readers might forward an "abbreviated" message to someone who won't be able to understand it easily. . . .

Keep Sentences Short, Simple, and Focused

A sentence that's too long is like a person who keeps talking without taking a breath:

> Attached is a revised version of the on-line catalogue which includes the feedback we have received from the reviewers who responded to our requests for comments, and even though we are expecting several additional reviews to arrive this week we thought it would be helpful for you to see what we have been able to do so far and this version should be close enough to complete for you to proceed with your own part of the project. (78 words)

Long sentences are very hard to read, especially on a computer screen. See how much easier the message is to grasp when it's broken down into shorter sentences, each conveying one primary thought.

> Attached is a revised version of the on-line catalogue. (9 words) This version includes the feedback we have received from the reviewers who responded to our requests for comments. (18 words) Even though we are expecting several additional reviews to arrive this week, we thought it would be helpful for you to see what we have been able to do so far. (31 words) This version should be close enough to complete for you to proceed with your own part of the project. (19 words)

Even though the revision is easier to read, however, it still contains two very long sentences, and the average sentence length is 19 words.

Here's one more version, edited to remove unnecessary words that add to the sentence length:

> Attached is a revised version of the on-line catalogue. (9 words) This version includes the feedback we received from reviewers. (9 words) Even though we expect several more reviews this week, we thought it would be helpful for you to see what we have done so far. (25 words) This version should be complete enough for you to proceed with your part of the project. (16 words)

Now the average sentence length is 15 words, and no sentence is longer than 25 words. The revision conveys the message quickly and clearly.

There are no hard-and-fast rules for sentence length. But here are some pointers:

- Keep the *average* length of your sentences to 15–20 words, with one primary thought per sentence. Avoid linking several thoughts with "and."
- Avoid writing sentences that are longer than 28–30 words. People often have to read overly long sentences more than once to grasp their meaning.
- Lower the average sentence length when you are conveying highly technical information and/or communicating with readers whose first language is not English.

Avoid Choppy Sentences

Sometimes the problem isn't sentences that are too long, it's sentences that are too short and choppy:

> We have received your invoice. The invoice does not include your employer I.D. number. Please send a new invoice with the employer I.D. number. Then we can process your payment.

Those four sentences can easily be combined so that the message reads more smoothly:

> The invoice you sent does not include your employer I.D. number. Please send a new invoice so we can process your payment.

Notice that each of the sentences in the revised message quickly and clearly conveys one complete thought. . . .

Make Sure That Subjects and Verbs Agree

When they are used together, a subject and verb must agree in number. If the subject is singular, the verb must be singular; if the subject is plural, the verb must be plural.

It's not always easy to identify the subject to which the verb relates. Read the sentences below. Which verb completes each sentence correctly?

1. It is the variety of customization options that (make/makes) their product so popular.
2. Neither of the designers (have/has) met the deadline.
3. Our management committee (approve/approves) the budget.

4. The public (is/are) likely to be interested in the results of our investigation.
5. E-mail is one of those technological advances that often (take/takes) more of our time than it saves.

Although it would be easy to assume that the subjects of those sentences are all plural, the subject of each sentence is actually singular. Thus, the verb must also be singular.

In these revisions, the singular subject is underlined and the singular verb appears in boldface.

1. It is the <u>variety</u> of customization options that **makes** their product so popular.
2. <u>Neither</u> of the designers **has** met the deadline.
3. Our <u>management committee</u> **approves** the budget.
4. The <u>public</u> **is** likely to be interested in the results of our investigation.
5. <u>E-mail</u> is one of those technological advances that often **takes** more of our time than it saves.

. . . Use Pronouns Correctly

The English language uses many different types of pronouns to replace nouns and other pronouns. Common pronoun problems include:

- Using "me" or "myself" when "I" is correct
- Confusing readers by using pronouns that do not clearly refer to a noun or another pronoun
- Using gender-related pronouns when both genders are represented

Me, Myself, or I?

Which pronoun—me or I—would you put on each blank line in these sentences?

1. Alice, Nathan, and _____ will fly to New York on Thursday.
2. While the IT manager is away, you can speak with Michael DeBrer or _____ about any problems with your system.

I or Me?

To decide whether "I" or "me" is correct, remove the other person from the sentence and say the sentence aloud to yourself:

You can deliver the plans to the Production Supervisor or _____.
You can deliver the plans to **me**.
You can deliver the plans to the Production Supervisor or **me**.

Here are some guidelines:

- When the pronoun is the **subject** of the action, use I

  Alice, Nathan, and I will fly to New York on Thursday.

- When the pronoun is the **object** of the action or the preposition, use **me**

  While the IT manager is away, you can speak with Michael DeBrer or **me** about any problems with your system.

A common mistake is to use **myself** in place of **I** or **me,** as in this example:

Incorrect The engineer and **myself** will take the prototype to the client's office.

Correct The engineer and **I** will take the prototype to the client's office.

There are really only three situations in which you might use **myself** in business writing.

- For added emphasis

  I **myself** would have preferred to study another option before making a decision.

- To make the point that you did something on your own

  I spoke with each of the team members **myself.**

- When referring to yourself as you would refer to another person

  I consider **myself** an excellent candidate for that position.

Who or That?

- Use **who** when you are referring to people
- Use **that** when you are referring to animals and things

The associate **who** handled the sale did an excellent job, and the order **that** resulted was larger than we had expected.

That or Which?

Generally, use **that,** without a comma, when pointing to something specific, and which, with a comma, when adding information that is not essential:

Please send me the manuscript **that** is ready to be typeset.

That points to a specific manuscript—the one that is ready to be typeset.

Please send me the manuscript, **which** I will give to the editor for review.
The manuscript, **which** is on the table, is ready to be typeset.

Which adds information—what will be done with the manuscript in the first example, and where the manuscript is in the second example.

. . . Make Sure Pronouns Have Clear References

Pronouns that do not clearly refer to a noun or another pronoun can confuse readers:

> The engine in the delivery truck was in such bad shape that the mechanics had to replace **it.**

What did the mechanics have to replace—the engine or the truck?

> Sandra Weisman told Mona Seiple that **she** would be responsible for leading the new team.

Who would be responsible for leading the team—Sandra or Mona?

> We have approval to hire three temporary workers, and new computers will be installed this week. **This** means that we are more likely to meet our deadline.

Why are we more likely to meet the deadline—because of the temporary workers or the new computers, or both?

In those sentences, readers have no way of knowing which interpretation is correct. Not only do unclear pronoun references like those lead to confusion, they sometimes result in serious mistakes. See how much clearer the meaning is when the sentences are rewritten:

> The mechanics replaced the delivery truck's engine because it was in such bad shape.
>
> Sandra Weisman told Mona Seiple that Mona would be responsible for leading the new team.
>
> We have approval to hire three temporary workers, and new computers will be installed this week. The extra staff and computers mean that we are more likely to meet our deadline.

. . . Use Gender-Neutral Language

Pick up almost any business communication written before 1969 and you'll find that the writer assumed that "he" was writing to an audience of men. Here's an example:

> The ideal applicant will have a degree in business from an accredited four-year college. In addition, he will have worked in our industry or a related field for at least three years. He should be able to demonstrate an ability to solve problems and think critically.

Today, it is no longer acceptable to use pronouns that imply you are speaking only about men when women are also represented. The 21st Century version of that paragraph would read:

> Ideally, applicants will have a degree in business from an accredited four-year college. In addition, they will have worked in our industry or a related field for at least three years. They should be able to demonstrate an ability to solve problems and think critically.

To keep your writing as gender-neutral as possible, use one of these techniques:

Use plural instead of singular pronouns. In English, plural pronouns such as "they" and "their" do not refer to gender. Notice that we used plural pronouns to revise the sentence in the example above.

Eliminate the pronoun altogether. Instead of, "The manager needs to send his recommendations for budget cuts to the team in February," write, "The manager needs to recommend budget cuts to the team in February."

Speak directly to your reader. Instead of, "Any client who purchases a package before the official release date will receive a discount on his invoice," write, "If you purchase a package before the official release date, you will receive a discount on your invoice."

Structure the sentence so you can use "who." Instead of, "If a new employee begins work during the last week of a month, he will attend the orientation session during the first week of the next month," write, "A new employee who begins work during the last week of a month will attend the orientation session during the first week of the next month."

Avoid He/She and S/He

Attempts to keep sentences gender-neutral sometimes result in very awkward constructions. We recommend using "he and she" or "he or she" instead, and only when you can't come up with a better alternative.

Evolving Grammar

Although still not technically correct, it's becoming increasingly acceptable to use the plural pronouns "their," "they," and "them" with singular indefinite pronouns such as "everyone." For example, instead of writing "Human Resources asked everyone to select his/her new benefits plan before March 1," most people think it would be okay to write, "Human Resources asked everyone to select their new benefits plan before March 1." Even better, you could say, "Human Resources asked all employees to select their new benefits plan before March 1."

... PUNCTUATE PROPERLY: USE COMMAS, SEMICOLONS, AND OTHER PUNCTUATION MARKS TO HELP CONVEY YOUR MESSAGE

Some people think that punctuation isn't important anymore. But they're wrong—punctuation always matters. Punctuation does for writing what the pauses, shifts in tone, emphasized words, and gestures do in a conversation—it helps make your meaning clear. Missing or incorrect punctuation can confuse readers; it can even change the meaning.

Notice how one little comma in the next example prevents an unpleasant implication and makes the meaning of the sentence clear:

Unclear	As you know nothing changed as a result of the investigation.
Clear	As you know, nothing changed as a result of the investigation.

The most common—and the most commonly misused—punctuation mark is the comma. Other important marks to think about are the semicolon, the colon, and the apostrophe.

!!!!!

E-mail writers seem to love exclamation marks! They're okay, but use them sparingly! They can make your writing seem very unprofessional!!!

Pointers for Using Commas*

Avoid comma splices. You can (and usually should) use a comma when you join two independent clauses (complete sentences) with a coordinating conjunction (and, but, for, nor, or, so, yet). But you can't use a comma by itself to splice the sentences together:

Wrong	We can expect some changes in the project timeline, we will do our best to meet your deadline.
Right	We can expect some changes in the project timeline, but we will do our best to meet your deadline.

Don't set off essential information with commas. You can (and usually should) use commas to set off information that could be removed without changing the meaning. But don't put commas around information that needs to be there:

No Commas Wanted	All clients who have not responded to the survey should be sent a follow-up e-mail.

Taking out the words "who have not responded to the survey" would change the meaning to say that *all* clients should be sent a follow-up e-mail, not just those who have not responded. So you would not separate those words from the rest of the sentence with commas:

Commas Needed	Linda Gomez, who has been our CFO for the past five years, is leaving the company to start a business of her own.

---

*Commas are so tricky that we've written an entire book on the subject. *Just Commas: Nine Basic Rules* is available on our Web site (www.writeitwell.com), where you'll also find comma tips available for free download.

The words "who has been our CFO for the past five years" convey extra information that could be removed without changing the meaning of the sentence. That's why you would set them off with commas.

"Do I Need to Use a Comma after the Last Item in a Series?"

The answer to this common question is, "Sometimes." Commas make it clear that each item in the series is a separate item. It's never wrong to use a comma before the "and" in a series, so we think that it's easier to use it all the time.

The budget includes funds for hiring new staff, providing computer software training for everyone on the development team, and expanding the library.

Don't use a comma to separate a group of words from the subject of the sentence. Be careful not to use a coma to separate a complete sentence from a dependent clause (a group of words that depends on the rest of the sentence for its meaning):

Wrong   The team leader is responsible for setting the agenda, and for scheduling the meeting.

The comma in that sentence creates a sentence fragment because the words "scheduling the meeting" don't stand on their own as a complete sentence:

Right   The team leader is responsible for setting the agenda and for scheduling the meeting.

As a rule, use a comma after an introductory clause. If the clause is short and there is absolutely no chance of misunderstanding, you could leave the comma out. But it's never wrong to use it:

Comma Needed   Geographically closer to Atlanta than Miami, Tallahassee is more like a traditional southern city than its neighbors to the south.

Comma Suggested   After a three-hour meeting(,) the directors adjourned without making a decision.

Pointers for Using Semicolons

Now, you should be warned that few adults do understand semicolons; other adults just think they understand.

—*"Between a Comma and a Period"*
The New York Times, *November 7, 1999*

The easiest way to think about semicolons are that they are more than commas but less than periods. They have two primary uses:

- To join two closely related independent clauses (sentences) without using a coordinating conjunction

    Tests of the reconfigured system will begin in two weeks; we hope for better results this time.

    We sent invitations to 500 of our most valued customers; however, only 173 people had responded by last Friday.

- To separate elements in a series that already contains commas

    The transition will be led by Amelia Margolis, manager of marketing; Dion Trang, manager of IT; and Milas Lupchek, director of customer relations.

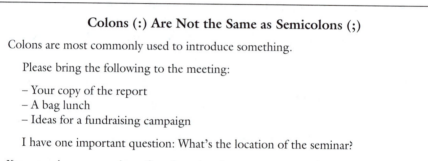

Colons (:) Are Not the Same as Semicolons (;)

Colons are most commonly used to introduce something.

Please bring the following to the meeting:

– Your copy of the report
– A bag lunch
– Ideas for a fundraising campaign

I have one important question: What's the location of the seminar?

You can also use a colon after the salutation on your e-mail, although these days a comma is also acceptable, and some people think it's more friendly to use a comma.

Dear Ms. Durphy:
Dear Ms. Durphy,

Pointers for Using Apostrophes

Another punctuation mark that writers often misuse is the apostrophe. Apostrophes are used to signify ownership and to replace the missing letter in contractions.

Apostrophes used to signify ownership. Notice that the apostrophe is placed differently, depending on whether the word that owns or possesses something is singular or plural. Generally, the apostrophe goes before the "s" when the word is singular and after the "s" when the word is plural:

    Our company's products
    The client's concerns
    Several employees' questions
    The computers' defects

But *be careful:* Some words are already plural in form. In those cases, the apostrophe goes *before* the "s":

> The children's books
> The women's careers

Also, these *possessive pronouns* do not need apostrophes because they already indicate ownership:

its	theirs	hers	his
whose	yours	ours	

Using Apostrophes with Plural Nouns Ending in "s"

To show possession with a plural noun that ends in "s," add only an apostrophe:

> The computers' keyboards
> The Williams' shipment

Apostrophes that replace the missing letter in contractions. Here are some common contractions:

It is = it's	does not = doesn't	we would = we'd
cannot = can't	do not = don't	who is = who's
I have = I've	should not = shouldn't	I will = I'll

You can use contractions to convey a casual, friendly tone:

Formal   It is a good idea to send hard copies of documents to anyone who cannot print them out before the meeting.

Casual   It's a good idea to send hard copies of documents to anyone who can't print them out before the meeting.

"Its" and "It's" Are Different Words

Be careful not to confuse the possessive pronoun "its," which needs no apostrophe, with the contraction "it's," which does.

Wrong   The taxi lost **it's** wheel coming around the curve.
Right   The taxi lost **its** wheel coming around the curve.

If you're not sure whether to use "its" or "it's," remember that an apostrophe in a contraction means something has been left out. The letter that has been left out of "it's" is the "i" in "it is." You wouldn't say, "The taxi lost **it is** wheel. . . ." So "its" must be the right form.

Pointers for Using Dashes and Parentheses

Dashes and parentheses make your writing more expressive. Both types of punctuation are used to set off information that is not essential to the meaning of a sentence. Dashes tend to highlight the information, while parentheses tend to minimize it or play it down. You can see how that works in the two versions of the same sentence that follow:

> We are pleased to welcome Dewey Freire—former CEO of Elements.com—to our advisory board.

> We are pleased to welcome Deway Freire (former CEO of Elements.com) to our advisory board.

Tips for Using Dashes

If the information you are setting off comes in the middle of a sentence, use a dash before *and after* the phrase or clause:

Use Hyphens to Create a Dash

If you can't easily create dashes on your e-mail system, it's okay to use two hyphens:

> Thanks for giving me your travel dates--I'll put them on my calendar.

Wrong	The ideal candidate will have good people skills—friendliness, patience, and an ability to listen, and the ability to learn new skills quickly.
Right	The ideal candidate will have good people skills—friendliness, patience, and an ability to listen—and the ability to learn new skills quickly.

If the information you are setting off comes at the end of a sentence, you need to use only one dash:

> The system was out of service for six full days this month—we need to replace it immediately.

Tips for Using Parentheses

Common errors include leaving out the closing mark, enclosing too much text in parentheses, and using parentheses too often.

Be sure to use both an opening *and* a closing parenthesis:

Wrong	As I mentioned, I'll be out of town (without access to e-mail or voice mail) from Nov. 12 to Nov. 23.
Right	As I mentioned, I'll be out of town (without access to e-mail or voice mail) from Nov. 12 to Nov. 23.

Don't overuse parentheses. The words in parentheses are usually extra information that you could either leave out, add to the main sentence, or put into a separate sentence. Notice how the parenthetical phrases interrupt this paragraph:

> Are you and Ari now viewing the Planning Guide (the one you drafted last spring) as one of the components (in addition to the Development Guide)? If I recall correctly from our last meeting (the one in the new offices on Spring Street), Ari mentioned the possibility of a separate component (the Planning Guide?).

Keep parenthetical statements short. They should never be longer than the sentence in which they are enclosed.

Use parentheses the first time you mention an acronym:

> Please send the course outline to me, care of the Education Development Department (EDD).

Part 4

Reports and Other Longer Documents

Nearly everyone in business and industry writes reports. The word *report* is really just a generic term for a variety of documents that vary in form and purpose. Some reports are purely informative; others are persuasive or argumentative. Reports can be simple check lists, or they can take the form of interoffice memos and e-mails, letters to clients, or more full-blown documents that are the results of weeks, if not months, of effort.

Business and technical writing practice sometimes distinguishes between formal and informal reports. Formal reports generally follow a multi-part format and are used primarily to present the results of a detailed project. Such a project often involves a considerable outlay of capital plus time and effort. The format for a formal report may mandate a cover letter or a memo of transmittal attached to a bound document consisting of an abstract, a table of contents, a glossary, an introduction, a detailed discussion of all aspects of the topic, a set of conclusions and recommendations, and pages and pages of attachments.

Informal reports tend to be shorter documents. Their formats are less complex, consisting only of such essential items as an introduction, a discussion, a set of conclusions, and, where appropriate, a list of recommendations.

To ensure that their reports are useful, writers should take the same kind of process approach that they would use when writing any other business or technical document. They should plan their reports carefully from the start so that they can clearly define and stick to their intended purpose whether that purpose be to provide information, to analyze information and draw conclusions, or to make recommendations.

Because they are often action-oriented, reports require writers to analyze their audience—or audiences—carefully. By virtue of their length, some reports can intimidate readers who need to know the information they contain.

A classification of the kinds of audiences that business and technical writing address, which Thomas Pearsall first delineated in his *Audience Analysis for Technical Writing* (1969), is especially relevant to the audiences for reports. Those audiences can include any, or all, of the following:

- the layperson
- the executive
- the expert
- the technician
- the operator.

Each brings a different background and a different set of needs to his or her reading of a report that writers must take into account if they hope to produce an effective document. In general, using abstracts and visual aids to supplement the contents of both formal and informal reports will make those reports more accessible to larger groups of readers of varying degrees of expertise.

Proposals are in some ways simply specialized reports, although they can be written in letter, memo, or e-mail form as well. The primary purpose of a proposal is to persuade readers to do something. Sales letters and requests for adjustments are two fairly simple proposals. Internal proposals, which are aimed at changing policies and procedures within an organization, and external proposals, such as documents seeking grants or funding, are more complex persuasive documents. Because proposals aim at convincing an audience to act in a way the writer wants, a process approach can help lead writers to a successful proposal.

This section of *Strategies* begins with an essay by J. C. Mathes and Dwight W. Stevenson that returns to a familiar theme in this anthology: the importance of audience analysis. Mathes and Stevenson offer a detailed examination of the problem of audience analysis and a solution to that problem designed to meet the needs of the several audiences that reports in particular, and other business and technical documents in general, address.

Underscoring the need for attention to audience, Robert W. Dodge reports on the results of a survey done at Westinghouse in the 1960s to determine the reading habits of busy managers. Those results still have relevance for business and technical writers today. Dodge concludes that, while the body of the report may have required the most time for the writer to produce, busy managers read this section of the report with considerably less frequency than they read the other sections, usually giving more attention to the abstract. Christian K. Arnold further underscores the importance of the abstract in meeting audience needs in his discussion of the key elements of an effective abstract. Vincent Vinci then offers what amounts to a checklist that business and technical report writers can use as a last step in an effective writing process.

Supposedly a picture is worth a thousand words. In creating and using illustrations to supplement written text, business and technical writers need to make sure that those thousand words are the ones they intended. Walter E. Oliu, Charles T. Brusaw, and Gerald J. Alred offer easy-to-follow instructions for the effective use of some of the most widely used visual aids.

David W. Ewing then discusses persuasive writing, addressing the wider issue of the different strategies needed to persuade rather than inform readers. Ewing's general comments are then complemented by those of Philip C. Kolin about how to write effective proposals. This section of *Strategies* ends with advice from Richard Johnson-Sheehan about how writers can use matters of style better to persuade an audience to accept a business or technical proposal.

J. C. Mathes and Dwight W. Stevenson

Audience Analysis: The Problem and a Solution

J. C. Mathes and Dwight W. Stevenson are both Professors Emeriti of Technical Communications in the College of Engineering at the University of Michigan.

Every communication situation involves three fundamental components: a writer, a message, and an audience. However, many report writers treat the communication situation as if there were only two components: a writer and his message. Writers often ignore their readers because writers are preoccupied with their own problems and with the subject matter of the communication. The consequence is a poorly designed, ineffective report.

As an example, a student related to the class her first communication experience on a design project during summer employment with an automobile company. After she had been working on her assignment for a few weeks, her supervisor asked her to jot him a memo explaining what she was doing. Not wanting to take much time away from her work and not thinking the report very important, she gave him a handwritten memo and continued her technical activities. Soon after, the department manager inquired on the progress of the project. The supervisor immediately responded that he had just had a progress report, and thereupon forwarded the engineer's brief memo. Needless to say, the engineer felt embarrassed when her undeveloped and inadequately explained memo became an official report to the organization. The engineer thought her memo was written just to her supervisor, who was quite familiar with her assignment. Due to her lack of experience with organizational behavior, she made several false assumptions about her report audience, and therefore about her report's purpose.

The inexperienced report writer often fails to design his report effectively because he makes several false assumptions about the report writing situation. If the writer would stop to analyze the audience component, he would realize that:

1. It is false to assume that the person addressed is the audience.
2. It is false to assume that the audience is a group of specialists in the field.
3. It is false to assume that the report has a finite period of use.
4. It is false to assume that the author and the audience always will be available for reference.
5. It is false to assume that the audience is familiar with the assignment.
6. It is false to assume that the audience has been involved in daily discussions of the material.
7. It is false to assume that the audience awaits the report.
8. It is false to assume that the audience has time to read the report.

Assumptions one and two indicate a writer's lack of awareness of the nature of his report audience. Assumptions three, four, and five indicate his lack of appreciation of the dynamic nature of the system. Assumptions six, seven, and eight indicate a writer's lack of consideration of the demands of day-by-day job activity.

A report has value only to the extent that it is useful to the organization. It is often used primarily by someone other than the person who requested it. Furthermore, the report may be responding to a variety of needs within the organization. These needs suggest that the persons who will use the report are not specialists or perhaps not even technically knowledgeable about the report's subject. The specialist is the engineer. Unless he is engaged in basic research, he usually must communicate with persons representing many different areas of operation in the organization.

In addition, the report is often useful over an extended period of time. Each written communication is filed in several offices. Last year's report can be incomprehensible if the writer did not anticipate and explain his purpose adequately. In these situations, even within the office where a report originated, the author as well as his supervisor will probably not be available to explain the report. Although organizational charts remain unchanged for years, personnel, assignments, and professional roles change constantly. Because of this dynamic process, even the immediate audience of a report sometimes is not familiar with the writer's technical assignment. Thus, the report writer usually must design his report for a dynamic situation.

Finally, the report writer must also be alert to the communication traps in relatively static situations. Not all readers will have heard the coffee break chats that fill in the details necessary to make even a routine recommendation convincing. A report can arrive at a time when the reader's mind is churning with other concerns. Even if it is expected, the report usually meets a reader who needs to act immediately. The reader usually does not have time to read through the whole report; he wants the useful information clearly and succinctly. To the reader, time probably is the most important commodity. Beginning report writers seldom realize they must design their reports to be used efficiently rather than read closely.

The sources of the false assumptions we have been discussing are not difficult to identify. The original source is the artificial communication a student is required to perform in college. In writing only for professors, a student learns to write for audiences of one, audiences who know more than the writer knows, and audiences who have no instrumental interests in what the report contains. The subsequent source, on the job, is the writer's natural attempt to simplify his task. The report writer, relying upon daily contact and familiarity, simply finds it easier to write a report for his own supervisor than to write for a supervisor in a different department. The writer also finds it easier to concentrate upon his own concerns than to consider the needs of his readers. He finds it difficult to address complex audiences and face the design problems they pose.

AUDIENCE COMPONENTS AND PROBLEMS THEY POSE

To write a report you must first understand how your audience poses a problem. Then you must analyze your audience in order to be able to design a report structure that provides an optimum solution. To explain the components of the report audience you must do more than just identify names, titles, and roles. You must determine who your audiences are as related to the purpose and content of your report. "Who" involves the specific operational functions of the persons who will read the report, as well as their educational and business backgrounds. These persons can be widely distributed, as is evident if you consider the operational relationships within a typical organization.

Classifying audiences only according to directions of communication flow along the paths delineated by the conventional organizational chart, we can identify three types of report audiences: *horizontal, vertical,* and *external.* For example, in the organization chart in Figure 1, *Part of Organization Chart for Naval Ship Engineering Center,*[1] horizontal audiences exist on each level. The Ship Concept Design Division and the Command and Surveillance Division form horizontal audiences for each other. Vertical audiences exist between levels. The Ship Concept Design Division and the Surface Ship Design Branch form vertical audiences for each other. External audiences exist when any unit interacts with a separate organization, such as when the Surface Ship Design Branch communicates with the Newport News Shipbuilding Company.

What the report writer first must realize is the separation between him and any of these three types of audiences. Few reports are written for horizontal audiences within the same unit, such as from one person in the Surface Ship Design Branch to another person or project group within the Surface Ship Design Branch itself. Instead, a report at least addresses horizontal audiences within a larger framework, such as from the Surface Ship Design Branch to the Systems Analysis Branch. Important reports usually have complex audiences, that is, vertical and horizontal, and sometimes external audiences as well.

An analysis of the problems generated by horizontal audiences—often assumed to pose few problems—illustrates the difficulties most writers face in all

---

[1]A reference in H. B. Benford and J. C. Mathes, *Your Future in Naval Architecture,* Richards Rosen, New York, 1968.

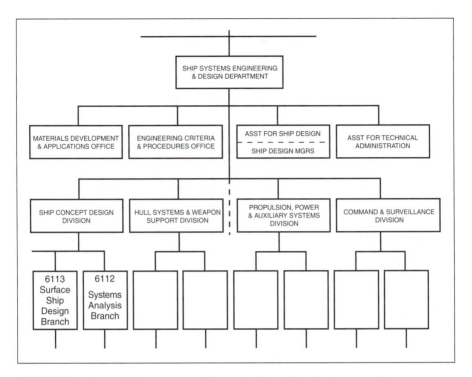

FIGURE 1   Part of Organization Chart for Naval Ship Engineering Center

report writing situations. A systems engineer in the Systems Analysis Branch has little technical education in common with the naval architect in the Surface Ship Design Branch. In most colleges he takes only a few of the same mathematics and engineering science courses. The systems engineer would not know the wave resistance theory familiar to the naval architect, although he could use the results of his analysis. In turn, the naval architect would not know stochastics and probability theory, although he could understand systems models. But the differences between these audiences and writers go well beyond differences in training. In addition to having different educational backgrounds, the audiences will have different concerns, such as budget, production, or contract obligations. The audiences will also be separated from the writer by organizational politics and competition, as well as by personality differences among the people concerned.

When the writer addresses a horizontal audience in another organizational unit, he usually addresses a person in an organizational role. When addressed to the role rather than the person, the report is aimed at a department or a group. This means the report will have audiences in addition to the person addressed. It may be read primarily by staff personnel and subordinates. The addressee ultimately may act on the basis of the information reported, but at times he serves only to transfer the report to persons in his department who will use it. Furthermore, the report may have audiences in addition to those in the department addressed. It may be forwarded to other persons elsewhere, such as lawyers and comptrollers. The report travels routinely throughout organizational paths, and will have unknown or unanticipated audiences as well.

Consequently, even when on the same horizontal organizational level, the writer and his audience have little in common beyond the fact of working for the same organization, of having the same "rank" and perhaps of having the same educational level of attainment. Educational backgrounds can be entirely different; more important, needs, values, and uses are different. The report writer may recommend the choice of one switch over another on the basis of cost-efficiency analysis; his audiences may be concerned for business relationships, distribution patterns, client preferences, and budgets. Therefore, the writer should not assume that his audience has technical competence in the field, familiarity with the technical assignment, knowledge of him or of personnel in his group, similar value perspectives, or even complementary motives. The differences between writer and audience are distinctive, and may even be irreconcilable.

The differences are magnified when the writer addresses vertical audiences. Reports directed at vertical audiences, that is, between levels of an organization chart, invariably have horizontal audience components also. These complex report writing situations pose significant communication problems for the writer. Differences between writer and audience are fundamental. The primary audiences for the reports, especially informal reports, must act or make decisions on the basis of the reports. The reports thus have only instrumental value, that is, value insofar as they can be used effectively. The writer must design his report primarily according to how it will be used.

In addition to horizontal audiences and to vertical audiences, many reports are also directed to external audiences. External audiences, whether they consist of a few or many persons, have the distinctive, dissimilar features of the complex vertical audience. With external audiences these features invariably are exaggerated, especially those involving need and value. An additional complication is that the external audience can judge an entire organization on the basis of the writer's report. And sometimes most important of all, concerns for tact and business relationships override technical concerns.

In actual practice the writer often finds audiences in different divisions of his own company to be "external" audiences. One engineer encountered this problem in his first position after graduation. He was sent to investigate the inconsistent test data being sent to his group from a different division of the company in another city. He found that the test procedures being used in that division were faulty. However, at his supervisor's direction he had to write a report that would not "step on any toes." He had to write the report in such a manner as to have the other division correct its test procedures while not implying that the division was in any way at fault. An engineer who assumes that the purpose of his report is just to explain a technical investigation is poorly prepared for professional practice.

Most of the important communication situations for an engineer during his first five years out of college occur when he reports to his supervisor, department head, and beyond. In these situations, his audiences are action-oriented line management who are uninterested in the technical details and may even be unfamiliar with the assignment. In addition, his audiences become acquainted with him professionally through his reports; therefore, it is more directly the report than the investigation that is important to the writer's career.

Audience components and the significant design problems they pose are well illustrated by the various audiences for a formal report written by an engineer on the development of a process to make a high purity chemical, as listed in Figure 2, *Complex Audience Components for a Formal Report by a Chemical Engineer on a Process to Make a High Purity Chemical.* The purpose of the report was to explain the process; others would make a feasibility study of the process and evaluate it in comparison to other processes.

The various audiences for this report, as you can determine just by reading their titles, would have had quite different roles, backgrounds, interests, values, needs, and uses for the report. The writer's brief analysis of the audiences yielded the following:

> He could not determine the nature of many of his audiences, who they were, or what the specifics of their roles were.

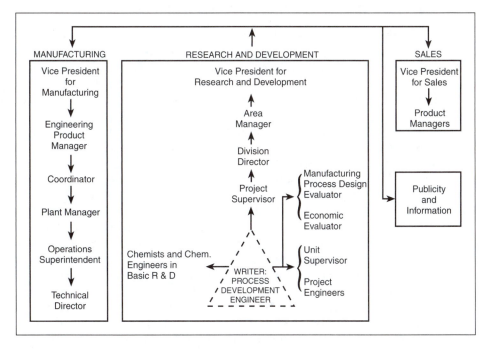

FIGURE 2 Complex Audience Components for a Formal Report by a Chemical Engineer on a Process to Make a High Purity Chemical

His audiences had little familiarity with his assignment.

His report would be used for information, for evaluation of the process, and for evaluation of the company's position in the field.

Some of his audiences would have from a minute to a half hour to glance at the report, some would take the report home to study it, and some would use it over extended periods of time for process analysis and for economic and manufacturing feasibility studies.

The useful lifetime of the report could be as long as twenty years.

The report would be used to evaluate the achievements of the writer's department.

The report would be used to evaluate the writing and technical proficiencies of the writer himself.

This report writer classified his audiences in terms of the conventional organization chart. Then to make them more than just names, titles, and roles he asked himself what they would know about his report and how they would use it. Even then he had only partially solved his audience problem and had just begun to clarify the design problems he faced. To do so he needed to analyze his audiences systematically.

A METHOD FOR SYSTEMATIC AUDIENCE ANALYSIS

To introduce the audience problem that report writers must face, we have used the conventional concept of the organization chart to classify audiences as *horizontal, vertical,* and *external.* However, when the writer comes to the task of performing an instrumentally useful audience analysis for a particular report, this concept of the organization and this classification system for report audiences are not very helpful.

First, the writer does not view from outside the total communication system modeled by the company organization chart. He is within the system himself, so his view is always relative. Second, the conventional outsider's view does not yield sufficiently detailed information about the report audiences. A single bloc on the organization chart looks just like any other bloc, but in fact each bloc represents one or several human beings with distinctive roles, backgrounds, and personal characteristics. Third, and most importantly, the outsider's view does not help much to clarify the specific routes of communication, as determined by audience needs, which an individual report will follow. The organization chart may describe the organization, but it does not describe how the organization functions. Thus many of the routes a report follows—and consequently the needs it addresses—will not be signaled by the company organization chart.

In short, the conventional concept of report audiences derived from organization charts is necessarily abstract and unspecific. For that reason a more effective method for audience analysis is needed. In the remaining portion of this . . . [selection], we will present a three-step procedure. The procedure calls for preparing an egocentric organization chart to identify individual report readers, characterizing these readers, and classifying them to establish priorities. Based upon an egocentric view of the organization and concerned primarily with what report readers need, this system should yield the information the writer must have if he is to design an individual report effectively.

Prepare an Egocentric Organization Chart

An egocentric organization chart differs from the conventional chart in two senses. First, it identifies specific individuals rather than complex organizational units. A bloc on the conventional chart may often represent a number of people, but insofar as possible the egocentric chart identifies particular individuals who are potential readers of reports a writer produces. Second, the egocentric chart categorizes people in terms of their proximity to the report writer rather than in terms of their hierarchical relationship to the report writer. Readers are not identified as organizationally superior, inferior, or equal to the writer but rather as near or distant from the writer. We find it effective to identify four different degrees of distance as is illustrated in Figure 3, *Egocentric Organization Chart*. In this figure, with the triangle representing the writer, each circle is an individual reader identified by his organizational title and by his primary operational concerns. The four degrees of distance are identified by the four concentric rings. The potential readers in the first ring are those people with whom the writer associates daily. They are typically those people in his same office or project group. The readers in the second ring are those people in other offices with whom the writer must normally interact in order to perform his job. Typically, these are persons in adjacent and management groups. The readers in the third ring are persons relatively more distant but still within the same organization. They are

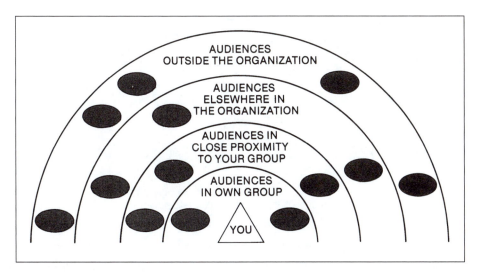

FIGURE 3 Egocentric Organization Chart

distant management, public relations, sales, legal department, production, purchasing, and so on. They are operationally dissimilar persons. The readers in the fourth ring are persons beyond the organization. They may work for the same company but in a division in another city. Or they may work for an entirely different organization.

Having prepared the egocentric organization chart, the report writer is able to see himself and his potential audiences from a useful perspective. Rather than seeing himself as an insignificantly small part of a complex structure—as he is apt to do with the conventional organizational chart—the writer sees himself as a center from which communication radiates throughout an organization. He sees his readers as individuals rather than as faceless blocs. And he sees that what he writes is addressed to people with varying and significant degrees of difference.

A good illustration of the perspective provided by the egocentric organization chart is the chart prepared by a chemical engineer working for a large corporation, Figure 4, *Actual Egocentric Organization Chart of an Engineer in a Large Corporation*. It is important to notice how the operational concerns of the persons even in close proximity vary considerably from those of the development engineer. What these people need from reports written by this engineer, then, has little to do with the processes by which he defined his technical problems.

The chemical engineer himself is concerned with the research and development of production processes and has little interest in, or knowledge of, budgetary matters. Some of the audiences in his group are chemists concerned with

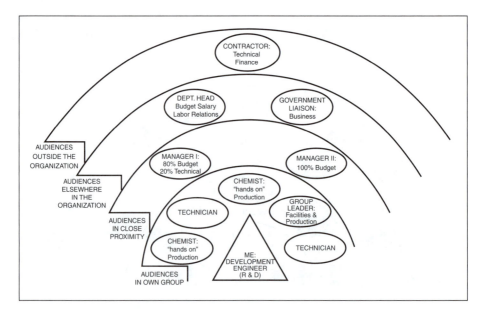

FIGURE 4 Actual Egocentric Organization Chart of an Engineer in a Large Corporation

production—not with research and development. Because of this they have, as he said, "lost familiarity with the technical background, and instead depend mostly on experience." Other audiences in his group are technicians concerned only with operations. With only two years of college, they have had no more than introductory chemistry courses and have had no engineering courses.

Still another audience in his group is his group leader. Rather than being concerned with development, this reader is concerned with facilities and production operations. Consequently, he too is "losing familiarity with the technical material." Particularly significant for the report writer is that his group leader in his professional capacity does not use his B.S.Ch.E. degree. His role is that of manager, so his needs have become administrative rather than technical.

The concerns of the chemical engineer/report writer's audiences in close proximity to his group change again. Instead of being concerned with development or production operations, these audiences are primarily concerned with the budget. They have little technical contact, and are described as "business oriented." Both Manager I and Manager II are older, and neither has a degree in engineering. One has a Ph.D. degree in chemistry, the other an M.S. degree in technology. Both have had technical experience in the lab, but neither can readily follow technical explanations. As the chemical engineer said, both would find it "difficult to return to the law."

The report writer's department head and other persons through whom the group communicates with audiences elsewhere in the organization, and beyond it, have additional concerns as well as different backgrounds. The department head is concerned with budget, personnel, and labor relations. The person in contact with outside funding units—in this case, a government agency—has business administration degrees and is entirely business oriented. The person in contact with subcontractors has both technical and financial concerns.

Notice that when this writer examined his audiences even in his own group as well as those in close proximity to him, he saw that the natures, the backgrounds, and especially the operational concerns of his audiences vary and differ considerably. As he widened the scope of his egocentric organization chart, he knew less and less about his audiences. However, he could assume they will vary even more than those of the audiences in close proximity.

Thus, in the process of examining the audience situation with an egocentric organization chart, a report writer can uncover not only the fact that audiences have functionally different interests, but also the nature of those functional differences. He can proceed to classify the audiences for each particular report in terms of audience needs.

Preparation of the egocentric organization chart is the first step of your procedure of systematic audience analysis. Notice that this step can be performed once to describe your typical report audience situation but must be particularized for each report to define the audiences for that report. Having prepared the egocentric chart once, the writer revises his chart for subsequent reports by adding or subtracting individual audiences.

Characterize the Individual Report Readers

In the process of preparing the egocentric organization chart, you immediately begin to think of your individual report readers in particular terms. In preparing the egocentric chart discussed above, the report writer mentioned such items as a reader's age, academic degrees, and background in the organization as well as his operational concerns. All of these particulars will come to mind when you think of your audiences as individuals. However, a systematic rather than piecemeal audience analysis will yield more useful information. The second step of audience analysis is, therefore, a systematic characterization of each person identified in the egocentric organization chart. A systematic characterization is made in terms of *operational, objective,* and *personal* characteristics.

The *operational characteristics* of your audiences are particularly important. As you identify the operational characteristics for a person affected by your report, try to identify significant differences between his or her role and yours. What are his professional values? How does he spend his time? That is, will his daily concerns and attitudes enable him to react to your report easily, or will they make it difficult for him to grasp what you are talking about? What does he know about your role, and in particular, what does he know or remember about your technical assignment and the organizational problem that occasioned your report to come to him? You should also consider carefully what he will need from your report. As you think over your entire technical investigation, ask yourself if that person will involve staff personnel in action on your report, of if he will in turn activate other persons elsewhere in the organization, when he receives the report. If he should, you must take their reactions into account when you write your report.

In addition, you should ask yourself, "How will my report affect his role?" A student engineer recently told us of an experience he had during summer employment when he was asked to evaluate the efficiency of the plant's waste treatment process. Armed with his fresh knowledge from advanced chemical engineering courses, to his surprise he found that, by making a simple change in the process, the company could save more than $200,000 a year. He fired off his report with great anticipation of glowing accolades—none came. How had his report affected the roles of some of his audiences? Although the writer had not considered the report's consequences when he wrote it, the supervisor, the manager, and related personnel now were faced with the problem of accounting for their waste of $200,000 a year. It should have been no surprise that they were less than elated over his discovery.

By *objective characteristics* we mean specific, relevant background data about the person. As you try to identify his or her educational background, you may note differences you might have otherwise neglected. Should his education seem to approximate yours, do not assume he knows what you know. Remember that the half-life of engineering education today is about five years. Thus, anyone five to ten years older than you, if you are recently out of college, probably will be only superficially familiar with the material and jargon of your advanced technical courses. If you can further identify his past professional experiences and roles, you might be able to anticipate his first-hand knowledge of your role and technical activities as well as to clarify any residual organizational commitments and value systems he might have. When you judge his knowledge of your technical area, ask yourself, "Could he participate in a professional conference in my field of specialization?"

For *personal characteristics,* when you identify a person by name, ask yourself how often the name changes in this organizational role. When you note his or her approximate age, remind yourself how differences in age can inhibit communication. Also note personal concerns that could influence his reactions to your report.

A convenient way to conduct the audience analysis we have been describing and to store the information it yields is to use an analysis form similar to the one in Figure 5, *Form for Characterizing Individual Report Readers* [p. 182]. It may be a little time-consuming to do this the first time around, but you can establish a file of audience characterizations. Then you can add to or subtract from this file as an individual communication situation requires.

One final point: This form is a means to an end rather than an end in itself. What is important for the report writer is that he thinks systematically about the questions this form raises. The novice usually has to force himself to analyze his audiences systematically. The experienced writer does this automatically.

Classify Audiences in Terms of How They Will Use Your Report

For each report you write, trace out the communication routes on your egocentric organization chart and add other routes not on the chart. Do not limit these routes to those specifically identified by the assignment and the addresses of the report. Rather, think through the total impacts of your report on the organization. That is, think in terms of the first, second, and even some third-order consequences of your report, and trace out the significant communication routes involved. All of these consequences define your actual communication.

When you think in terms of consequences, primarily you think in terms of the uses to which your report will be put. No longer are you concerned with your technical investigation itself. In fact, when you consider how readers will use your report, you realize that very few of your potential readers will have any real interest in the details of your technical investigation. Instead, they want to know the answers to such questions as "Why was this investigation made? What is the significance of the problem it addresses? What am I supposed to do with the results of this investigation? What will it cost? What are the implications—for sales, for production, for the unions? What happens next? Who does it? Who is responsible?"

It is precisely this audience concern for nontechnical questions that causes so much trouble for young practicing engineers. Professionally, much of what the engineer spends his time doing is, at most, of only marginal concern to many of his audiences. His audiences ask questions about things which perhaps never entered his thoughts during his own technical activities when he received the assignment, defined the problem, and performed his investigation. These questions, however, must enter into his considerations when he writes his report.

NAME: TITLE:

A.   OPERATIONAL CHARACTERISTICS:
     1. His role within the organization and consequent value system:

     2. His daily concerns and attitudes:

     3. His knowledge of your technical responsibilities and assignment:

     4. What he will need from your report:

     5. What staff and other persons will be activated by your report
        through him:

     6. How your report could affect his role:

B.   OBJECTIVE CHARACTERISTICS:
     1. His education—levels, fields, and years:

     2. His past professional experiences and roles:

     3. His knowledge of your technical area:

C.   PERSONAL CHARACTERISTICS:
     Personal characteristics that could influence his reactions—age,
     attitudes, pet concerns, etc.

FIGURE 5   Form for Characterizing Individual Report Readers

Having defined the communication routes for a report you now know what audiences you will have and what questions they will want answered. The final step in our method of audience analysis is to assign priorities to your audiences. Classify them in terms of how they will use your report. In order of their importance to you (not in terms of their proximity to you), classify your audiences by these three categories:

- *Primary audiences*—who make decisions or act on the basis of the information a report contains.
- *Secondary audiences*—who are affected by the decisions and actions.
- *Immediate audiences*—who route the report or transmit the information it contains.

The *primary audience* for a report consists of those persons who will make decisions or act on the basis of the information provided by the report. The report overall should be designed to meet the needs of these users. The primary audience can consist of one person who will act in an official capacity, or it can consist of several persons representing several offices using the report. The important point here is that the primary audience for a report can consist of persons from any ring on the egocentric organization chart. They may be distant or in close proximity to the writer. They may be his organizational superiors, inferiors, or equals. They are simply those readers for whom the report is primarily intended. They are the top priority users.

In theory at least, primary audiences act in terms of their organizational roles rather than as individuals with distinctive idiosyncrasies, predilections, and values. Your audience analysis should indicate when these personal concerns are likely to override organizational concerns. A typical primary audience is the decision maker, but his actual decisions are often determined by the evaluations and recommendations of staff personnel. Thus the report whose primary audience is a decision maker with line responsibility actually has an audience of staff personnel. Another type of primary audience is the production superintendent, but again his actions are often contingent upon the reactions of others.

In addition, because the report enters into a system, in time both the line and staff personnel will change; roles rather than individuals provide continuity. For this reason, it is helpful to remember the words of one engineer when he said, "A complete change of personnel could occur over the lifetime of my report." The report remains in the file. The report writer must not assume that his primary audience will be familiar with the technical assignment. He must design the report so that it contains adequate information concerning the reasons for the assignment, details of the procedures used, the results of the investigation, and conclusions and recommendations. This information is needed so that any future component of his primary audience will be able to use the report confidently.

The *secondary audience* for a report consists of those persons other than primary decision makers or users who are affected by the information the report transmits into the system. These are the people whose activities are affected when a primary audience makes a decision, such as when production supervision has to adjust to management decisions. They must respond appropriately when a primary audience acts, such as when personnel and labor relations have to accommodate production line changes. The report writer must not neglect the needs of his secondary audiences. In tracing out his communication routes, he will identify several secondary audiences. Analysis of their needs will reveal what additional information the report should contain. This information is often omitted by writers who do not classify their audiences sufficiently.

The *immediate audience* for a report are those persons who route the report or transmit the information it contains. It is essential for the report writer to identify his immediate audiences and not to confuse them with his primary audiences. The immediate audience might be the report writer's supervisor or another middle management person. Yet usually his role will be to transmit information rather than to use the information directly. An information system has numerous persons who transmit reports but who may not act upon the information or who may not be affected by the information in ways of concern to the report writers. Often, a report is addressed to the writer's supervisor, but except for an incidental memo report, the supervisor serves only to transmit and expedite the information flow throughout the organizational system.

A word of caution: at times the immediate audience is also part of the primary audience; at other times the immediate audience is part of the secondary audience. For each report you write, you must distinguish those among your readers who will function as conduits to the primary audience.

As an example of these distinctions between categories of report audiences, consider how audiences identified on the egocentric organization chart, Figure 4 [p. 178], can be categorized. Assume that the chemical engineer writes a report on a particular process improvement he has designed. The immediate audience might be his Group Leader. Another would be Manager I, transmitting the report to Manager II. The primary audiences might be Manager II and the Department Head; they would ask a barrage of nontechnical questions similar to those we mentioned a moment ago. They will decide whether or not the organization will implement the improvement recommended by the writer. The Department Head also could be part of the secondary audience by asking questions relating to labor relations and union contracts. Other secondary audiences, each asking different questions of the report, could be:

> The person in contact with the funding agency, who will be concerned with budget and contract implications.

> The person in contact with subcontractors, determining how they are affected.

> The Group Leader, whose activities will be changed.

> The "hands on" chemist, whose production responsibilities will be affected.

> The technicians, whose job descriptions will change.

In addition to the secondary audiences on the egocentric organization chart, the report will have other secondary audiences throughout the organization— technical service and development, for example, or perhaps waste treatment.

At some length we have been discussing a fairly detailed method for systematic audience analysis. The method may have seemed more complicated than it actually is. Reduced to its basic ingredients, the method requires you, first, to identify all the individuals who will read the report, second, to characterize them, and third, to classify them. The *Matrix for Audience Analysis*, Figure 6, is a convenient device for characterizing and classifying your readers once you have identified them. At a glance, the matrix reveals what information you have and what information you still need to generate. Above all, the matrix forces you to think systematically. If you are able to fill in a good deal of specific information in each cell (particularly in the first six cells), you have gone a long way towards seeing how the needs of your audiences will determine the design of your report.

Characteristics / Types of audiences	Operational	Objective	Personal
Primary	①	④	⑦
Secondary	②	⑤	⑧
Immediate	③	⑥	⑨

FIGURE 6   Matrix for Audience Analysis

We have not introduced a systematic method for audience analysis with the expectation that it will make your communication task easy. We have introduced you to the problems you must account for when you design your reports— problems you otherwise might ignore. You should, at least, appreciate the complexity of a report audience. Thus, when you come to write a report, you are less likely to make false assumptions about your audience. To develop this attitude is perhaps as important as to acquire the specific information the analysis yields. On the basis of this attitude, you now are ready to determine the specific purpose of your report.

Richard W. Dodge

What to Report

When he wrote this article in 1962, Richard W. Dodge was the Editor of Westinghouse Engineer.

Technical reports *can* be a useful tool for management—not only as a source of general information, but, more importantly, as a valuable aid to decision making. But to be effective for these purposes, a technical report must be geared to the needs of management.

Considerable effort is being spent today to upgrade the effectiveness of the technical report. Much of this is directed toward improving the writing abilities of engineers and scientists; or toward systems of organization, or format, for reports. This effort has had some rewarding effects in producing better written reports.

But one basic factor in achieving better reports seems to have received comparatively little attention. This is the question of audience needs. Or, expressed another way, "*What does management want in reports?*" This is an extremely basic question, and yet it seems to have had less attention than have the mechanics of putting words on paper.

A recent study conducted at Westinghouse sheds considerable light on this subject. While the results are for one company, probably most of them would apply equally to many other companies or organizations.

The study was made by an independent consultant with considerable experience in the field of technical report writing. It consisted of interviews with Westinghouse men at every level of management, carefully selected to present an accurate cross section. The list of questions asked is shown in Table 1 [p. 188]. The results were compiled and analyzed, and from the report several conclusions are apparent. In addition, some suggestions for report writers follow as a natural consequence.

TABLE I   Questions Asked of Managers

 1. What types of reports are submitted to you?
 2. What do you look for *first* in the reports submitted to you?
 3. What do you want from these reports?
 4. To what depth do you want to follow any one particular idea?
 5. At what level (how technical and how detailed) should the various reports be written?
 6. What do you want emphasized in the reports submitted to you? (Facts, interpretations, recommendations, implications, etc.)
 7. What types of decisions are you called upon to make or to participate in?
 8. What type of information do you need in order to make these decisions?
 9. What types of information do you receive that you don't want?
10. What types of information do you want but not receive?
11. How much of a typical or average report you receive is useful?
12. What types of reports do you write?
13. What do you think your boss wants in the reports you send him?
14. What percentage of the reports you receive do you think desirable or useful? (In kind or frequency.)
15. What percentage of the reports you write do you think desirable or useful? (In kind or frequency.)
16. What particular weaknesses have you found in reports?

WHAT MANAGEMENT LOOKS FOR IN ENGINEERING REPORTS

When a manager reads a report, he looks for pertinent facts and competent opinions that will aid him in decision making. He wants to know right away whether he should read the report, route it, or skip it.

To determine this, he wants answers fast to some or all of the following questions:

- What's the report about and who wrote it?
- What does it contribute?
- What are the conclusions and recommendations?
- What are their importance and significance?
- What's the implication to the Company?
- What actions are suggested? Short range? Long range?
- Why? By whom? When? How?

The manager wants this information in brief, concise, and meaningful terms. He wants it at the beginning of the report and all in one piece.

For example, if a summary is to convey information efficiently, it should contain three kinds of facts:

1. What the report is about;
2. The significance and implications of the work; and
3. The action called for.

To give an intelligent idea of what the report is about, first of all the problem must be defined, then the objectives of the project set forth. Next, the reasons for doing the work must be given. Following this should come the conclusions. And finally, the recommendations.

Such summaries are informative and useful, and should be placed at the beginning of the report.

The kind of information a manager wants in a report is determined by his management responsibilities, but how he wants this information presented is determined largely by his reading habits. This study indicates that management report reading habits are surprisingly similar. Every manager interviewed said he read the *summary* or abstract; a bare majority said they read the *introduction* and *background* sections as well as the *conclusions* and *recommendations;* only a few managers read the *body* of the report or the *appendix* material.

The managers who read the *background* section, or the *conclusions* and *recommendations,* said they did so ". . . to gain a better perspective of the material being reported and to find an answer to the all-important question: What do we do next?" Those who read the *body* of the report gave one of the following reasons:

1. Especially interested in subject
2. Deeply involved in the project
3. Urgency of problem requires it
4. Skeptical of conclusions drawn.

And those few managers who read the *appendix* material did so to evaluate further the work being reported. To the report writer, this can mean but one thing: If a report is to convey useful information efficiently, the structure must fit the manager's reading habits.

The frequency of reading chart in Figure 1 [p. 190] suggests how a report should be structured if it is to be useful to management readers.

SUBJECT MATTER INTEREST

In addition to what facts a manager looks for in a report and how he reads reports, the study indicated that he is interested in five broad technological areas. These are:

1. Technical problems
2. New projects and products
3. Experiments and tests
4. Materials and processes
5. Field troubles.

Managers want to know a number of things about each of these areas. These are listed in Table 2 [p. 191]. Each of the sets of questions can serve as an effective checklist for report writers.

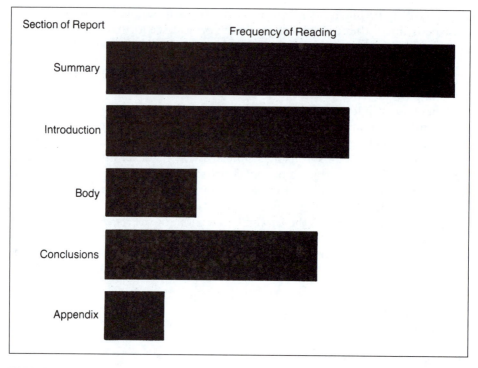

FIGURE I   How Managers Read Reports

In addition to these subjects, a manager must also consider market factors and organization problems. Although these are not the primary concern of the engineer, he should furnish information to management whenever technical aspects provide special evidence or insight into the problem being considered. For example, here are some of the questions about marketing matters a manager will want answered:

- What are the chances for success?
- What are the possible rewards? Monetary? Technological?
- What are the possible risks? Monetary? Technological?
- Can we be competitive? Price? Delivery?
- Is there a market? Must one be developed?
- When will the product be available?

And, here are some of the questions about organization problems a manager must have answered before he can make a decision:

- Is it the type of work Westinghouse should do?
- What changes will be required? Organization? Manpower? Facilities? Equipment?
- Is it an expanding or contracting program?
- What suffers if we concentrate on this?

TABLE 2 What Managers Want to Know

Problems	Tests and Experiments
What is it?	What tested or investigated?
Why undertaken?	Why? How?
Magnitude and importance?	What did it show?
What is being done? By whom?	Better ways?
Approaches used?	Conclusions? Recommendations?
Thorough and complete?	Implications to Company?
Suggested solutions? Best? Consider others?	**Material and Processes**
What now?	Properties, characteristics, capabilities?
Who does it?	Limitations?
Time factors?	Use requirements and environment?
New Projects and Products	Areas and scope of application?
Potential?	Cost factors?
Risks?	Availability and sources?
Scope of application?	What else will do it?
Commercial implications?	Problems in using?
Competition?	Significance of application to Company?
Importance to Company?	**Field Troubles and Special**
More work to be done? Any problems?	**Design Problems**
Required manpower, facilities and equipment?	Specific equipment involved?
Relative importance to other projects and	What trouble developed? Any trouble history?
products?	How much involved?
Life of project or product line?	Responsibility? Others? Westinghouse?
Effect on Westinghouse technical position?	What is needed?
Priorities required?	Special requirements and environment?
Proposed schedule?	Who does it? Time factors?
Target date?	Most practical solution?
	Recommended action?
	Suggested product design changes?

These are the kinds of questions Westinghouse management wants answered about projects in these five broad technological areas. The report writer should answer them whenever possible.

LEVEL OF PRESENTATION

Trite as it may sound, the technical and detail level at which a report should be written depends upon the reader and his use of the material. Most readers—certainly this is true for management readers—are interested in the significant material and in the general concepts that grow out of detail. Consequently, there is seldom real justification for a highly technical and detailed presentation.

Usually the management reader has an educational and experience background different from that of the writer. *Never* does the management reader have the same knowledge of and familiarity with the specific problem being reported that the writer has.

Therefore, the writer of a report for management should write at a technical level suitable for a reader whose educational and experience background is in a field different from his own. For example, if the report writer is an electrical engineer, he should write his reports for a person educated and trained in a field such as chemical engineering, or mechanical engineering, or metallurgical engineering.

All parts of the report *should preferably* be written on this basis. The highly technical, mathematical, and detailed material—if necessary at all—can and should be placed in the appendix.

MANAGEMENT RESPONSIBILITIES

The information presented thus far is primarily of interest to the report writer. In addition, however, management itself has definite responsibilities in the reporting process. These can be summed up as follows:

1. Define the project and the required reports;
2. Provide proper perspective for the project and the required reporting;
3. See that effective reports are submitted on time; and
4. See that the reports are properly distributed.

An engineering report, like any engineered product, has to be designed to fill a particular need and to achieve a particular purpose within the specific situation. Making sure that the writer knows what his report is to do, how it is to be used, and who is going to use it—all these things are the responsibilities of management. Purpose, use, and reader are the design factors in communications, and unless the writer knows these things, he is in no position to design an effective instrument of communication—be it a report, a memorandum, or what have you.

Four conferences at selected times can help a manager control the writing of those he supervises and will help him get the kind of reports he wants, when he wants them.

Step 1—At the beginning of the project. The purpose of this conference is to define the project, make sure the engineer involved knows what it is he's supposed to do, and specify the required reporting that is going to be expected of him as the project continues. What kind of decisions, for example, hinge upon his report? What is the relation of his work to the decision making process of management? These are the kinds of questions to clear up at this conference.

If the project is an involved one that could easily be misunderstood, the manager may want to check the effectiveness of the conference by asking the engineer to write a memorandum stating in his own words his understanding of the project, how he plans to handle it, and the reporting requirements. This can ensure a mutual understanding of the project from the very outset.

Step 2—At the completion of the investigation. When the engineer has finished the project assignment—but before he has reported on it—the manager should have him come in and talk over the results of his work. What did he find out? What conclusions has he reached? What is the main supporting evidence for these conclusions? What recommendations does he make? Should any future action be suggested? What is the value of the work to the Company?

The broader perspective of the manager, plus his extensive knowledge of the Company and its activities, puts him in the position of being able to give the engineer a much better picture of the value and implications of the project.

The mechanism for getting into the report the kind of information needed for decision making is a relatively simple one. As the manager goes over the material with the engineer, he picks out points that need to be emphasized, and those that can be left out. This is a formative process that aids the engineer in the selection of material and evidence to support his material.

Knowing in advance that he has a review session with his supervisor, the chances are the engineer will do some thinking beforehand about the project and the results. Consequently, he will have formed some opinions about the significance of the work, and will, therefore, make a more coherent and intelligent presentation of the project and the results of his investigation.

This review will do something for the manager, too. The material will give him an insight into the value of the work that will enable him to converse intelligently and convincingly about the project to others. Such a preview may, therefore, expedite decisions influencing the project in one way or another.

Step 3—After the report is outlined. The manager should schedule a third conference after the report is outlined. At this session, the manager and the author should review the report outline step-by-step. If the manager is satisfied with the outline, he should tell the author so and tell him to proceed with the report.

If, however, the manager is not satisfied with the outline and believes it will have to be reorganized before the kind of report wanted can be written, he must make this fact known to the author. One way he can do this is to have the author tell him why the outline is structured the way it is. This usually discloses the organizational weakness to the author and consequently he will be the one to suggest a change. This, of course, is the ideal situation. However, if the indirect approach doesn't work, the more direct approach must be used.

Regardless of the method used to develop a satisfactory outline, the thing the manager must keep in mind is this: It's much easier to win the author's consent to structural changes at the outline stage, i.e., *before* the report is written than afterward. Writing is a personal thing; therefore, when changes are suggested in organization or approach, these are all too frequently considered personal attacks and strained relations result.

Step 4—After the report is written. The fourth interview calls for a review and approval by the manager of the finished report and the preparation of a distribution list. During this review, the manager may find some sections of the report that need changing. While this is to be expected, he should limit the extent of these changes. The true test of any piece of writing is the clarity of the statement. If it's clear and does the job, the manager should leave it alone.

This four-step conference mechanism will save the manager valuable time, and it will save the engineer valuable time. Also, it will ensure meaningful and useful project reports—not an insignificant accomplishment itself. In addition, the process is an educative one. It places in the manager's hands another tool he can use to develop and broaden the viewpoint of the engineer. By eliminating misunderstanding and wasted effort, the review process creates a more helpful and effective working atmosphere. It acknowledges the professional status of the engineer and recognizes his importance as a member of the engineering department.

Christian K. Arnold

The Writing of Abstracts

The late Christian K. Arnold was an instructor in English and a technical writer and editor before becoming Associate Executive Secretary of the Association of State Universities and Land-Grant Colleges.

The most important section of your technical report or paper is the abstract. Some people will read your report from cover to cover; others will skim many parts, reading carefully only those parts that interest them for one reason or another; some will read only the introduction, results, and conclusions; but everyone who picks it up will read the abstract. In fact, the percentage of those who read beyond the abstract is probably related directly to the skill with which the abstract is written. The first significant impression of your report is formed in the reader's mind by the abstract; and the sympathy with which it is read, if it is read at all, is often determined by this first impression. Further, the people your organization wants most to impress with your report are the very people who will probably read no more than the abstract and certainly no more than the abstract, introduction, conclusions, and recommendations. And the people you should want most to read your paper are the ones for whose free time you have the most competition.

Despite its importance, you are apt to throw your abstract together as fast as possible. Its construction is the last step of an arduous job that you would rather have avoided in the first place. It's a real relief to be rid of the thing, and almost anything will satisfy you. But a little time spent in learning the "rules" that govern the construction of good abstracts and in practicing how to apply them will pay material dividends to both you and your organization.

The abstract—or summary, foreword, or whatever you call the initial thumbnail sketch of your report or paper—has two purposes: (1) it provides the specialist in the field with enough information about the report to permit him to decide whether he could read it with profit, and (2) it provides the administrator or executive with enough knowledge about what has been done in the study or project and with what results to satisfy most of his administrative needs.

It might seem that the design specifications would depend upon the purpose for which the abstract is written. To satisfy the first purpose, for instance, the abstract needs only to give an accurate indication of the subject matter and scope of the study; but, to satisfy the second, the abstract must summarize the results and conclusions and give enough background information to make the results understandable. The abstract designed for the first purpose can tolerate any technical language or symbolic shortcuts understood at large by the subject-matter group; the abstract designed for the second purpose should contain no terms not generally understood in a semitechnical community. The abstract for the first purpose is called a *descriptive abstract;* that for the second, an *informative abstract.*

The following abstract, prepared by a professional technical abstracter in the Library of Congress, clearly gives the subject-matter specialist all the help he needs to decide whether he should read the article it describes:

> Results are presented of a series of cold-room tests on a Dodge diesel engine to determine the effects on starting time of (1) fuel quantity delivered at cranking speed and (2) type of fuel-injection pump used. The tests were made at a temperature of $-10°F$ with engine and accessories chilled at $-10°F$ at least 8 hours before starting.

Regardless of however useful this abstract might be on a library card or in an index or an annotated bibliography, it does not give an executive enough information. Nor does it encourage everyone to read the article. In fact, this abstract is useless to everyone except the specialist looking for sources of information. The descriptive abstract, in other words, cannot satisfy the requirement of the informative abstract.

But is the reverse also true? Let's have a look at an informative abstract written for the same article:

> A series of tests was made to determine the effect on diesel-engine starting characteristics at low temperatures of (1) the amount of fuel injected and (2) the type of injection pump used. All tests were conducted in a cold room maintained at $-10°F$ on a commercial Dodge engine. The engine and all accessories were "cold-soaked" in the test chamber for at least 8 hours before each test. Best starting was obtained with 116 cu mm of fuel, 85 per cent more than that required for maximum power. Very poor starting was obtained with the lean setting of 34.7 cu mm. Tests with two different pumps indicated that, for best starting characteristics, the pump must deliver fuel evenly to all cylinders even at low cranking speeds so that each cylinder contributes its maximum potential power.

This abstract is not perfect. With just a few more words, for instance, the abstracter could have clarified the data about the amount of fuel delivered: do the figures give flow rates (what is the unit of time?) or total amount of fuel injected (over how long a period?)? He could easily have defined "best" starting. He could have been more specific about at least the more satisfactory type of pump: what is the type that delivers the fuel more evenly? Clarification of these points would not have increased the length of the abstract significantly.

The important point, however, is not the deficiencies of the illustration. In fact, it is almost impossible to find a perfect, or even near perfect, abstract, quite possibly because the abstract is the most difficult part of the report to write. This difficulty stems from the severe limitations imposed on its length, its importance to the over-all acceptance of the report or paper, and, with informative abstracts, the requirement for simplicity and general understandability.

The important point, rather, is that the informative abstract gives everything that is included in the descriptive one. The informative abstract, that is, satisfies not only its own purpose but also that of the descriptive abstract. Since values are obtained from the informative abstract that are not obtained from the descriptive, it is almost always worthwhile to take the extra time and effort necessary to produce a good informative abstract for your report or memo. Viewed from the standpoint of either the total time and effort expended on the writing job as a whole or the extra benefits that accrue to you and your organization, the additional effort is inconsequential.

It is impossible to lay down guidelines that will lead always to the construction of an effective abstract, simply because each reporting job, and consequently each abstract, is unique. However, general "rules" can be established that, if practiced conscientiously and applied intelligently, will eliminate most of the bugs from your abstracts.

1. *Your abstract must include enough specific information about the project or study to satisfy most of the administrative needs of a busy executive.* This means that the more important results, conclusions, and recommendations, together with enough additional information to make them understandable, must be included. This additional information will most certainly include an accurate statement of the problem and the limitations placed on it. It will probably include an interpretation of the results and the principal facts upon which the analysis was made, along with an indication of how they were obtained. Again, *specific* information must be given. One of the most common faults of abstracts in technical reports is that the information given is too general to be useful.

2. *Your abstract must be a self-contained unit, a complete report-in-miniature.* Sooner or later, most abstracts are separated from the parent report, and the abstract that cannot stand on its own feet independently must then either be rewritten or will fail to perform its job. And the rewriting, if it is done, will be done by someone not nearly as sympathetic with your study as you are. Even if it is not separated from the report, the abstract must be written as a complete, independent unit if it is to be of the most help possible to the executive. This rule automatically eliminates the common deadwood phrases like "this report contains . . ." or "this report is a report on . . ." that clutter up many abstracts. It also eliminates all references to sections, figures, tables, or anything else contained in the report paper.

3. *Your abstract must be short.* Length in an abstract defeats every purpose for which it is written. However, no one can tell you just how short it must be. Some authorities have attempted to establish arbitrary lengths, usually in terms of a certain percentage of the report, the figure given normally falling between three and ten per cent. Such artificial guides are unrealistic. The abstract for a 30-page report must necessarily be longer, percentagewise, than the abstract for a 300-page report, since there is certain basic information that must be given regardless of the length of the report. In addition, the information given in some reports can be summarized much more briefly than can that given in other reports of the same over-all dimensions. Definite advantages, psychological as well as material, are obtained if the abstract is short enough to be printed entirely on one page so that the reader doesn't even have to turn a page to get the total picture that the abstract provides. Certainly, it should be longer than the interest span of an only mildly interested and very busy executive. About the best practical advice that can be given in a vacuum is to make your abstract as short as possible without cutting out essential information or doing violence to its accuracy. With practice, you might be surprised to learn how much information you can crowd into a few words. It helps, too, to learn to blue-pencil unessential information. It is perhaps important to document that "a meeting was held at the Bureau of Ordnance on Tuesday, October 3, 1961, at 2:30 P.M." somewhere, but such information is just excess baggage in your abstract: it helps neither the research worker looking for source material nor the administrator looking for a status or information summary. Someone is supposed to have once said, "I would have written a shorter letter if I had had more time." Take the time to make your abstracts shorter; the results are worth it. But be careful not to distort the facts in the condensing.

4. *Your abstract must be written in fluent, easy-to-read prose.* The odds are heavily against your reader's being an expert in the subject covered by your report or paper. In fact, the odds that he is an expert in your field are probably no greater than the odds that he has only a smattering of training in any technical or scientific discipline. And even if he were perfectly capable of following the most obscure, tortured technical jargon, he will appreciate your sparing him the necessity for doing it. T. O. Richards, head of the Laboratory Control Department, and R. A. Richardson, head of the Technical Data Department, both of the General Motors Corporation, have written that their experience shows the abstract cannot be made too elementary. "We never had [an abstract] . . . in which the explanations and terms were too simple." This requirement immediately eliminated the "telegraphic" writing often found in abstracts. Save footage by sound practices of economy and not by cutting out the articles and the transitional devices needed for smoothness and fluency. It also eliminates those obscure terms that you defend on the basis of "that's the way it's always said."

5. *Your abstract must be consistent in tone and emphases with the report paper, but it does not need to follow the arrangement, wording, or proportion of the original.* Data, information, and ideas introduced into the abstract must also appear in the report or paper. And they must appear with the same emphases. A conclusion or recommendation that is qualified in the report paper must not turn up without the qualification in the abstract. After all, someone might read both the abstract and the report. If this reader spots an inconsistency or is confused, you've lost a reader.

6.   *Your abstract should make the widest possible use of abbreviations and numerals, but it must not contain any tables or illustrations.* Because of the space limitations imposed on abstracts, the rules governing the use of abbreviations and numerals are relaxed for it. In fact, all figures except those standing at the beginning of sentences should be written as numerals, and all abbreviations generally accepted by such standard sources as the American Standards Association and "Webster's Dictionary" should be used.

By now you must surely see why the abstract is the toughest part of your report to write. A good abstract is well worth the time and effort necessary to write it and is one of the most important parts of your report. And abstract writing probably contributes more to the acquisition of sound expository skills than does any other prose discipline.

Vincent Vinci

Ten Report Writing Pitfalls: How to Avoid Them

Vincent Vinci was Director of Public Relations for Lockheed Electronics when he wrote this article.

The advancement of science moves on a pavement of communications. Chemists, electrical engineers, botanists, geologists, atomic physicists, and other scientists are not only practitioners but interpreters of science. As such, the justification, the recognition and the rewards within their fields result from their published materials.

Included in the vast field of communications is the report, a frequently used medium for paving the way to understanding and action. The engineering manager whose function is the direction of people and programs receives and writes many reports in his career. And therefore the need for technical reports that communicate effectively has been internationally recognized.

Since scientific writing is complicated by specialized terminology, a need for precision and the field's leaping advancement, the author of an engineering report can be overwhelmed by its contents. The proper handling of contents and communication of a report's purpose can be enhanced if the writer can avoid the following 10 pitfalls.

PITFALL 1: IGNORING YOUR AUDIENCE

In all the forms of communications, ignoring your audience in the preparation of a report is perhaps the greatest transgression. Why? All other forms of communication, such as instruction manuals, speeches, books and brochures, are directed to an indefinable or only partially definable audience. The report, on the other hand, is usually directed to a specific person or group and has a specific purpose.

So, it would certainly seem that if one knows both the "who" and the "why," then a report writer should not be trapped by this pitfall.

But it is not enough to know the who and why, you need to know "how." To get to the how, let's assume that the reader is your boss and has asked you to write a trip report. You are to visit several plants and report on capital equipment requirements. Before you write the first word, you will have to find out what your boss already knows about these requirements. It is obvious that he wants a new assessment of the facilities' needs. But, was he unsatisfied with a recent assessment and wants another point of view, or is a new analysis required because the previous report is outdated—or does he feel that now is the time to make the investment in facilities so that production can be increased over the next five years? That's a lot of questions, but they define both the who and why of your trip and, more importantly, your report.

By this time you may get the feeling that I am suggesting you give him exactly what he wants to read. The answer is yes and no. No, I don't mean play up to your boss's likes and dislikes. I do mean, however, that you give him all the information he needs to make a decision—the pros and the cons.

I mean also that the information be presented in a way that he is acclimated to in making judgments. For example, usually a production-oriented manager or executive (even the chief executive) will think in terms of his specialty. The president of a company who climbed the marketing ladder selling solvents will think better in marketing terms. Therefore, perhaps the marketing aspects of additional equipment and facilities should be stressed. You should also be aware that if you happen to be the finance director, your boss will expect to see cost/investment factors too.

A simple method for remembering, rather than ignoring, your audience is to place a sheet of paper in front of you when you start to write your report. On the paper have written in bold letters **WHO, WHY** and **HOW,** with the answers clearly and cogently defined. Keep it in front of you throughout the preparation of your report.

PITFALL 2: WRITING TO IMPRESS

Nothing turns a reader off faster than writing to impress. Very often reports written to leave a lasting scholarly impression on top management actually hinder communication.

Generally, when a word is used to impress, the report writer assumes that the reader either knows its meaning or will take the trouble to look it up. Don't assume that a word familiar to you is easily recognized by your reader. I recall a few years ago, there was a word "serendipity" which became a fashionable word to impress your reader with. And there was "fulsome" and "pejorative," and more. All are good words, but they're often misused or misapplied. They were shoved into reports to impress, completely disregarding the reader. Your objective is that your reader comprehend your thoughts, and there should be a minimum of impediments to understanding—understanding with first reading, and no deciphering.

Unfortunately, writing to impress is not merely restricted to use of obscure words but also includes unnecessary detail and technical trivia. Perhaps the scientist, chemist, chemical engineer and others become so intrigued with technical fine points that the meaningful (to your audience) elements of a report are buried. And quite often the fault is not so much a lack of removing the chaff from the grain but an attempt to technically impress the reader. Of course there exist reports that are full of technical detail because the nature of the communication is to impart a new chemical process, compound or technique. Even when writing this kind of report, you should eliminate any esoteric technical facts that do not contribute to communication, even though you may be tempted to include them to exhibit your degree of knowledge in the field.

PITFALL 3: HAVING MORE THAN ONE AIM

A report is a missile targeted to hit a point or achieve a mission. It is not a barrage of shotgun pellets that scatter across a target indiscriminately.

Have you ever, while reading a report, wondered where or what it was leading to—and even when you are finished you weren't quite sure? The writer probably had more than one aim, thereby preventing you from knowing where the report was heading.

Having more than one aim is usually the sign of a novice writer, but the pitfall can also trip up an experienced engineer if he does not organize the report toward one objective.

It is too easy to say that your report is being written to communicate, to a specific audience, information about your research, tests, visit, meeting, conference, field trip, progress or any other one of a range of activities that may be the subject. If you look at the first part of the sentence, you will see that "specific audience" and "information" are the key words that have to be modified to arrive at the goal of your report. For instance, you must define the specific audience such as the "members of the research council," "the finance committee" or "the chief process engineer and his staff."

Secondly, you need to characterize the information, such as "analysis of a new catalytic process," "new methods of atomic absorption testing" or "progress on waste treatment programs." You should be able to state the specific purpose of your report in one sentence: e.g., "The use of fibrous material improves scrubber efficiency and life—a report to the product improvement committee."

When you have arrived at such a definition of your purpose and audience, you can then focus both the test results and analysis toward that purpose, tempered with your readers in mind.

The usual error made in writing reports is to follow the chronology of the research in the body of the report with a summary of a set of conclusions and recommendations attached. The proper procedure to follow is to write (while focusing on your report goal) the analysis first (supported by test essentials or any other details), then your introduction or summary—sort of reverse chronology. But be sure that your goal and audience are clearly known because they become the basis of organizing your report.

PITFALL 4: BEING INCONSISTENT

If you work for an international chemical firm, you may be well aware of problems in communicating with plant managers and engineers of foreign installations or branches. And I'm not referring to language barriers, because for the most part these hurdles are immediately recognized and taken care of. What is more significant is units of measure. This problem is becoming more apparent as the United States slowly decides whether or not to adopt the metric system. Until it is adopted your best bet is to stick to one measurement system throughout the report. Preferably, the system chosen should be that familiar to your audience. If the audience is mixed, you should use both systems with one (always the same one) in parentheses. Obviously, don't mix units of measure because you will confuse or annoy your readers.

Consistency is not limited to measurements but encompasses terms, equations, derivations, numbers, symbols, abbreviations, acronyms, hyphenation, capitalization and punctuation. In other words, consistency in the mechanics of style will avoid work for your reader and smooth his path toward understanding and appreciating the content of the report.

If your company neither has a style guide nor follows the general trends of good editorial practice, perhaps you could suggest instituting a guide. In addition to the U.S. Government Printing Office Style Manual, many scientific and engineering societies have set up guides which could be used.

PITFALL 5: OVERQUALIFYING

Chemical engineers, astronomers, geologists, electrical engineers, and scientists of any other discipline have been educated and trained to be precise. As a result, they strive for precision, accuracy, and detail. That tends to work against the scientist when it comes to writing. Add to that the limited training received in the arts, and you realize why written expression does not come easily.

Most reports, therefore, have too many modifiers—adjectives, clauses, phrases, adverbs and other qualifiers. Consider some examples: the single-stage, isolated double-cooled refractory process breakdown, or the angle of the single-rotor dc hysteresis motor rotor winding. To avoid such difficult-to-comprehend phrases, you could in the first example write "the breakdown of the process in single-stage, isolated double-cooled refractories," and in the second, "the angle of the rotor winding in single-rotor dc hysteresis motors can cause . . . ," and so on. This eliminates the string of modifiers and makes the phrase easier to understand.

Better still, if your report allows you to say at the beginning that the following descriptions are only related to "single-stage, isolated double-cooled refractories" or "single-rotor dc hysteresis motors," you can remove the cumbersome nomenclature entirely.

In short, to avoid obscuring facts and ideas, eliminate excessive modifiers. Try to state your idea or main point first and follow with your qualifying phrases.

PITFALL 6: NOT DEFINING

Dwell, lake, and barn, all are common words. Right? Right and wrong. Yes, they are common to the nonscientist. To the mechanical engineer, dwell is the period a cam follower stays at maximum lift; to a chemical engineer, lake is a dye compound; and to an atomic physicist, a barn is an atomic cross-sectional area (10^{-24} cm^2).

These three words indicate two points: first, common words are used in science with other than their common meanings; and second, terms need to be defined.

In defining terms you use in a report, you must consider what to define and how to define. Of the two, I consider what to define a more difficult task and suggest that you review carefully just which terms you need defined. If you analyze the purpose, the scope, the direction and your audience (reader/user), you will probably get a good handle on such terms.

"How to define" ranges from the simple substitution of a common term for an uncommon one, to an extended or amplified explanation. But whatever the term, or method of definition, you need to slant it both to the reader and to the report purpose.

PITFALL 7: MISINTRODUCING

Introductions, summaries, abstracts and forewords—whatever you use to lead your reader into your report, it should not read like an exposition of a table of contents. If it does, you might as well let your audience read the table of contents.

The introduction, which should be written after the body of the report, should state the subject, purpose, scope, and the plan of the report. In many cases, an introduction will include a summary of the findings or conclusions. If a report is a progress report, the introduction should relate the current report to previous reports. Introductions, then, not only tell the sequence or plan of the report, but tell the what, how and why of the subject as well.

PITFALL 8: DAZZLING WITH DATA

Someone once said that a good painter not only knows what to put in a painting, but more importantly he knows what to leave out. It's much the same with report writing. If you dazzle your reader with tons of data, he may be moved by the weight of the report but may get no more out of it than that.

The usual error occurs in supportive material that many engineers and scientists feel is unnecessary to give a report scientific importance. The truth is that successful scientific writing (which includes reports) is heavily grounded in reality, simplicity, and understanding—not quantity.

The simplest way to evaluate the relevancy of information is to ask yourself after writing a paragraph, "What can I remove from this paragraph without destroying its meaning and its relationship to what precedes and what will follow?" Then, ask another question, "Does my reader require all that data to comprehend, evaluate or make a decision with?" If you find you can do without excess words, excess description and excessive supportive data, you will end up with a tighter, better and more informative report.

These principles should also be used to evaluate graphs, photographs, diagrams and other illustrations. Remember, illustrations should support or aid comprehension rather than being a crutch on which your report leans. The same should be kept in mind when determining just how much you should append to your report. There is no need to copy all your lab notes to show that detailed experimentation was performed to substantiate the results. A statement that the notes exist and are available will suffice.

PITFALL 9: NOT HIGHLIGHTING

Again, I believe the analogy of the painter applies. A good painter also knows what to highlight and what to subdue in a portrait or scene.

If you don't accent the significant elements, findings, illustrations, data, tests, facts, trends, procedures, precedents, or experiments pertinent to the subject and object of your report, you place the burden of doing so on your reader. As a result, he may consider the report a failure, draw his own conclusions, or hit upon the significant elements by chance. In any event, don't leave it up to your reader to search out the major points of your report.

Highlighting is one step past knowing what goes into your report and what to leave out (see Pitfall 8 above). All the key points of your report should define and focus on the purpose of your report. They must be included in your summary or conclusions, but these sections are not the only places to highlight. Attention should be called to key elements needed for the understanding of your material throughout the body of the report. Several methods may be used: you can underline an important statement or conclusion, you can simply point out that a particular illustration is the proof of the results of an experiment, or, as most professional writers do, you can make the key sentence the first or last sentence of a paragraph.

PITFALL 10: NOT REWRITING

Did you ever hear of an actor who hadn't rehearsed his lines before stepping before an audience? An actor wouldn't chance it—his reputation and his next role depend on his performance. The engineer shouldn't chance it either. Don't expect the draft of your report to be ready for final typing and reproduction without rewriting.

Once you have judged what your report will contain and how it will be organized, just charge ahead and write the first draft. Don't worry about choosing the precise word, turning that meaningful phrase, or covering all the facts in one paragraph or section. Once you have written your first draft (and the quicker you accomplish this the more time you will have to perfect the text), you are in a better position to analyze, tailor, and refine the report as a whole. Now you are also able to focus all the elements toward your purpose and your audience.

As you begin the rewriting process simply pick up each page of your draft, scan it, and ask yourself what role the material on that page plays in the fulfillment of the report's objective and understanding. You will find that this will enable you to delete, add, change and rearrange your material very quickly.

After you have completed this process, then rewrite paragraph by paragraph, sentence by sentence, and word by word. Your final step is to repeat the procedure of examining each page's contents. When you are satisfied with its flow and cohesion, then you will have a good report, one you know will be well received and acted upon.

Walter E. Oliu, Charles T. Brusaw, and Gerald J. Alred

Creating Visuals

Walter E. Oliu is a technical writer with the U.S. Nuclear Regulatory Commission. Charles T. Brusaw is retired from NCR Corporation where he was a senior program instructor. Gerald J. Alred is Professor of English at the University of Wisconsin–Milwaukee.

DESIGNING AND INTEGRATING VISUALS WITH TEXT

To make the most effective use of visuals and to integrate them smoothly with the text of your document, consider your graphics requirements even before you begin to write. Plan your visuals—tables, graphs, drawings, charts, maps, or photographs—when you're planning the scope and organization of your final work, whether it's a report, newsletter, brochure, or Web site. Make graphics an integral part of your outline, noting approximately where each should appear throughout the text. At each place where you plan to include a visual, either make a rough sketch of the visual or write "illustration of . . ." and enclose each suggestion in a box in your outline. If you are working on your computer, you can copy and paste graphics directly into your outline at the appropriate places, using your computer's clipboard feature. Like other information in a working outline, these boxes and sketches can be moved, revised, or deleted as required.

The following guidelines, which apply to most visual materials you might use to supplement or clarify the information in your text, will help you create and incorporate your visual materials within your documents effectively. . . .

- **Why include your visual?** Explain in the text why you've included the illustration. The description for each visual will vary with its complexity and its importance. Consider your audience: nonexperts requite lengthier explanations than experts do, as a rule.

- **Is the information in your visual accurate?** Gather the information from reliable sources.
- **Is your visual focused?** Include only information necessary to the discussion in the text and eliminate unnecessary labels, arrows, boxes, and lines.
- **Are terms and symbols in your visual defined and consistent?** Define all acronyms in the text, figure, or table. If any symbols are not self-explanatory, include a listing (known as a *key*) that defines them. Keep terminology consistent. Do not refer to something as a "proportion" in the text and as a "percentage" in the illustration.
- **Does your visual specify measurements and distances?** Specify the units of measurement used or include a scale of relative distances, when appropriate. Ensure that relative sizes are clear or indicate distance by a scale, as on a map.
- **Is the lettering readable?** Position any explanatory text or labels horizontally for ease or reading, if possible.
- **Is the caption clear?** Give each illustration a concise caption that clearly describes its contents.
- **Is there a figure or table number?** Assign a figure or table number for documents containing five or more illustrations. The figure or table number precedes the title:

    Figure 1. Projected sales for 2007–2010

    Note that graphics (photographs, drawings, maps, and so on) are generically labeled "figures," while tables (data organized in rows and columns) are labeled "tables."

- **Is a list of figures or tables needed?** In documents with more than five illustrations, list the illustrations by title, together with figure and page numbers, or table and page numbers, following the table of contents. Title figures "List of Figures" and tables "List of Tables."
- **Are figure or table numbers referred to in your text?** Refer to each illustration by its figure or table number in the text of your document.
- **Are visuals appropriately placed?** Place illustrations as close as possible and following the text where they are discussed. Place lengthy and detailed illustrations in an appendix and refer to them in the text of your document.
- **Do visuals stand out from surrounding text?** Allow adequate white space on the page around and within each illustration.

A discussion of specific types of visuals commonly used in on-the-job writing follows. Your topic will ordinarily determine the best type of visual to use.

Tables

A table is useful for showing large numbers of specific, related data in a brief space. The data may be numerical, as in Figure 1 [p. 210], or verbal. . . . Because a table displays information in rows and columns, your readers can easily compare data and see their significance more clearly than if they were presented in the text.

Following are the elements of a typical table, as shown in Figure 1, with guidelines:

- *Table number.* Number each table sequentially throughout the text.
- *Table title.* Create a title that describes concisely what the table represents; place it above the table.
- *Boxhead.* In the boxhead (beneath the title) provide column headings that are brief but descriptive. Include units of measurement either as part of the heading or enclosed in parentheses beneath it. Standard abbreviations and symbols are acceptable. Avoid vertical or diagonal lettering.
- *Stub.* In the left-hand vertical column of a table, called the stub, list all the items to be shown in the body of the table.
- *Body.* Provide data in the body of your table below the column headings and to the right of the stub. (Each datum element is located in a *cell.*) Within the body, arrange columns so that the terms to be compared appear in adjacent rows and columns. Where no information exists for a specific cell, substitute a row of dots or a dash to acknowledge the gap. If you substitute the abbreviation "N/A" for missing data in a cell, add a footnote to clarify whether it means "not available" or "not applicable."
- *Rules.* Use rules (lines) to separate your table into its various parts. Include horizontal rules below the title, below the body of the table, and between the column headings and the body of the table. You may include vertical rules to separate the columns, but do not use rules to enclose the sides of the table.
- *Source line.* Below the table, include a source line to identify where you obtained the data (when appropriate). Many organizations place the source line below the footnotes.
- *Footnotes.* Include a footnote when you need to explain an item in the table. Use symbols (*, †) or lowercase letters (sometimes in parentheses) rather than numbers to make it clear that the notes are not part of the data or the main text.
- *Continuing tables.* When you must divide your table to continue it on another page, repeat the column headings and give the table number at the head of each new page with a "continued" label ("Table 3, continued"), as shown in Figure 2 [p. 211].

To list relatively few items that would be easier for the reader to grasp in tabular form, use an informal table.

The sound-intensity levels (decibels) for the three frequency bands (in hertz) were determined to be the following:

Frequency Band (Hz)	Decibels
600–1,200	68
1,200–2,400	62
2,400–4,800	53

Table number → | Boxhead → | Stub → | Rule →

Table title ← | Column headings ← | Body ←

Table 1   U.S. Population, Employment, and Product-by-Sector Projections 1980–2010*

	1980	1990	2000	2010	% Change 1980–2010*
Population (Thousands)					
Total	226,549	249,402	276,241	300,431	32.6
Under 18	63,755	64,156	71,789	73,617	15.5
18–64	137,240	154,011	169,131	186,709	36.0
65 and Over	25,549	31,235	35,322	40,104	57.0
Employment (Thousands of Jobs)					
All-Industry Total	113,726	138,981	157,656	176,164	54.9
Manufacturing	20,777	19,756	18,890	18,850	(9.3)[†]
Retail Trade	17,853	23,020	26,402	29,450	65.0
Services	24,558	38,188	49,474	59,379	141.8
Government	18,796	21,203	23,065	24,750	31.7
Other	31,743	36,815	39,825	43,737	37.8
Gross State Product (Millions of 1987 Dollars)					
All-Industry Total	3,697,140	4,888,324	6,025,600	7,219,400	95.3
Manufacturing	725,428	928,483	1,105,409	1,279,410	76.4
Retail Trade	320,134	478,080	603,150	726,774	127.0
Services	609,012	869,360	1,103,812	1,371,477	125.2
Government	494,431	564,163	617,863	686,916	38.9
Other	1,548,135	2,048,238	2,595,367	3,154,824	103.8

Source line → | Footnote →

Source: U.S. Small Business Administration, Office of Advocacy, from data provided by the Bureau of Economic Analysis.
*Represents percent change
[†]Figures in parenthesis are negative

FIGURE I   Table

Although you need not include titles or table numbers to identify informal tables, you do need to include headings that describe the information provided and columns and rows that are properly aligned. You may also need to acknowledge the source of the information, as discussed in the [following] Ethics Note. . . .

ETHICS NOTE

Whether you reprint a preexisting image or use published information to create your own graph or table, you must acknowledge your borrowings in a source or credit line. Place source information below the caption for a figure and below any footnotes at the bottom of a table. (See Figure 1 for an example.) If you wish to use illustrations found on the Web or in printed sources, obtain written permission from the copyright holder. Material that is not copyrighted (generally limited to publications of the U.S. federal government) can be reproduced without permission, but you must still acknowledge the source in a credit line.

Graphs

Graphs, also called charts, present numerical data in visual form, showing trends, movements, distributions, and cycles more readily than tables do. Although

Table 4

Assessment of Electronic Media and Format Standards in Federal Agencies: Number, Percentage, and Basis for Use by Agency

| Format | Standard for each format used | | | | | | | |
| | Agency mandated | | Common agency practice | | Other | | None | |
	Number	Percent	Number	Percent	Number	Percent	Number	Percent
Database								
Oracle	7	38.9	8	44.4	1	5.6	1	5.6
WAIS	1	4.3	21	91.3	0	0.0	0	0.0
MARC	1	33.3	1	33.3	0	0.0	0	0.0
Sybase	0	0.0	4	80.0	0	0.0	0	0.0
dBase	0	0.0	8	80.0	0	0.0	0	0.0
Other	2	4.4	21	46.7	12	26.7	9	20.0
Spreadsheet								
Lotus 1-2-3	6	25.0	9	37.5	3	12.5	5	20.8
Excel	4	11.8	24	70.6	0	0.0	4	11.8
Other	0	0.0	0	0.0	0	0.0	2	50.0
Tagged markup								
HTML	21	13.3	114	72.2	6	3.8	15	9.5
SGML	2	13.3	9	60.0	2	13.3	1	6.7
XML	0	0.0	1	33.3	0	0.0	1	33.3
Other	0	0.0	7	53.8	1	7.7	4	30.8
Image								

Table 4, continued

Assessment of Electronic Media and Format Standards in Federal Agencies: Number, Percentage, and Basis for Use by Agency

| Format | Standard for each format used | | | | | | | |
| | Agency mandated | | Common agency practice | | Other | | None | |
	Number	Percent	Number	Percent	Number	Percent	Number	Percent
Audio								
WAV	2	15.4	8	61.5	0	0.0	2	15.4
AU	4	66.7	0	0.0	0	0.0	1	16.7
AIFF	1	50.0	0	0.0	0	0.0	0	0.0
Other	0	0.0	1	20.0	2	40.0	1	20.0
Video								
MOV	0	0.0	5	62.5	0	0.0	2	25.0
MPEG	1	10.0	5	50.0	1	10.0	2	20.0
AVI	1	20.0	3	60.0	0	0.0	0	0.0
Other	0	0.0	0	0.0	1	50.0	0	0.0
Text								
ASCII	21	13.3	87	70.7	6	4.9	15	9.5

FIGURE 2   Divided Table (Continued on a Second Page)

Designing Your Document: Creating Tables

- Use tables to present data that you want readers to quickly evaluate and compare, and that would be difficult or tedious to present in your main text.
- Identify each table with a concise, descriptive title and a unique table number.
- Use horizontal lettering, if possible.
- Do not enclose the left and right sides with vertical rules.
- Include a source line when necessary to identify where you obtained your data.
- For tables continued on another page, repeat the table number (followed by "continued"), title, and column headings.
- Use informal tables—those without a title or number—when there are only a few items to categorize.

---

graphs present statistics in a format that is easy to understand, they are less accurate than tables. For this reason, they are often accompanied by tables that give exact numbers. (Note the difference between the graph and the table showing the same data in Figure 3.) To solve the problem of showing only approximate data in graphs, you can include the exact data for each column—if this will not clutter your graph—giving the reader both a quick overview of the data and accurate numbers. (See Figures 4 [p. 214] and 8 [p. 217].) The most commonly used graphs are line graphs, bar graphs, pie graphs, and picture graphs, all of which you can easily render on your computer once you have entered your data into a spreadsheet or database application.

Line Graphs

The line graph shows the relationship between two or more sets of figures. The graph is composed of a vertical axis and a horizontal axis that intersect at right angles, each representing one set of data. The relationship between the two sets is readily indicated by points plotted along appropriate intersections of the two axes that are then connected to form a continuous line. The line graph's vertical axis usually represents amounts, and its horizontal axis usually represents increments of time (Figure 4). Line graphs with more than one plotted line allow for comparisons between two sets of data. In creating such graphs, label each plotted line, as shown in Figure 5 [p. 214]. You can emphasize the difference between the two lines by shading the space that separates them. The following guidelines apply to most line graphs:

- Give your graph a title that describes the data clearly and concisely.
- Indicate the zero point (where the two axes meet). If the range of data makes it inconvenient to begin at zero, insert a break in the scale (Figure 6 [p. 215]); otherwise, the graph would show a large area with no data.

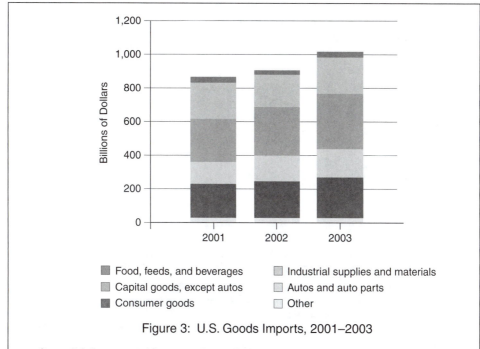

Figure 3: U.S. Goods Imports, 2001–2003

Source: U.S. Department of Commerce, Census Basis.

Table 3: U.S. Goods Imports, 2001–2003

	2001	2002	2003	02–03	98–03	96–03
Imports	*Billions of Dollars*			*Percent Change*		
Total (BOP Basis)*	876.4	917.2	1,030.2	12.3	54.1	92.0
Food, feeds, and beverages	39.7	41.2	43.6	5.7	40.6	57.9
Industrial supplies and materials	213.8	200.1	222.6	10.7	36.7	59.9
Capital goods, except autos	253.3	269.6	296.9	10.1	61.0	121.0
Autos and auto parts	139.8	149.1	179.5	20.4	51.7	95.6
Consumer goods	193.8	216.5	239.6	10.7	63.8	95.3
Other	29.3	35.4	43.9	24.1	106.1	148.0

Source: U.S. Department of Commerce.
*Balance of Payment Basis for Total. Census Basis for Sectors.

FIGURE 3   Graph and Table Showing the Same Data

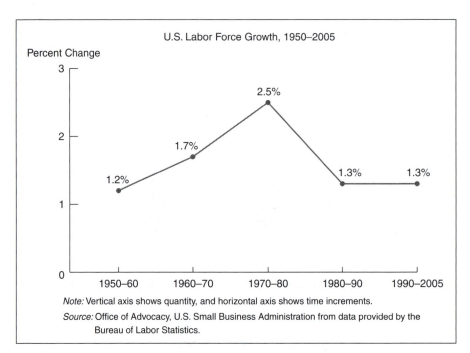

FIGURE 4 Single-Line Graph

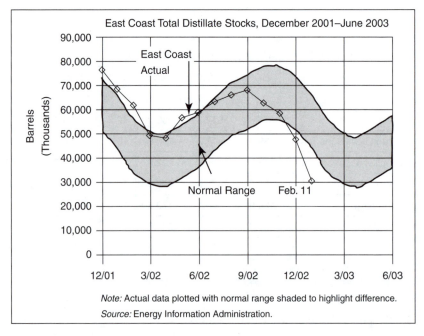

FIGURE 5 Double-Line Graph with Shading

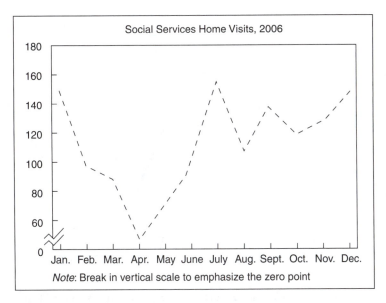

FIGURE 6 Line Graph with Vertical Axis Broken

- Divide the vertical axis into equal portions, from the least amount at the bottom to the greatest amount at the top. The caption for this scale may be placed at the upper left (as in Figure 5) or, as is more often the case, vertically along the vertical axis (as in Figure 6).
- Divide the horizontal axis into equal units from left to right, and label them to show what values each represents.
- Include enough points to plot—accurately depict—the data; too few data points will distort depiction of the trends (Figure 7 [p. 216]).
- Keep grid lines to a minimum so that the curved lines stand out. Detailed grid lines are unnecessary because precise values are usually shown either on the graph or in an accompanying table.
- Include a label or a key when necessary to define symbols or visual cues to the data, such as in the box below the data in Figure 8 [p. 217].
- Include a source line under the graph at the lower left, indicating where you obtained the data (see Figures 4, 5, and 8).
- Present all type horizontally if possible, although the type for the vertical-axis caption is usually presented vertically (see Figure 6).

Be sure to proportion the vertical and horizontal scales so that they present data precisely and without visual distortion—to do otherwise is inaccurate and potentially misleading. In Figure 7, the graph on the left gives the appearance of a slight decline followed by steady increase in investment returns because the scale is compressed, with some of the years selectively omitted. The graph on the right represents the trend more accurately because the years are evenly distributed without omissions.

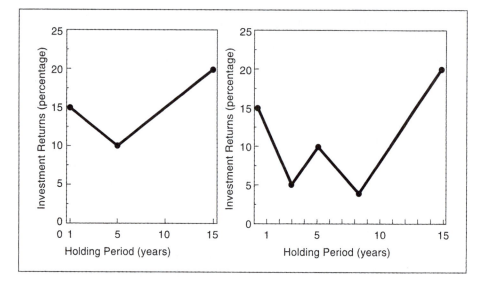

FIGURE 7   Distorted (*left*) and Distortion-Free (*right*) Expressions of Data

---

ETHICS NOTE

Be careful not to omit or distort the data in your visuals. At the least, misleading visuals call into question the credibility of you and your organization—and they are unethical. Using misleading visuals can also subject you and your organization to lawsuits.

Bar Graphs

Bar graphs consist of horizontal or vertical bars of equal width but scaled in length or height to represent some quantity. They are commonly used to show the following proportional relations:

- Different types of information during different periods of time (Figure 8)
- Quantities of the same kind of information at different periods of time (Figure 9 [p. 218])
- Quantities of different information during a fixed period of time (Figure 10 [p. 219])
- Quantities of the different parts that make up a whole (Figure 11 [p. 220])

Bar graphs can also indicate what proportion of a whole the various component parts represent. In such a graph, the bar, which is theoretically equivalent to 100 percent, is divided according to the proportion of the whole that each item sampled represents. (Compare the displays of the same data in Figures 10 and 12 [p. 220].) In some bar graphs, the completed bar does not represent 100 percent because not all parts of the whole have been included or not all are pertinent in the sample (Figure 11). Bar graphs are also used to track project schedules, where each bar represents the time allotted for each task of a project. . . .

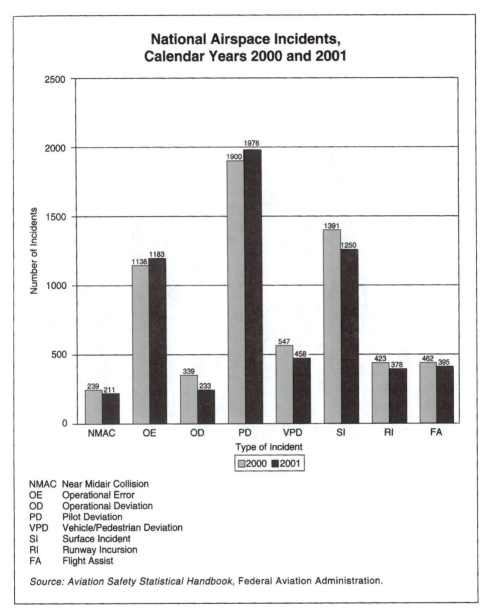

FIGURE 8 Bar Graph of Different Types of Information during Different Time Periods

Note that in Figure 14 [p. 222], a type of bar graph showing travel frequency, the exact quantities appear at the end of each picture column, eliminating the need to have an accompanying table giving the percentages. If the bars are not labeled, as in Figures 11 and 12, the different portions must be clearly indicated by shading, cross-hatching, or other devices. Include a key that represents the various subdivisions.

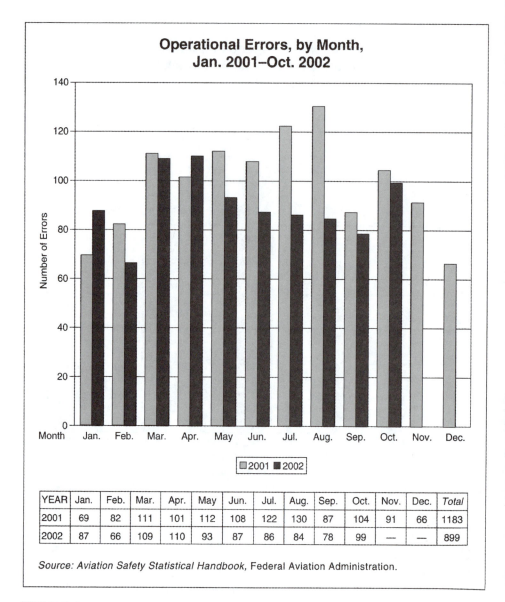

Operational Errors, by Month, Jan. 2001–Oct. 2002

YEAR	Jan.	Feb.	Mar.	Apr.	May	Jun.	Jul.	Aug.	Sep.	Oct.	Nov.	Dec.	Total
2001	69	82	111	101	112	108	122	130	87	104	91	66	1183
2002	87	66	109	110	93	87	86	84	78	99	—	—	899

Source: Aviation Safety Statistical Handbook, Federal Aviation Administration.

FIGURE 9 Bar Graph Showing Quantities of the Same Kind of Information at Different Periods of Time

Pie Graphs

A pie graph presents data as wedge-shaped sections of a circle. The circle equals 100 percent, or the whole, of some quantity (a tax dollar, a bus fare, the hours of a working day), with the wedges representing the various parts into which the whole is divided. In Figure 13 [p. 221], for example, the circle stands for a city tax dollar and is divided into units equivalent to the percentages of the tax dollar

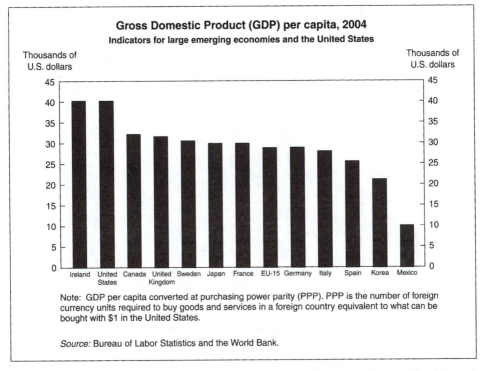

Gross Domestic Product (GDP) per capita, 2004
Indicators for large emerging economies and the United States

Note: GDP per capita converted at purchasing power parity (PPP). PPP is the number of foreign currency units required to buy goods and services in a foreign country equivalent to what can be bought with $1 in the United States.

Source: Bureau of Labor Statistics and the World Bank.

FIGURE 10   Bar Graph Showing Quantities of Different Information during a Fixed Period of Time

spent on various city services. Note that the slice representing salaries is slightly offset (exploded) from the others to emphasize that data. This feature is commonly available on spreadsheet and database software.

The relationships among the various statistics presented in a pie graph are easy to grasp, but the information is often general. For this reason, a pie graph is often accompanied by a table that presents the actual figures on which the percentages in the graph are based.

Following are guidelines for constructing pie graphs:

- Keep in mind that the complete 360° circle is equivalent to 100 percent.
- When possible, begin at the 12 o'clock position and sequence the wedges clockwise, from largest to smallest. (This is not always possible because the default setting for some charting software sequences the data counterclockwise.)
- Apply a distinctive pattern or various shades of gray for each wedge.
- Keep all labels horizontal and, most important, provide the percentage value of each wedge.
- Check to see that all wedges and their respective percentages add up to 100 percent.

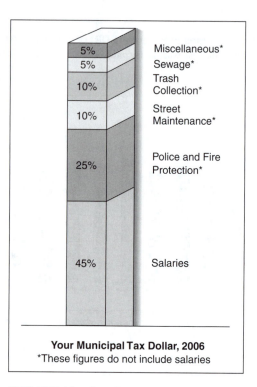

FIGURE 11   Bar Graph Showing Different
Quantities of Different Parts of a Whole

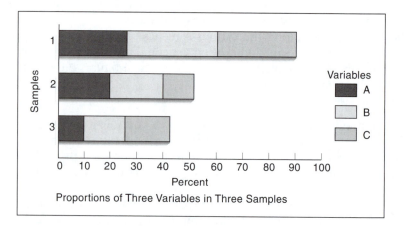

FIGURE 12   Bar Graph in Which Not All Parts of a Whole Have
Been Included

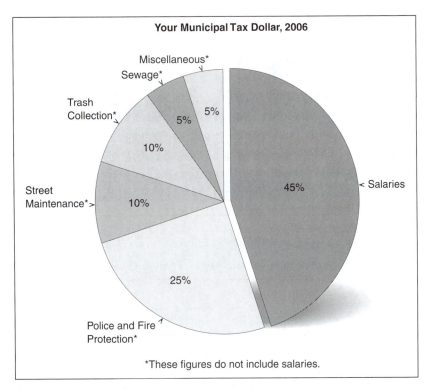

FIGURE 13   Exploded Pie Graph (with Same Data as in Figure 11)

Although pie graphs have a strong visual impact, they also have drawbacks. If more than five or six items of information are presented, the graph looks cluttered and, unless percentages are labeled on each section, the reader cannot compare the values of the sections as accurately as on a bar graph.

Picture Graphs

Picture graphs (also called *pictograms*) are modified bar graphs that use picture symbols to represent the item for which data are presented. Each symbol corresponds to a specified quantity of the item, as shown in Figure 14 [p. 222]. Note that exact numbers are also included because the picture symbol can indicate only approximate figures. Pictograms usually work well for nonexpert audiences because they make the data more vivid and easier to remember. They are also popular in presentations because they add an element of entertainment to the data. Here are some tips on preparing picture graphs:

- Use symbols that are self-explanatory.
- Have each symbol represent a specific number of units and be sure to include accurate numerical quantities for each row of data.
- Show larger quantities by increasing the number of symbols rather than by creating a larger symbol.

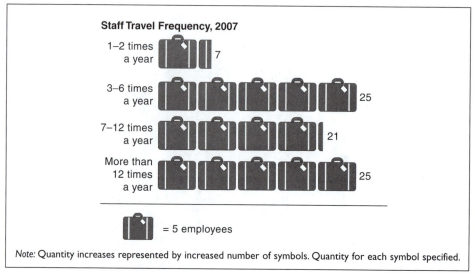

FIGURE 14   Picture Graph

Dimensional-Column Graphs

Consider a common on-the-job reporting requirement—tracking a series of expenses over a given period of time. Assume that you wish to show your company's expenses over a three-month period for security, courier, mail, and custodial services. Once you enter the data for these services into a spreadsheet program, you can display them in a variety of graph styles. As you select from among the options available, keep in mind your reader's need to interpret the data accurately and quickly, so keep the graph style as simple as possible for the information shown.

Graphs that depict columns as three-dimensional pillars are popular—they give the data a solid, three-dimensional, building-block appearance. They can, however, obscure rather than clarify the information, depending on how they are displayed. Consider the graph in Figure 15. Although the data are accurate, they cannot be interpreted as shown. The axis showing expenditures cannot be correlated with most columns representing the various services. The graph also obscures the columns for courier and guard-force services. Finally, this graph style does not allow readers to spot trends for expenditures over the three-month period, which is key information to decision-makers.

The graph in Figure 16 [p. 224] presents more clearly the same data as that shown in Figure 15. The trends of expenditures are easy to spot, and all the data are at least visible. Yet this graph is not ideal. To interpret the information, the reader would need to put a ruler on the page and align the tops of the columns with the axis showing expenditures.

The three-dimensional appearance can also cause confusion: Is the front or back of each column the correct data point? Also somewhat confusing is that, at first glance, the reader is tricked into interpreting the spaces between the column clusters as columns because they are of equal width.

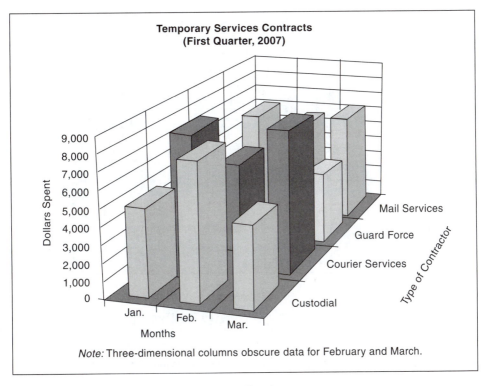

FIGURE 15 Three-Dimensional Column Graph

The graphs in Figures 17 [p. 225] and 18 [p. 226] would best represent the data, depending on your intent. Figure 17 avoids the ambiguity of the graph in Figure 16 by showing the data in two dimensions. It also displays the horizontal lines for expenditures on the vertical axis, thus making the data for each column easier to interpret. If you wished to show relative expenses among the four variables for a given quarter, this graph would be ideal. However, if you wished to show trends for the entire three-month period in contract expenditures at a glance, the graph in Figure 18 . . . is preferable.

When precise dollar amounts for each service are equally important, you can provide a table showing that information. As Figures 15 through 18 show, the more complicated a graph looks, the harder it is to interpret. On balance, simpler is better for the reader. Use this principle when you review your computer graphics on-screen in several styles and consider your reader's needs before deciding which style to use.

Drawings

A drawing is useful when your reader needs an impression of an object's general appearance or an overview of a series of steps or directions. . . . Drawings are the best choice when you need to focus on details or relationships that a photograph cannot capture. A drawing can emphasize the significant place of a mechanism, or its function, and omit what is not significant—for example, a cutaway drawing can show the

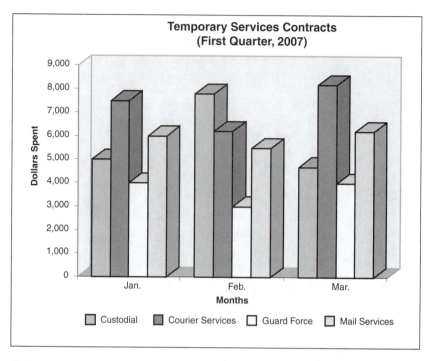

FIGURE 16 Three-Dimensional Column Graph (Same Data as in Figure 15)

internal parts of a piece of equipment in such a way that their relationship to the overall equipment is clear (Figure 19 [p. 226]). An exploded-view drawing can show the proper sequence in which parts fit together or the details of each individual part (Figure 20 [p. 227]).

Drawings are also the best option for illustrating simple objects or tasks that do not require photography (Figure 21 [p. 228]). However, if the actual appearance of an object (a dented fender) or a phenomenon (an aircraft wind-tunnel experiment) is necessary to your document, a photograph is essential. For drawings that require a high degree of accuracy and precision, seek the help of a graphics specialist.

Many organizations have their own format specifications for drawings. In the absence of such specifications, the following guidelines should be helpful:

- Show the equipment from the point of view of the person who will use it.
- When illustrating part of a system, show its relationship to the largest system of which it is a part.
- Draw the different parts of an object in proportion to one another, unless you indicate that certain parts are enlarged.
- For drawings used to illustrate a process, arrange them from left to right and from top to bottom.
- Label important parts of each drawing so that text references to them are clear and consistent.
- Depending on the complexity of what is shown, label the parts themselves— see Figure 20—or use a letter or number key.

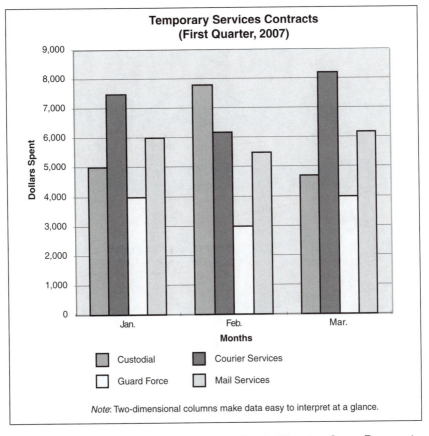

FIGURE 17   Two-Dimensional Column Graph (Showing Same Data as in Figures 15 and 16)

For general-interest images needed to illustrate newsletters, brochures, or presentation slides, use the clip-art libraries provided with word-processing programs, graphics programs, and the numerous image libraries available on the Web. These sources contain thousands of noncopyrighted symbols, shapes, and images of people, equipment, furniture, building, and the like. An added advantage is that you can adjust their size according to your document's layout and fit your space requirements. . . .

Flowcharts

A flowchart is a diagram that shows the stages of a process from beginning to end; it presents an overview that allows readers to grasp essential steps quickly and easily. Flowcharts can illustrate a variety of processes ranging from the stages required to refine bauxite ore into aluminum to the steps required to prepare a manuscript for publication.

Flowcharts can take several forms to represent the steps in a process: labeled blocks . . . pictorial representations . . . or standardized symbols. . . . The items in

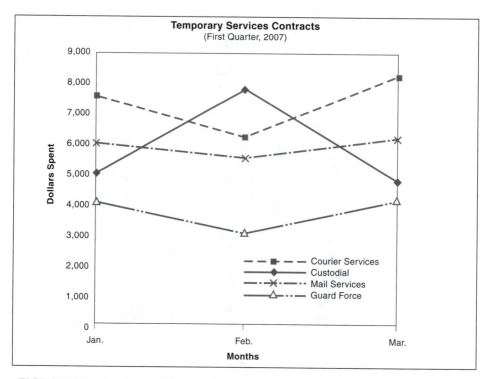

FIGURE 18   Line Graph (Showing Same Data as in Figures 15 through 17)

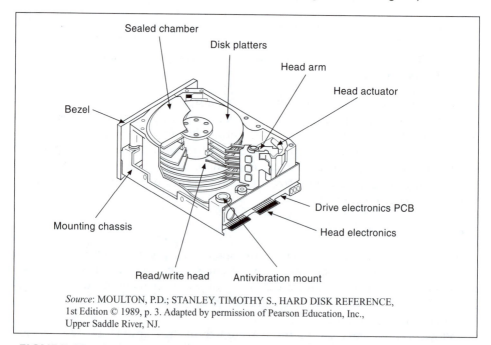

Source: MOULTON, P.D.; STANLEY, TIMOTHY S., HARD DISK REFERENCE,
1st Edition © 1989, p. 3. Adapted by permission of Pearson Education, Inc.,
Upper Saddle River, NJ.

FIGURE 19   Cutaway Drawing of a Hard Disk Drive

Installation

As you unpack the WorkCentre, familiarize yourself with its contents. After the WorkCentre is installed, and the Ready Indicator is lit, the WorkCentre is ready to make copies.

IMPORTANT: Save the carton and packing materials. They should be used to repack the WorkCentre if it has to be shipped for servicing or in case you move.

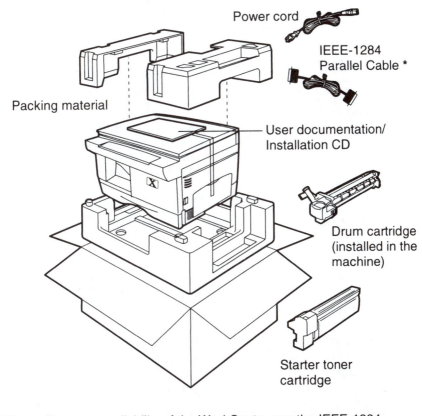

Power cord

IEEE-1284
Parallel Cable *

Packing material

User documentation/
Installation CD

Drum cartridge
(installed in the
machine)

Starter toner
cartridge

* Note: To ensure reliability of the WorkCentre use the IEEE-1284 compliant parallel cable that is supplied with the machine. Only cables labeled "IEEE-1284" can be used with your WorkCentre.

Source: © Xerox Corporation. Used with permission.

Note: Exploded view shows parts of a mechanism, packing container, and part names.

FIGURE 20   Exploded-View Drawing

Prevent Repetitive-Motion Injuries

Before beginning keying and during breaks throughout the day, take time to do the stretches as shown.

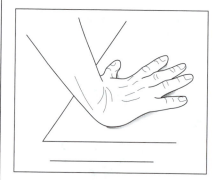

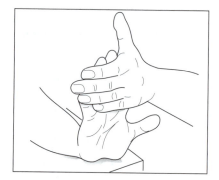

Gently press the hand against a firm flat surface, stretching the fingers and wrist. Hold for five seconds.

Rest the forearm on the edge of a table. Grasp the fingers of one hand and gently bend back the wrist, stretching the hands and wrist. Hold for five seconds.

FIGURE 21   Drawing

any flowchart are always connected according to the sequence in which the steps occur and typically flow left to right or top to bottom. When the flow is otherwise, indicate it with arrows. Flowcharts that document computer programs and other information-processing procedures use standardized symbols set forth in *Information Processing: Documentation Symbols and Conventions for Data, Program, and System Flowcharts, Program Network Charts, and System Resources Charts,* ISO publication 1985 (E).

Follow these guidelines when creating a flowchart:

- With labeled blocks and standardized symbols, use arrows to show the direction of flow, especially if the flow is opposite to the normal direction. With pictorial representations, use arrows to show the direction of all flow.
- Label each step in the process, or identify it with a conventional symbol. Steps can also be represented pictorially or by captioned blocks.
- Include a key if the flowchart contains symbols that your audience may not understand.
- Leave adequate white space on the page. Do not crowd the steps and directional arrows too closely together.

Organizational Charts

An organizational chart shows how the various components of an organization are related to one another. Such an illustration is useful when you want to give readers an overview of an organization or indicate the lines of authority within the organization (Figure 22).

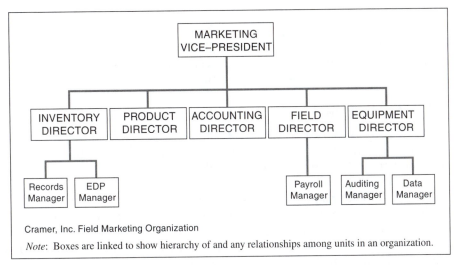

Cramer, Inc. Field Marketing Organization

Note: Boxes are linked to show hierarchy of and any relationships among units in an organization.

FIGURE 22   Organizational Chart

The title of each organizational component (office, section, division) is placed in a separate box. These boxes are then linked to a central authority. If useful to your readers, include the name of the person occupying the position identified in each box. As with all illustrations, place the organizational chart as close as possible to the text that refers to it.

Maps

Maps can be used to show the specific geographic features of an area (roads, mountains, rivers) or to show information according to geographic distribution (population, housing, manufacturing centers, and so forth). . . .

Keep in mind the following points as you create maps for use with your text:

1. Clearly identify all boundaries within the map. Eliminate those that are unnecessary to the area you want to show.
2. Eliminate unnecessary information from your map. For example, if population is the focal point, do not include mountains, roads, rivers, and so on.
3. Include a scale of miles or feet, or of kilometers or meters, to give your reader an indication of the map's proportions.
4. Indicate which direction is north.

5. Emphasize key features by using shading, dots, crosshatching, or appropriate symbols, and include a key telling what the different colors, shadings, or symbols represent. . . .

Photographs

Photographs are vital to show the surface appearance of an object or to record an event or the development of a phenomenon over a period of time. Not all representations, however, call for photographs. They cannot depict the internal workings of a mechanism or below-the-surface details of objects or structures. Such details are better shown in drawings or diagrams.

Highlighting Photographic Objects

If you are taking the photo, stand close enough to the object so that it fills your picture frame. A camera will photograph only what it is aimed at; accordingly, select important details and the camera angles that will record these details. To show the relative size of an unfamiliar object, place a familiar object—such as a ruler, a book, a tool, or a person—near the object that is to be photographed. . . .

Ask the printer reproducing your publication for special handling requirements if you use glossy photographs. If you use digital photos, ask about the preferred resolution of the images before you shoot. The higher the resolution, the better the quality. You can digitize film photos using scanners and digital cameras, so ensure that the equipment has the necessary memory for the resolution required by the printer.

Using Color

Color is in many cases the only way to communicate crucial information. In medical, chemical, geological, and botanical publications, for example, readers often need to know exactly what an object or a phenomenon looks like to accurately interpret it. In these circumstances, color reproduction is the only legitimate option available.

Posting color images at a Web site is no more complicated or expensive than posting black-and-white images. For publications, however, preparing and printing color photographs are complex technical tasks performed by graphics and printing professionals. If you are planning to use color photographs in your publication, discuss with these professionals the type, quality, and number of photographs required.

Be mindful that color reproduction is significantly more expensive than black-and-white reproduction. Color can also be tricky to reproduce accurately without losing contrast and vividness. For this reason, the original photographs must be sharply focused and rich in contrast.

David W. Ewing

Strategies of Persuasion

David W. Ewing was Executive Editor-Planning of the Harvard Business Review *and a member of the faculty of the Harvard Business School.*

When we review reports, letters, and memoranda that get the intended results, we find a fascinating diversity of approaches. Some are gentle in approach, taking readers by the hand and leading them to a certain finding or recommendation. Others are brisk and abrupt. Some are objective in approach, carefully examining both sides of an idea, like a judge writing a difficult decision—at least until the end. Others burst with impatience to explain one side, and only one side, of a proposal or argument. Some flow swimmingly; others erupt like Mount Vesuvius.

How is it possible that communications taking such different approaches can all be effective? It is not sufficient to answer glibly, "It depends on the situation," or "Communications should mirror the personal style of the communicator." How does a written communication depend on the situation? Which style of the communicator should be reflected in a given communication? Examine the writing of most business executives, professionals, and public leaders, and you will find not one but *many* styles of exposition.

RULES EVERY PERSUADER SHOULD KNOW

The explanation lies in a set of relationships among the communicator, the reader, the message, and the time-space environment. These relationships work in predictable ways and are an important part of the knowledge of every good business and professional writer. They come into play in the planning stages of writing, when the writer is considering how he or she will proceed, and in the main body of the presentation, to accomplish what he or she has promised in the opening paragraphs. The relationships affect some of the most important decisions a writer makes—the choice of ideas to use, the comparative emphasis to be given various arguments pro and con, the types of reasons and supporting material used, the establishment of credibility, and other matters.

It is convenient but self-defeating to follow fixed prescriptions for persuasion, such as to put your strongest arguments first or last or to identify with the readers. Such nostrums were fine for the age of patent medicines and snake-oil peddlers, but not for the age of diagnostic medicine. What is more, they are belittling. They assume that writers are witless. Good writers vary their approaches in response to their readings of different situations. Just as a good golfer plays an approach to a green differently depending on the wind, so a good writer uses different strategies depending on the crosscurrents of mood and feeling.

How should you choose your approach to a group of readers? What elements of the approach should be tailored to the situation? These are the topics of this . . . [selection]. Let us assume that the substance of the intended message is clear in your mind. . . .

1. Consider Whether Your Views Will Make Problems for Readers.

J. C. Mathes and Dwight W. Stevenson tell of the student engineer who was asked to evaluate the efficiency of the employer plant's waste-treatment process.[1] He found that, by making a simple change, the company could save more than $200,000 a year. Anticipating an enthusiastic response, he wrote up and delivered his report. Although he waited with great expectations, no accolades came. Why? What he hadn't counted on was that now his supervisors would have to explain to their bosses why they had allowed a waste of $200,000 a year. They were far from elated to read his report.

"If you want to make a man your enemy," Henry C. Link once said, "tell him simply, 'You are wrong.' This method works every time." Under the illusion that their sciences are "hard," physicists, biologists, and others may assume that it is necessary only to worry about setting the facts forth accurately. From posterity's standpoint, perhaps yes—but not from the standpoint of writing for current results. As Kenneth Boulding, head of the American Association for the Advancement of Science, has pointed out, the so-called "hard" sciences are in many ways "soft," and vice versa. "You can knock forever on a deaf man's door," said Zorba the Greek, and the deaf man can be a physicist as well as a marketing manager or public official.

If your views are bad news for readers, you proceed to report them, but with empathy and tact and an effort to put yourself in the readers' shoes. You work as carefully as if you were licking honey off a thorn.

---

[1]J. C. Mathes and Dwight W. Stevenson, *Designing Technical Reports* (Indianapolis: Bobbs-Merrill, 1976), pp. 18–19.

2. Don't Offer New Ideas, Directives, or Recommendations for Change until Your Readers Are Prepared for Them.

"Should I state my surprising findings at the very beginning of my memorandum?" a writer asks. "Should I go slow with my heretical proposal and hold the reader's hand?" asks another.

Generally speaking, the answer to all such questions depends on the extent of your audience's resistance to change, the amount of change you are asking for, the uncertainty in readers' minds as to your understanding of their situation, and what psychologists call the "perceived threat" of your communication, that is, how much it seems (to readers) to upset their values and interests. The more change, uncertainty, and/or threat, the slower you should proceed, the more carefully you should prepare your readers.

For instance, if your boss is enthusiastic about a new promotion scheme that he (or she) has paid a consultant $25,000 to devise, naturally you will want to go slow in shooting it down (at least if you want to stay in his good graces). In fact, any written criticism of the scheme is probably out of order until you have had a chance to talk with him and get a feel for the proper timing of any forthcoming criticism. When you do commit yourself to writing, you should probably review the arguments for the new scheme as fairly as possible, making it crystal clear to him that you understand them. Only then does it become timely to turn to the facts or conditions that, in your opinion, raise serious questions about the plan.

On the other hand, suppose the faulty promotion scheme is of little personal interest to your boss—it is not his (or her) "baby." Now the situation is different. You can launch right into the shortcomings, throwing your heaviest objections first. The fact is that the boss may not even want to know about the lesser objections, much less the supporting arguments once advanced for the plan; the main things he should know are that (a) the plan is in trouble and (b) the major, most compelling reasons why.

Clearly, this strategy is plain, everyday common sense—what you would normally do in communicating orally instead of in writing. Only in writing you must be more explicit and thorough, because a document lacks the expressiveness and visual advantages of a spoken dialogue.

Now consider another type of situation. Suppose it is your unhappy task to write a department manager that the extra appropriations he (or she) was promised have been cancelled. First of all, how would you handle it if you saw him often and could talk with him personally? . . . You would not (we hope) pussyfoot around trying to withhold the bad news from him. Also, once you had indicated the main message, you would probably backtrack a little and make it clear that the step is being taken with reluctance.

"Joe, it looks as if we're not going to be able to give you the extra budget we promised," you might say, getting down to business. That is the message in a nutshell—now for the review of common ground. "We know how much you have counted on getting those people and funds. There's no doubt you could manage them well and put them to good use. And we know that the morale of your people is involved in this, too. But the fact is that the sales we counted on are not coming in. We've got to cut somewhere, and, frankly, we feel it's got to be your department because. . . ."

If you are communicating by letter or memorandum, the strategy is exactly the same (only "Joe" may now be "Mr. Wyncoop"). After your lead, you review the main needs as he and you understand them, perhaps spelling them out more than you would have in a face-to-face meeting but choosing the same ones. Only then do you turn to the new conditions that make it necessary to do an about-face.

3. Your Credibitility with Readers Affects Your Strategy.

In general, communication research indicates that the chances of opinion change vary with the communicator's authority with his or her readers. In their succinct summary of the field, *Persuasion,* social scientists Marvin Karlins and Herbert I. Abelson point out that credibility itself is a variable; that is, it can be influenced by the words of the communicator.[2] Above all, as psychologists repeatedly emphasize, credibility lies in the eye of the beholder. . . .[3]

In written communications there are two types of credibility. It may be given or it may be acquired. Between the two lies a world of practical difference.

Given credibility may result from your position in an organization. If, let us say, you are the boss writing directions to a subordinate, your credibility is likely to be high. Given credibility also may result from reputation—a well-known chemist has more credibility in communications about polymers than a good industrial engineer has, but the latter would possess more credibility in communications about time-and-motion studies. It may result from the individuals and groups the writer is associated with—if he or she is a member of the same trade union the reader belongs to, a union held in high esteem by both, he or she has more credibility in a memorandum on grievance procedures than a member of the board of directors would have.

Though you may be high in given credibility, you may yet need to remind some readers of the fact. In the case of an obvious credential, such as a position in an important organization, a letterhead may be enough to do the trick. Another device is to insert a few lines of biographical data at the head of a report or brochure. If you have had experience or associations that carry weight with the reader, perhaps you can interject them early in the message. "During a visit I had last week with Zach Jarvis," you might say, knowing that Dr. Zachary P. Jarvis is a magic name with your reader, or "The Executive Committee of the Aberjona Basin Association asked me to join their meeting on Monday . . ." knowing that group carries a triple-A rating in the mind of the reader.

---

[2]Marvin Karlins and Herbert I. Abelson. *Persuasion* (New York: Springer, 1970); see pp. 107–132.

[3]See, for example, Ralph L. Rosnow and Edward J. Robinson, Eds. *Experiments in Persuasion* (New York: Academic, 1967).

Of course, you do not want to overplay your hand at such name dropping. A report to a fairly diverse audience by a famous black organization began simply:

> The National Association for the Advancement of Colored People has for many years been dedicated to the task of defending the economic, social and political rights and interests of black Americans. The growing national debate about energy has led us to examine the question to ascertain the implications for black Americans.[4]

Again, an attorney of the American Civil Liberties Union, in a letter to members of the organization soliciting donations, began:

> My dear friend:

> I am the ACLU lawyer who went into court last April to defend freedom of speech in Skokie, Illinois, for a handful of people calling themselves "nazis."

> The case has had an enormous impact on my life.

> It has also gravely injured the ACLU financially. . . .[5]

Acquired credibility, on the other hand, is earned by thoughts and facts in the written message. I may not know you from Adam. Yet if you send me a letter or report that carefully, helpfully describes something I am interested in, you gain credibility in my estimation.

Some studies suggest that, if you are low in given credibility and seek to acquire it with an audience, a useful technique is to cite ideas or evidence that support the reader's existing views.[6] As Disraeli once said, "My idea of an agreeable person is a person who agrees with me." The very fact that you feel confident and knowledgeable enough to articulate these views is likely to lift you several notches in the reader's estimation.

Still another approach is that old standby of persuaders—identifying yourself, in an early section, with the goals and interests of the audience. Possibly the most famous example of this strategy is the opening of Marc Antony's funeral oration, in Shakespeare's *Julius Caesar:* "I come to bury Caesar, not to praise him. . . ."

Finally, you can acquire credibility by citing authorities who rate highly with your intended audience, or by exhibiting documentary evidence that, because of its source, lends prestige and authority to your proposals or ideas.

"For success in negotiation," says C. Northcote Parkinson and Nigel Rowe, "it is vitally important that people will believe what you say and assume that any promise you make will be kept. But it is no good saying: 'Trust me. Rely on my word.' Only politicians say that."[7]

---

[4] *The Wall Street Journal,* January 12, 1978.
[5] David Goldberger, letter dated March 20, 1978.
[6] Karlins and Abelson, op. cit., pp. 115–119.
[7] C. Northcote Parkinson and Nigel Rowe, "Better Communcation: Business's Best Defense," *The McKinsey Quarterly,* Winter 1978, p. 26.

Even if you have prestigious credentials, you cannot take too much for granted. In an age of television sets, radios, cassettes, and record players in every home, credibility—at least with the public—may come quicker for the singer or comedian than for the judge, business executive, or medical researcher. In fact, because of an association with an organization or profession, you may be stereotyped as a member of "them" or "the establishment." It may behoove you to establish that you are a person with a name, a personality, certain interests, certain experiences—not just a nameless representative.

4. If Your Audience Disagrees with Your Ideas or Is Uncertain about Them, Present Both Sides of the Argument.

Behavioral scientists generally find that, if an audience is friendly to a persuader, or has no contrary views on the topic and will get none in the future, a one-sided presentation of a controversial question is most effective.[8] For instance, if your point is that sales of product X in the St. Louis territory could be doubled and you are writing to enthusiastic salespeople of product X, your best course is to concentrate on facts and examples showing the enormous potential of product X. There is no shortage of evidence showing that people generally prefer reading material that confirms their beliefs, and that they develop resistance to material that repudiates their beliefs. (As we shall see presently, however, this does not mean you cannot change their minds.)

But suppose your audience has not made up its mind, so far as you know? In this case you would do well to deal with *both* sides of the argument (or all sides, if there are more than two). Follow the same approach if the reader disagrees with you at the outset. For one thing, a two-sided presentation suggests to an uncertain or hostile audience that you possess objectivity. For another, it helps the reader remember your view by putting the pros and cons in relationship to one another. Also, it meets the reader's need to be treated as a mature, informed individual. As Karlins and Abelson point out:

> Conspicuously underlying your presentation is the assumption that the audience would be on your side if they only knew the truth. The other points of view should be presented with the attitude "it would be natural for you to have this idea if you don't know all the facts, but when you know all the facts, you will be convinced."[9]

Karlins and Abelson tested the reactions of audiences in postwar Germany to Voice of America broadcasts. They found that the most persuasive programs were those that included admissions of shortcomings in United States living conditions.[10]

---

[8]Experiments supporting this conclusion are reported by Carl I. Hovland, Arthur A. Lumsdaine, and F. Sheffield in *Experiments on Mass Communiation* (Princeton: Princeton University Press, 1949). Cited by Karlins and Abelson, op. cit., p. 22.
[9]Karlins and Abelson, op. cit., p. 26.
[10]See *Factors Affecting Credibility in Psychological Warfare Communications* (Washington, D.C.: Human Resources Research Office, George Washington University, 1956).

Again, observation of businessmen's reactions to scores of *Harvard Business Review* articles advocating controversial measures convinces me that the most influential articles have been those that have acknowledged the shortcomings, weaknesses, and limitations of their arguments. When an author wants to sell a new idea to a sophisticated audience, he or she should be candid about the soft spots in his argument.

5. Win Respect by Making Your Opinion or Recommendation Clear.

Although strategy may call for a two-sided argument, this does not mean you should be timid in setting forth your conclusions or proposals at the end. We assume here that you have definite views and seek to persuade your audience to adopt them. The two-sided approach is a *means* to that end; it does not imply compromising or obfuscating your conclusions. An official at Armour & Company once criticized many reports from subordinates to bosses on the ground that, after presenting much data, they concluded, in effect, "Here is what I found out and maybe we should do this or maybe we should do that." The typical response of a boss to such a memorandum, he noted, was to do nothing. Hence, the time taken both in writing and reading was wasted.[11]

6. Put Your Strongest Points Last if the Audience Is Very Interested in the Argument, First if It Is Not so Interested.

This question is referred to by social scientists as the "primacy-recency" issue in persuasion. The argument presented first is said to have primacy; the argument presented last, recency. Although studies of the question have produced inconsistent findings and no firm rules can be drawn, it appears that if your audience is deeply concerned with your subject you can afford to lead it along from the weakest points to the strongest. The audience's great interest will keep it reading, and putting the weaker points at the start tends to create rising reader expectations about what is coming. When you end with your strongest punch, therefore, you do not let readers down.

If your audience is not so concerned with the topic, on the other hand, it may be best to use the opposite approach. Now you cannot risk leading readers along a winding path. They may drop out before you reach the end. So grab their attention right at the beginning with your strongest argument or idea.

In any case, put the recommendation, facts, or arguments you most want the reader to *remember* first or last. Although experiments by social scientists on the primacy-recency issue are inconclusive, there is a firm pattern on the question of recall. The ideas you state first or last have a better chance of being remembered than the ideas stated in the middle of your appeal or case.

---

[11]John Ball and Cecil B. Williams, *Report Writing* (New York: Ronald, 1955).

7. Don't Count on Changing Attitudes by Offering Information Alone.

"People are hostile to big business because they don't know enough facts about it," businesspeople are heard to say. Or, "If customers knew the truth about our costs, they would not object to our prices." Companies have poured large sums into advertising and public relations campaigns on this assumption; civic organizations have often based their hopes on it.

"The trouble with the assumption," states Karlins and Abelson, "is that it is almost never valid. There is a substantial body of research findings indicating that cognition—knowing something new—increasing information—is effective as an attitude change agent only under very specialized conditions."[12]

Social scientists do concede, however, that presentations of facts alone may strengthen the opinions of people who already agree with the persuader. The information reassures them and helps them defend themselves in discussions with others.

8. "Testimonials" Are Most Likely to Be Persuasive if Drawn from People with Whom Readers Associate.

It is well known that a person's attitudes and opinions are strongly influenced by the groups to which he or she belongs or wants to belong—work units in a company, labor unions, bowling teams, social clubs, church associations, ethnic associations and so on. To muster third-party support for your proposal or idea, therefore, you would do well to cite the behavior, findings, or beliefs of groups to which your readers belong. In so doing, you allay any feelings of isolation readers might have if tempted to follow your ideas. You suggest that they are not alone with you, that there is group support for the points being made.

As every school child learns, the predominant attitudes of a group toward individuals or regarding standards of behavior, performance, or status influence an individual member's perceptions. For instance, a study of boys at a camp demonstrated that their ratings of various individuals' performances at shooting and canoeing were biased by their knowledge of the status of the rated individuals in the camp society. Thus a boy generally regarded as a leader was seen as performing better with the rifle or canoe than was a boy generally regarded as a follower, even though the first boy's performance was not actually superior.[13]

In addition, it seems fair to say that as modern television, radio, records, and cassettes have brought national celebrities into the home and automobile, these people, too, have been stamped with approval or disapproval by millions of groups across the country.

---

[12]Karlins and Abelson, op. cit., p. 33.
[13]Ibid., p. 50.

Accordingly, if your readers are young, dissident, or "long hairs," refer to a Richard Dreyfuss or a Joan Baez for supporting statements, not to a Gerald Ford or an Arnold Palmer. If your readers are electrical engineers, quote well-regarded scientific sources as your authority, not star salespeople or public relations people. Take into account also that the more deeply attached your readers are to a group, the greater the influence of the group norms on them. For instance, one experiment by social scientists showed that the opinions of Catholic students who took their religion seriously were less influenced by the answers of nonserious Catholics than were the opinions of Catholic students who placed little value on their church membership.[14]

9. Be Wary of Using Extreme or "Sensational" Claims and Facts.

Both research in behavioral science and common sense confirm this rule.[15] Do not be misled by the fact that flashy journalists make successful use of extreme and bizarre cases to dramatize a story. The situation in business and professional writing is different from that in journalism.

When you seek the confidence and cooperation of your readers—and typically you do in the kinds of communications we deal with in this book—it is best to write in terms of the real world as you and they perceive it. Observable, believable, realistic statements carry more weight than any other kind. Although you want reader attention, you do not want to shock your audience with outlandish examples or arguments. These may help you to succeed in making the reader sit up—but they will also provoke distrust and suspicion.

Examples are common in the letters sections of newspapers. A writer who identified himself as a former vice-president of a well-known bank opposed a large power company's plan to build a new plant in a rural area near his town. His letter began as follows: "A great many of us . . . are profoundly disturbed by the proposal now being considered to disrupt and destroy the marvelous little valley southwest of [name of town], in order to build bigger and better power plants. This would be a devastating blow to the last unspoiled bit of country left in Connecticut. . . ."[16]

Like a batter who hits the first two pitches foul and quickly gets two strikes against him, this writer managed to distort the first two sentences he wrote. The proposed plant, though a very large one, would not "disrupt and destroy" the valley—only a small section of the valley area would be affected. Moreover, the valley was not "the last unspoiled bit of country" in the state—it was only a small parcel of the state's beautiful countryside. These exaggerations might have

---

[14]See H. Kelley, "Salience of Membership and Resistance to Change of Group-Anchored Attitudes," *Human Relations,* August 1955, pp. 255–289. Cited in Karlins and Abelson, op. cit., p. 58.

[15]See, for example, *Building Opposition to the Excess Profits Tax* (Princeton: Opinion Research Corporation, August 1952), and R. Weiss, "Conscious Technique for the Variation of Source Credibility," *Psychological Reports,* Vol. 20, 1969, p. 1159. Both cited in Karlins and Abelson, op. cit., pp. 36–37.

[16]*Lakeville Journal,* April 2, 1970, p. 11.

drawn cheers from rabid foes of the project, but the writer wasn't interested in appealing to them; he wanted to win uncommitted readers. At the very beginning, however, he antagonized them with hyperbole.

10. Tailor Your Presentation to the Reasons for Readers' Attitudes, if You Know Them.

Your chances of persuading readers are better if you can plan your appeal or argument to meet the main feelings, prejudices, or reasons for their beliefs. For instance, if reader beliefs are the result of their wanting to go along with certain groups they like or associate with, your best bet (as indicated earlier) is to show the acceptability of your point to these groups. If their attitudes reflect personal biases, such as an old grudge against someone in power, it is best to tailor your presentation to that prejudice. And so on.

Summarizing the implications of several behavioral studies, Karlins and Abelson present the example of three people who say they are against private ownership of industry. How should their reasons for this position influence one's choice of strategy or persuasion? The authors explain:

> One of them feels that way because he has only been exposed to one side of the story and has nothing else on which to base his opinions. The way to change this man's opinion may be to expose him to facts, take him to visit some factories, meet some workers and supervisors. A second person is against private ownership because that is the prevailing norm or social climate in the circles in which he finds himself. His attitudes are caused by his being a part of a group and conforming to its standards. You cannot change this fellow just by showing him facts. The facts must be presented in an atmosphere which suggests a social reward for changing his opinion. Some kind of status appeal might be a start in that direction. A third person may have negative attitudes toward private industry because by making business the scapegoat for all his troubles, he can unload his pent-up feelings of bitterness and disappointment at the world for not giving him a better break. . . . Trying to change this third person with facts may actually do more harm than good. The more evidence shows how wrong he is, the more he looks for reasons to support his beliefs. This kind of person can sometimes be influenced by helping him to understand why he has a particular attitude.[17]

11. Never Mention Other People without Considering Their Possible Effect on the Reader.

Other people may, as we saw earlier, be introduced for the sake of "testimonials." More commonly, however, other people's names are mentioned in the course of explaining a situation, narrating an event, or completing the format of a message. This use of names, too, may affect the power of your message.

A reference to the actions of another person—however simple and unobtrusive it may seem to you the writer—may alter your relationship with

---

[17]Karlins and Abelson, op. cit., p. 92.

readers. If readers consider that person a friend or enemy, their natural reaction is to begin thinking of the possible bearing of your communication on their friendship or antagonism. This reaction can have significant implications for your approach.

To illustrate, a doctoral student who had failed to meet his school's program requirements tried to muster faculty opinion in support of his petition for readmission by appearing daily at the entrance to the dining hall and handing out leaflets to faculty members. One such leaflet contained these words: "I am very unhappy about the strain my case has created for Professor [name of the program director]. I am distressed if last Friday's handout . . . created the impression that I was harping on his mistakes. I have told him and I tell you that I could understand his actions and decisions. . . ." The leaflet went on at some length to explain the doctoral student's feelings about the problem.

What this writer did not realize was the impact of the professor's name on his communication strategy. Almost everyone who received the leaflet was a colleague of the professor in question. Therefore the leaflet made it necessary for them to think of their relationship with the professor when they made up their minds about the petition. And their relationship with the professor was more important to them than their relationship with the doctoral student.

If the doctoral student considered it essential to mention the professor, he could have elected to: (1) try to win readers over while convincing them that their relations with the professor would not be affected, or (2) show that the professor was so far off base that readers were morally bound to risk their relationship with him. In the latter case, the leaflet should have contained ready-to-use arguments that readers could draw on in explaining to the professor why they sympathized with the doctoral student. Since the leaflet did neither of these things, it was a failure in persuasion.

Don't overlook the possible effect of distribution. Letters often go to third parties, with "cc" typed at the bottom followed by the names of those people. A memorandum often contains the names of several addressees in the "To" line at the top. Covering letters with reports may indicate several groups of readers. All this may affect your strategy. The background information that you could omit if writing only to Jones may be quite necessary if Brown, too, is an important reader; and the rather offhand treatment you give to a certain test or episode if writing to Jones and Brown might not be fitting at all if Larabee also is an intended reader. Many times the wise manager or professional rewrites part of a letter or memo after deciding to send a copy of it to an additional person who was not considered when the first draft was made.

Many people have strong feelings about "blind copies," that is, copies sent to persons other than those indicated after "cc" at the end of a letter or in the "To" line of a memorandum. Some people feel that blind copies never should be sent. Others feel that since a letter or memo is the property of the writer, he or she can distribute it at will. Although the latter view is legally correct, only an obtuse writer will distribute copies thoughtlessly if the content is in any way confidential, personal, or politically sensitive.

SIZING UP YOUR READERS

We have a tendency to abstract written communications from real life, to act as if the customary ground rules of influence and persuasion don't apply to a message that is in writing. We act with a naiveté almost unheard of in our face-to-face relationships. Not seeing readers, we act as if they weren't real people. "If we write the information clearly, accurately, and correctly," we think wishfully to ourselves, "surely that satisfies the requirements of a piece of paper." But Josh Billings's puckish maxim, "As scarce as truth is, the supply has always been in excess of the demand," applies to truth on paper as well as truth in conversation.

Think of your intended readers as the real people they will be when they take your letter or report out of the "in-box." Only then can you decide intelligently what information and ideas to emphasize and in what order to present them. To help you think of readers as three-dimensional people, ask yourself some questions about their situation and relationships with you. Are they:

- Deeply or only mildly interested in the subject of your communication?
- Familiar or unfamiliar with your views, competence, and feelings about them?
- Knowledgeable or ignorant of your authority in the area discussed, your status, and your associations of possible importance to them?
- Committed or uncommitted to a viewpoint, opinion, or course of action other than the one you favor in your letter, report, or other document?
- Likely or unlikely to find your proposal, idea, finding, or conclusion threatening or requiring considerable change in their thought or behavior?
- Inclined or uninclined to think and feel the way they do about the subject because of identifiable reasons, prejudices, or experiences?
- Associated formally or informally with groups or organizations involved in some way with the idea or proposal you deal with?

With answers to questions like these in mind, you will not see your readers as shadows on the wall. They will sit across from you. You can write as if talking *with* them, not talking to them.

Philip C. Kolin

Proposals

Philip C. Kolin, who has published widely in the fields of literature and communications, is Professor of English at the University of Southern Mississippi.

GUIDELINES FOR WRITING A SUCCESSFUL PROPOSAL

Regardless of the type of proposal you are called on to write, the following guidelines will help you to persuade your audience to approve your plan. Refer to these guidelines both before and while you formulate your plan.

1. **Approach writing a proposal as a problem-solving activity.** Your goal is to solve a problem that affects the reader. Do not lose sight of the problem as you plan and write your proposal. Everything in your proposal should relate to the problem, and the organization of your proposal should reflect your ability as a problem-solver. Psychologically, make the reader feel confident that you can solve the problem.

2. **Regard your audience as skeptical readers.** Even though you offer a plan that you think will benefit readers, do not be overconfident that they will automatically accept it as the best and only way to proceed. To determine the feasibility of your plan, readers will question everything you say. They will withhold their approval if your proposal contains errors, omissions, or inconsistencies. Consequently, try to examine your proposal from the readers' point of view. . . .

3. **Research your proposal carefully.** A winning proposal is not based only on a few well-meaning, general suggestions. All your good intentions and enthusiasm will not substitute for the hard facts readers will demand. Concrete examples persuade readers; unsupported generalizations do not. To make your proposal complete and accurate, you will have to do a lot of homework; for example, reading

previous correspondence or research about the problem, doing comparative shopping for the best prices, verifying schedules and timetables, interviewing customers and/or employees, making site visits. . . .

4. **Prove that your proposal is workable.** The bottom-line question from your reader is "Will this plan work?" Your proposal should be well thought out. It should contain no statements that say "Let's see what happens if we do X or Y." By analyzing and, when possible, by testing each part of your proposal in advance, you can eliminate any quirks and revise the proposal appropriately before readers evaluate it. What you propose should be consistent with the organization and capabilities of the company. It would be foolish to recommend, for example, that a small company (fifty employees) triple its workforce to accomplish your plan.

5. **Be sure that your proposal is financially realistic.** This point is closely associated with and follows from guideline 4. "Is it worth the money?" is another bottom-line question you can expect from readers. Do not submit a proposal that would require an unnecessarily large amount of money to implement. For example, it would be unrealistic to recommend that your company spend $20,000 to solve a $2,000 problem that might not ever recur. Study the ecomonic climate, too—are you in an economic slump or in a boom time?

6. **Package your proposal attractively.** Make sure your proposal is letter-perfect, inviting, and easy to read (e.g., use plenty of headings and other visual devices). . . . The appearance as well as the content of your proposal can determine whether it is accepted or rejected. Remember that readers, especially those unfamiliar with your work, will evaluate your proposal as evidence of the type of work you want to do for them. Take advantage of any software programs dealing with desktop publishing . . . that may be available to you. This software will allow you to prepare a proposal that looks as if it were done by a professional printer.

INTERNAL PROPOSALS

The primary purpose of an internal proposal, such as the one included in Figure 1, is to offer a realistic and constructive plan to help your company run its business more efficiently and economically. On your job you may discover a better way of doing something or a more efficient way to correct a problem. In the world of work, typical problems for which proposals are written focus on money, personnel, outdated technology, health concerns, and organizational communications. You believe that your proposed change will save your employer time, money, or further trouble. (Tina Escobar and Oliver Jabur in Figure 1 have identified and researched a more effective and less costly way for Community Federal Bank to do business and to satisfy its customers.) You decide to notify your department head, manager, or supervisor, or your employer may ask you for specific suggestions to solve a problem he or she has already identified. . . .

COMMUNITY FEDERAL BANK

FDIC/DIFM

Powell	*Monroe*	*Langston*
584-5200	*413-6000*	*796-3009*

TO: Michael L. Sappington, Executive Vice President
      Dorothy Woo, Langston Regional Manager

FROM: Tina Escobar, Oliver Jabur, ATM Services

DATE: June 2, 1994

RE: A Proposal to Install an ATM at the Mayfield Park Branch

PURPOSE

Clearly states why proposal is being sent

We are writing to propose a cost-effective solution to what is a growing problem at the Mayfield Park branch in Langston: inefficient servicing of customer needs and rising personnel costs. We recommend that you approve the purchase and installation, within the next three to four months, of an ATM at Mayfield. Such action is consistent with Community's goals of expanding electronic banking services and promoting our image as a self-serve yet customer-oriented institution.

THE PROBLEM WITH CURRENT SERVICES AT MAYFIELD PARK

Identifies problem by giving reader necessary background information

Currently, we employ four tellers at Mayfield. However, we are spending too much on personnel/salary for routine customer transactions. In fact, as determined by teller activity reports, nearly 25 percent of the 4 tellers' time each week is devoted to routine activities easily accommodated by ATMs. Here is a breakdown of teller activity for the month of May:

Teller #	Total Transactions	Routine Transactions
1	6,205	1,551
2	5,989	1,383
3	6,345	1,522
4	6,072	1,518
	24,611	5,974

FIGURE 1   An Internal Unsolicited Proposal

Divides problem into parts— volume, financial, personnel, customer service

Clearly, we are not fully using our tellers' sales abilities when they are kept busy with routine activities. To compound the problem, we expect business to increase by at least 25 percent at Mayfield in the next few months, as projected by this year's market survey. If we do not install an ATM, we will need to hire a fifth teller, at an annual cost of $20,800 ($15,500 base pay plus approximately 30 percent for fringes), for the additional 6,000 transactions we project.

Verifies that problem is widespread

Most important, customer needs are not being met efficiently at Mayfield. Recent surveys done for Community Federal by Watson-Perry demonstrate that customers are dissatisfied about not having the convenience of an ATM at Mayfield. Seventy-seven percent of the respondents to the Watson-Perry questionnaire pointed to the lack of an ATM as Mayfield's biggest drawback. Customers are unhappy about long waits in line to do simple banking business, such as deposits, withdrawals, and loan payments, and about having to drive to other branches to do their after-hours banking. Conversations we had with manager Rachael Harris-Ignara at Mayfield confirm that customers regularly complain to tellers and loan officers about not having an ATM.

Ultimately, the lack of an ATM at Mayfield Park hurts Community's image. With ATMs available to Mayfield residents at local stores and other banks, our institution risks having customers and potential customers disassociate Community from their banking needs. We not only miss the opportunity of selling them on our other services but also risk losing their business entirely.

A SOLUTION TO THE PROBLEM

Relates solution to individual parts of the problem

The purchase and installation of an ATM at Mayfield Park will initially result in significant savings in personnel costs and time. We will not have to hire a fifth teller at $20,800 for the anticipated increase in transactions. Thanks to an ATM, we will also be able to allocate more effectively the talents of the four existing tellers at Mayfield. These tellers will then be able to assist customers with questions and transactions that cannot be handled through an ATM, such as purchase of savings bonds, traveler's checks, or foreign currency. Mayfield tellers, therefore, will have more opportunities for greater involvement in customer services and can spend more time cross-selling our services. As a result, we will get additional personnel achievements without reducing the level of customer service. In fact, we will be improving that service.

The increase in teller availability will inevitably lead to greater customer satisfaction. An ATM will allow customers the option of meeting their banking needs electronically or through a teller. Customers in a hurry can

FIGURE 1 *(Continued)*

easily make a withdrawal with their ATM cards, while those who need more personal attention can take advantage of window service. As customer frustration is eased, so too will be the stress on tellers because of shorter lines and fewer complaints.

Shows problem can be solved and stresses how

It is feasible to install an ATM at Mayfield. This location does not pose the difficulties we faced at some older branches. Mayfield offers ample room to install a drive-up ATM in the stubbed-out fourth drive-up lane. This location is away from the heavily congested area in front of the bank, yet it is easily accessible from the main driveway and the side drive facing Cornith Avenue.

Judging from the ATM vendor's past work, the ATM could be installed and operational within three to four months. That is the amount of time it took to install ATMs at the first two locations in Powell and for Archer Avenue in Langston. Moreover, by authorizing the expenditure at Mayfield within the next month, you will ensure that ATM service is available before the Christmas season.

COSTS

Itemizes costs

The costs of implementing our proposal are as follows:

Diebold Drive-up ATM	$28,000.00
Installation fee	2,000.00
Maintenance (1 year)	1,500.00
	$31,500.00

Interprets costs for reader

This $31,500.00, however, does not truly reflect into our annual costs. We would be able to amortize, for tax purposes, the cost of installation of the ATM over five years. Our annual expenses would, therefore, look like this:

$30,000 (28,000 + 2,000) ÷ 5 years =
$6,000.00 + 1,500 (maintenance) or $7,500 per year.

Compared with the $20,800 a year the bank would have to expend for an additional fifth teller at Mayfield, the annual depreciated cost for the ATM ($7,500) reduces by nearly two-thirds the amount of money the bank will have to spend for much more efficient customer service.

FIGURE I *(Continued)*

CONCLUSION

Stresses benefits
for reader and
bank as a whole

Authorizing an ATM for the Mayfield Park branch is both feasible and
cost-effective. Adoption of this proposal will save our bank more than
$13,000.00 in teller services annually, reduce customer complaints, and
increase customer satisfaction and approval. We will be happy to discuss
this proposal with you anytime at your convenience.

FIGURE 1   *(Continued)*

Generally speaking, your proposal will be an informal, in-house message, so a
brief (usually one- or two-page) memo should be appropriate.

An internal proposal can be written about a variety of topics, such as:

- purchasing new or more advanced equipment: word processors, transducers,
  automobiles
- hiring new employees or training current ones to learn a new technique or
  process
- eliminating a dangerous condition or reducing an environmental risk to prevent
  accidents—for employees, customers, or the community at large
- improving communication within or between departments of a company or
  agency
- revising a policy to improve customer relations (eliminating an inconvenience,
  speeding up a delivery) or employees' morale (offering vanpooling; adding more
  options for a schedule).

As this list shows, internal proposals cover almost every activity or policy that
affects the day-to-day operation of a company or agency.

Your Audience and Office Politics

Writing an internal proposal requires you to be aware of and sensitive to office
politics. To be successful, your internal proposal should be written with the needs
and likes of your audience in mind. Remember that your boss will expect you to
be very convincing about both the problem you say exists and the solution you
propose to correct it. You cannot assume that your reader will automatically agree
with you that there is a problem or that your plan is the only way to solve it.

Your reader in fact may feel threatened by your plan. After all, you are advocating a change. Some managers regard such changes as a challenge to their
administration of an office or department. Or your reader may be indifferent, not
even wanting to give your work serious consideration. Or your manager-reader
may have certain "pet" projects or ways of doing things that you must take into

account. To surmount these and other obstacles, show that the change you propose is in everyone's best interest. Do not overlook the possibility that your boss may have to take your proposal up the organizational ladder for commentary and, eventually, approval. . . . Writing a proposal may mean working as a team with your boss—putting his or her name on the document, too, to get notice and credit.

Before you write an internal proposal, consider the implications of your plan for your boss and for other offices or sections in your company. A change you propose for your department or office (transfers; new budgets or schedules) may have sweeping and potentially disruptive implications for another office or division within your company. It is wise to discuss your plan with your boss before you put it in writing. Then you might provide your boss with a draft, asking for his or her revisions or feedback. . . .

Never submit an internal proposal that offers an idea you think will work but relies on someone else to supply the specific details on *how* it will work. For example, do not write an internal proposal that says the payroll, community relations, maintenance, or advertising department can give the reader the details and costs he or she needs about your proposal. That pushes the responsibility onto someone else, and your proposal could be rejected for lack of concrete evidence.

The Organization of an Internal Proposal

A short internal proposal follows a relatively straightforward plan of organization, from identifying the problem to solving it. Internal proposals usually contain four parts, as shown in Figure 1. Refer to this proposal as you read the following discussion.

The Introduction

Begin your proposal with a brief statement of why you are writing to your boss: "I propose that. . . ." State why you think a specific change is necessary now. Then succinctly define the problem and emphasize that your plan, if approved by the reader, will solve that problem. Where necessary, stress the urgency to act—within the next week? month?

Background of the Problem

In this section prove that a problem does exist by documenting its importance for your boss and your company. As a matter of fact, the more you show how the problem affects the boss's work (and area of supervision), the more likely you are to persuade him or her to act. And the more concrete evidence you cite, the easier it will be to convince the reader that the problem is significant and that action needs to be taken now.

Avoid vague (and unsupported) generalizations such as "we're losing money each day with this procedure (or piece of equipment)"; "costs continue to escalate";

"the trouble occurs frequently in a number of places"; "numerous complaints have come in"; "if something isn't done soon, more difficulty will result."

Instead, provide readers with quantifiable details about the number of dollars a company is actually losing per day, week, or month. Emphasize the financial trouble so that you can show how your plan (described in the next section) offers an efficient and workable solution. Indicate how many employees (or work-hours) are involved or how many customers are inconvenienced or endangered by a procedure or condition. Notice how Escobar and Jabur include this information in their proposal in Figure 1. Verify how widespread a problem is or how frequently it occurs by citing specific occasions. Rather than just saying that a new word processor would save the company "a lot of money," document how many work-hours are lost using other equipment for the routine jobs your company now has employees perform.

The Solution or Plan

In this section describe the change you want approved. Tie your solution (the change) directly to the problem you have just documented. Each part of your plan should help eliminate the problem or should help increase the productivity, efficiency, or safety you think is possible.

Your reader will again expect to find factual evidence. Do not give merely the outline of a plan or say that details can be worked out later. Supply details that answer the following questions: (1) Is the plan workable—can it be accomplished here in our office or plant? and (2) Is it cost-effective—will it really save us money in the long run or will it lead to even greater expenses?

To get the boss to say yes to both questions, supply the facts that you have gathered as a result of your research. For example, if you propose that your firm buy a new piece of equipment, do the necessary homework to locate the most efficient and cost-effective model available, as Tina Escobar and Oliver Jabur do in Figure 1. Supply the dealer's name, the costs, major conditions of service and training contracts, and warranties. Describe how your firm could use this equipment to obtain better results in the future. Cite specific tasks the new equipment can perform more efficiently at a lower cost than the equipment now in use.

It is also wise to raise alternative solutions, before the reader does, and to discuss their disadvantages. Notice how Tina Escobar and Oliver Jabur do this in Figure 1, in their discussion of the solution to the bank's customer-service problem. They show why installing an ATM is more necessary than hiring a fifth teller. Their discussion is based on a strong persuasive strategy, demonstrating to their readers that they have examined all alternatives and chosen the best.

If you are proposing that your company hire or reassign employees, indicate where these employees will come from, when they will start, what they will be paid, what skills they must have, where they will work and for how long. If you propose to assign current employees to a job, keep in mind that their salary will still have to be paid by your company. Just because they are coworkers does not mean they will work for nothing. Note again in Figure 1 how Tina Escobar and

Oliver Jabur raise and resolve the problem of new and profitable responsibilities for the tellers at Community Federal Bank.

A proposal to change a procedure must include the following details:

1. How the new (or revised) procedure will work
2. How many employees or customers will be affected by it
3. When it will go into operation
4. How much it will cost the employer to change procedures
5. What delays or losses in business might be expected while the company switches from one procedure to another
6. What employees, equipment, or locations are available to accomplish this change.

As these questions indicate, your boss will be concerned about schedules, working conditions, employees, methods, locations, equipment, and the costs involved in your plan for change. The costs, in fact, will be of utmost importance. Make sure that you supply a careful and accurate budget so that your reader will know what the change is going to cost. Moreover, make those costs attractive by emphasizing how inexpensive they are as compared to the cost of not making the change, as Escobar and Jabur do in the section labeled "Costs." Double-check your math.

The Conclusion

The conclusion to your internal proposal should be short—a paragraph or two at the most. Your intention is to remind the reader that the problem is serious, that the reason for change is justified, and that you think the reader needs to take action. Select the most important benefits and emphasize them again. In Figure 1, Escobar and Jabur again emphasize the savings that the bank will see by following their plan as well as the increase in customer satisfaction. Also indicate that you are willing to discuss your plan with your reader.

SALES PROPOSALS

A sales proposal is the most common type of external proposal. Its purpose is to sell your company's products or services for a set fee. Whether short or long, a sales proposal is a marketing tool that includes a sales pitch as well as a detailed description of the work you propose to do. Figures 2 and 3 [pp. 252–256] contain sample sales proposals.

The Audience and Its Needs

Your audience will usually be one or more business executives who have the power to approve or reject a proposal. Unlike readers of an internal proposal, your audience for a sales proposal may be even more skeptical since they may not know you or your work. Your proposal may also be evaluated by experts in

REYNOLDS INTERIORS
250 Commerce Avenue, S.W.
Portland, OR 97204–2129

503–555–8733   FAX 503–555–1629

January 21, 1994

Mr. Floyd Tompkins, Manager
General Purpose Appliances
Highway 41, South
Portland, OR 97222

Dear Mr. Tompkins:

In response to your Request 7521 for bids for an appropriate floor covering at your new showroom, Reynolds Interiors is pleased to submit the following proposal. After carefully reviewing your specifications for a floor covering and inspecting your new facility, we believe that the Armstrong Classic Corlon 900 is the most suitable choice. I am enclosing a sample of the Corlon 900 so that you can see how it looks.

Corlon's Advantages

Guaranteed against defects for a full three years, Corlon is one of the finest and most durable floor coverings manufactured by Armstrong. It is a heavy-duty commercial floors 0.085-inch thick for protection. Twenty-five percent of each tile consists of interface backing; the other 75 percent is an inlaid wear layer that offers exceptionally high resistance to everyday traffic. Traffic tests conducted by the Independent Floor Covering Institute repeatedly proved the superiority of Corlon's construction and resistance.

Another important feature of Corlon is the size of its rolls. Unlike other leading brands of similar commercial flooring—Remington or Treadmaster—Corlon comes in 12-foot-wide rather than 6-foot-wide rolls. This extra width will significantly reduce the number of seams in your floor, thus increasing its attractiveness and reducing the danger of tile split.

Installation Procedures

The Classic Corlon requires that we use the inlaid seaming process, a technical procedure requiring the services of a trained floor mechanic. Herman Goshen, our floor mechanic, has over fifteen years of experience working with the inlaid seam process. His professional work has been consistently praised by our customers.

Installation Schedule

We can install the Classic Corlon on your showroom floor during the first week of March, which fits the timetable specified in your request. The tile will take three and one-half days to install and will be ready to walk on immediately. We recommend, though, that you not move equipment onto the floor for twenty-four hours after installation.

FIGURE 2 An External Solicited Proposal

Costs

The following costs include the Classic Corlon tile, labor, equipment, and tax:

750 sq. yards of Classic Corlon at $12.50/yard	$ 9,375.00
Labor (28 hr. @ $15.00/hr.)	420.00
Sealing fluid (10 gal. @ 10.00/gal.)	100.00
Total	9,895.00
Tax (5 percent)	494.75
GRAND TOTAL	**$10,389.75**

Our costs are $250.00 under those you specified in your request.

Reynolds' Qualifications

Reynolds Interiors has been in business for more than twenty-eight years. In that time, we have installed many commercial floors in Portland and its suburbs. In the last year, we have served more than sixty customers, including the new multi-purpose Tradex plant in Portland.

Conclusion

Thank you for the opportunity to submit this proposal. We believe you will have a great deal of success with an Armstrong floor. If we can provide you with any further information, please call us.

Sincerely yours,

Sharon Scovill

Sharon Scovill
Sales Manager

Jack Rosen

Jack Rosen
Installation Supervisor

FIGURE 2 *(Continued)*

N B E

National Business Equipment

470 Rodgers Rd.   ■   Camden, NJ 08104–0826   ■   (201) 555–1100

September 20, 1994

Ms. Denise Taylor
Business Manager
Madison Tool and Die Company
3400 Veterans Boulevard
Camden, NJ 08104

Dear Ms. Taylor:

While we were servicing your Piko 2500 copier last week, it occurred to me that you might be interested in purchasing a newer model that will give you state-of-the-art features to make your copying work more efficient and reliable. In the past three years since you purchased your 2500, copier technology has advanced tremendously, giving users many benefits at surprisingly low cost.

Based on our assessment of Madison Tool and Die Company's needs for the most reliable office equipment available, we recommend that you purchase the Piko 4000. We believe that this copier will satisfactorily meet all of your copying needs for the present and well into the future.

ADVANTAGES OF THE PIKO 4000

With the Piko 4000, your cost per single copy will be reduced to only 1 1/2 cents, compared with approximately 3 cents per copy with your present machine. The 4000 model is designed to make as many as **75,000 copies per month,** almost double the recommended work load of your current copier. The following description of the main features of the Piko 4000 will show you its other advantages over your current copier.

Printing Technology

The Piko 4000's **printing rate of 80 copies per minute** will save you valuable time. The automatic duplexing capability allows you to turn single-sided documents into double-sided ones, thus reducing paper costs. In addition to the standard paper sizes—8 1/2 × 11, 8 1/2 × 14, and 11 × 17—**the 4000 functions effectively with other paper sizes and shapes,** including heavyweight sheet stock suitable for specifications and drawings. The Piko 4000 has original-size sensing, which can copy varying sizes automatically.

FIGURE 3   An External Unsolicited Proposal

Copier Control Security System

The Piko 4000's **electronic ID keypad** allows only authorized users, with individual entry codes, to run the copier. This keypad can control access, limit copies, and provide tabulated records of use for your accounting purposes.

Feeding/Sorting Capacity

The 4000 includes as standard equipment both a recirculating automatic document feeder and an automatic sorter/stapler. These features **can reduce the time spent on large copying jobs by as much as 50 percent.**

Printing Quality

The new Piko 4000 offers top-quality, high resolution reproduction **in four colors** in addition to black and white. Since seeing is believing, I am enclosing a copy of this proposal made on the 4000 as well as a copy of a blueprint to show you how both copied documents look.

Reduction/Enlargement Function

The Piko 4000's zoom capability allows you to increase or decrease the size of your copy within a range of 50–200 percent. Since these changes are made in increments of 1 percent, your copy is always the exact size you want.

Size

The Piko 4000 measures 4' × 3' × 4', **occupying one-quarter less space** than your present copier. Once the machine is installed, you will find this extra space useful for your storage needs.

INSTALLATION, TRAINING, AND SERVICE

We will deliver and install your new Piko 4000 within one week of receiving your purchase order. The installation requires approximately forty-five minutes.

Our local sales representative, Darlene Simpson, is available to instruct your employees in the operation and routine maintenance of the 4000. A phone call will allow us to set up a mutually convenient date for such instruction. If a problem occurs, **we offer customers the latest in remote diagnostics.** A modem installed in the Piko 4000 allows us to diagnose specific problems (and needed parts) before dispatching a service representative. We save time, so you save time. In addition, our "hot line" is available to you on business days from 8:00 A.M. to 5:00

FIGURE 3 *(Continued)*

P.M. We guarantee that a service representative will arrive at your office **within two business hours** from the time of your call.

COSTS

Below is the price of the Piko 4000 and appropriate initial supplies:

Piko 4000	$18,295.00
Toner (tube)	23.00
Paper (5,000 sheets of medium grade 8 1/2 × 11)	25.00
Service contract—optional	
(per year, parts and labor included)	2,300.00
Total	$20,643.00

We install your new Piko 4000 **free of charge.**

NATIONAL'S REPUTATION

Over the past 22 years, National has supplied copiers to more than 90 companies in the greater Camden area, including Biscayne Industries, Northeast Manufacturing, and Teunissen Accounting, Inc. In addition to providing quality products, we are **dedicated to giving our customers fast and efficient service long after a sale.**

We appreciate your using **National Business Equipment** for your repair needs and hope that you will decide to purchase the new Piko 4000. Please call me if you have any questions about the Piko 4000 or National.

If this proposal is acceptable, please sign and return a copy of this letter.

Sincerely yours,

Marion Copely

Marion Copely

Encls. Copier samples

I accept this proposal made by National Business Equipment.

_____

for Madison Tool and Die Company

FIGURE 3   *(Continued)*

other fields employed by your prospective customer. Readers of a sales proposal will evaluate your work according to (1) how well it meets their needs and (2) how well it compares with the proposals submitted by your competitors. Your proposal must convince readers that you can provide the most appropriate work or service and that your company is more reliable and efficient than any other firm.

The key to success is incorporating the "you attitude" . . . throughout your proposal. Relate your product, service, or personnel to the reader's exact needs as stated in the RFP [request for proposal] for a solicited proposal or through your own investigations for an unsolicited proposal. You cannot submit the same proposal for every job you want to win and expect to be awarded a contract. Different firms have different needs. The most important question the reader will raise, therefore, about your work is "How does this proposal meet our company's special requirements?" Some other fairly common questions readers will have as they evaluate your work include the following:

- Does the writer's firm understand our problem?
- Can the writer's firm deliver what it promises?
- Can the job be completed on time?
- What assurances does the writer offer that the job will be done exactly as proposed?

Answer each of these questions by demonstrating how your product or service is tailored to the customer's needs.

Organizing a Sales Proposal

A sales proposal can have the following parts: introduction, description of the proposed product or service, timetable, costs, qualifications of your company, and conclusion.

Introduction

The introduction to your sales proposal can be a single paragraph in a short sales proposal or several pages in a more complex one. Basically, the introduction should prepare readers for everything that follows in your proposal. The introduction itself may contain the following sections, which sometimes may be combined.

1. **Statement of purpose and subject of proposal.** Tell readers why you are writing and identify the specific subject of your work. If you are responding to an RFP, use specific code numbers or cite application dates, as the proposal in Figure 2 does. If your proposal is unsolicited, indicate how you learned of the problem, as Figure 3 does. Briefly define the solution you propose.
2. **Background of the problem you propose to solve.** Show readers that you are familiar with their problem and that you have a firm grasp of the importance and implications of the problem. In a solicited proposal, such as the one shown in

Figure 2, a section outlining the problem is usually unnecessary, because the potential client has already identified the problem and just wants to know how you would handle it. In that case, just point out how your company would solve the problem, mentioning your superiority over your competitors (see the third paragraph of Figure 2). In an unsolicited proposal, you need to describe the problem in convincing detail, identifying the specific trouble areas. However, if it is an external proposal to a current customer, such as the one in Figure 3, it would be unwise to point out past problems your client may have had with your company's services or products. In Figure 3, note how the problems of the Piko 2500 model are described mainly as a way to sell the advantages of the Piko 4000.

Description of the Proposed Product or Service

This section is the heart of your proposal. Before spending their money, customers will demand hard, factual evidence of what you claim can and should be done. Here are some points that you should cover.

1. **Carefully show your potential customers that your product or service is right for them.** Stress specific benefits of your product or service most relevant to your reader. Blend sales talk with descriptions of hardware.
2. **Describe your work in suitable detail—what it looks like, what it does, and how consistently and well it will perform in the readers' office, plant, hospital, or agency.** You might include a brochure, picture, or, as the writers of the proposals in Figures 2 and 3 do, a sample of your product for customers to study. Convince readers that your product is the most up-to-date and efficient one they could select. Note how the proposals in Figures 2 and 3 emphasize this.
3. **Stress any special features, maintenance advantages, warranties, or service benefits.** Highlight features that show the quality, consistency, or security of your work. For a service, emphasize the procedures you use, the terms of that service, and even the kinds of tools you use, especially any "state-of-the-art" equipment. Be sure to provide a step-by-step outline of what will happen and why each step is beneficial for readers.

Timetable

A carefully planned timetable shows readers that you know your job and that you can accomplish it in the right amount of time. Your dates should match any listed in an RFP. Provide specific dates when the work will begin, how long it will continue, and when you will be finished installing equipment, testing equipment, or training employees to use equipment. For proposals offering a service, specify how many times—by the hour, week, month—customers can expect to receive your help; for example, spraying three times a month for an exterminating service or making deliveries by 10:00 A.M. for a trucking company. Indicate whether follow-up visits or service calls will be provided.

Costs

Make your budget accurate, complete, and convincing. Accepted by both parties, a proposal is a binding legal agreement. Don't underestimate costs in the hope that a low bid will win you the job. You may get the job but lose money doing it, for the customer will rightfully hold you to your unrealistic figures. Neither should you inflate prices; competitors will beat you in the bidding.

Give customers more than the bottom-line cost. Show exactly what readers are getting for their money so they can determine if everything they need is included. Itemize costs for specific services, equipment, labor (by the hour or by the job), transportation, travel, or training you propose to supply. If something is not included or is considered optional, say so—additional hours of training, replacement of parts, and the like. If you anticipate a price increase, let the customer know how long current prices will stay in effect. That information may spur them to act favorably now.

Qualifications of Your Company

Emphasize your company's accomplishments and expertise in using relevant services and equipment. You might list previous work you have done that is identical or similar to the type of work you are proposing to do for the customer. You may even want to mention the names of a few local firms for whom you have worked that would be pleased to recommend you. Never misrepresent your qualifications or those of the individuals who work with or for you. Your prospective client may verify if you have in fact worked on similar jobs for the last five to six years.

Conclusion

This is the "call to action" section of your proposal. As with the conclusion in an internal proposal, encourage your reader to approve your plan. Stress the major benefits your plan has for the customer. Notice how the last paragraphs of Figures 2 and 3 do this effectively. Offer to answer any questions the reader may have. Some proposals end by asking readers to sign and return a copy of the proposal thus indicating their acceptance of it, as the proposal in Figure 3 does.

Richard Johnson-Sheehan

Writing Proposals with Style

Richard Johnson-Sheehan teaches classical rhetoric and professional writing at Purdue University where he also is Director of Composition.

WHAT IS STYLE?

If you are like most people, style is a rather murky concept. Some documents just seem to have a good style, while others do not. Consequently, you may have come to believe that good style is something some people have from birth. Good style might even seem accidental. You aren't sure how it happened, but your co-workers tell you that something you wrote is very readable, even eloquent.

What is style? Style works at a few different levels in a proposal. On the sentence level, good style might involve choosing the right words or forming sentences that are easy to read. On the paragraph level, style could involve weaving sentences together in ways that emphasize your main points and lead the readers comfortably through your ideas. At the document level, style involves setting an appropriate tone and weaving themes into your work that appeal to your readers' emotions and values. In his 1998 book *Technical Writing Style,* Dan Jones defines style the following way:

> Style affects or influences almost all other elements of writing. Style is your choices of words, phrases, clauses, and sentences, and how you connect these sentences. Style is the unity and coherence of your paragraphs and larger segments. Style is your tone—your attitude toward your subject, your audience, and yourself—in what you write. Style is who you are and how you reflect who you are, intentionally or unintentionally, in what you write (3).

Style does more than make the content easier to read and more persuasive. In many ways, it illustrates your clear-headedness, your emphasis on quality, and your willingness to communicate and work with the readers.

Style is not embellishment or ornamentation. Some people mistakenly believe that style is the spice sprinkled over a proposal to make the content more palatable to the readers. These writers throw in some extra adjectives and an occasional metaphor to perk up the bland parts of the proposal. But, just as spices are most flavorful when they are cooked into food, style needs to be carefully worked into the proposal. Indeed, a proposal that uses stylistic devices to embellish the content is merely hinting to the readers that the proposal lacks substance. Style enhances and amplifies content, but it should never be used to artificially embellish or hide a lack of content.

Classical rhetoricians like Cicero and Augustine discussed style in three levels: plain style, middle style, and grand style. The plain style is for instruction and demonstration, allowing the writer to lay out the facts or describe something in simple terms. The middle style is for persuading people to take action. When using the middle style, the writer highlights the benefits of taking action or doing something a particular way. The grand style is for motivating people to do something they already know they should do. For example, Winston Churchill, Martin Luther King, Jr., and John F. Kennedy regularly used the grand style to motivate their listeners to do what was right, even if people were reluctant to do it.

In proposals, there is little opportunity to use the grand style, because they tend to focus on instructing and persuading the readers. Proposals that use the grand style often seem too fanatical and excessive. So, in this . . . [essay] we will concentrate on using the plain and persuasive styles in proposals. A proposal that properly combines the plain and persuasive style will be both informative and moving for the readers.

WRITING PLAIN SENTENCES

In a proposal, the plain style tends to be used when the writers need to instruct the readers about a situation or process. Specifically, the plain style is used mostly in the Situation section, where you are describing a problem or opportunity for the readers, and the Qualifications section, where you are describing your background and experience. In some cases, the Plan section might also be written in plain style, especially when your plain is rather straightforward.

As a student, you were more than likely advised to "write clearly" or "write in concrete language" as though simply making up your mind to write clearly or concretely was all it took. In reality, writing plainly is a skill that requires practice and concentration. Fortunately, once a few simple guidelines have been learned and mastered, style writing will soon become a natural strength in your writing.

To start, let us consider the parts of a basic sentence. From your grammar classes, you learned that a sentence typically has three main parts: a subject, a verb, and a comment. The subject is what the sentence is about. The verb is what the subject is doing. And, the comment says something about the subject. For example, consider these three variations of the same sentence.

Subject	Verb	Comment
The Institute	provides	the government with accurate crime statistics.

Subject	Verb	Comment
The government	is provided	with accurate crime statistics by the Institute.

Subject	Verb	Comment
Crime statistics	were provided	to the government by the Institute.

The content in these sentences has not changed. Nevertheless, the emphasis in each sentence changes as we replace the subject slot with different nouns. Sentence A is *about* the "Institute." Sentence B is *about* the "government." Sentence C is *about* the "crime statistics." By changing the subject of the sentence, we essentially shift the focus of the sentence, drawing our readers' attention to different issues.

This simple understanding of the different parts of a sentence is the basis for eight guidelines that can be used to write plainer sentences in proposals, as shown in Figure 1. We will discuss these sentence guidelines in more detail in the following pages.

Guideline 1: The subject should be what the sentence is about.
Guideline 2: Make the "doer" the subject. Subject is the "doer."
Guideline 3: State the action in the verb.
Guideline 4: Put the subject early in the sentence.
Guideline 5: Eliminate nominalizations.
Guideline 6: Avoid excessive prepositional phrases.
Guideline 7: Eliminate redundancy.
Guideline 8: Make sentences "breathing length."

FIGURE I Sentence Guidelines

Guideline 1: The subject should be what the sentence is about

At a very simple level, weak style often occurs when the readers cannot easily identify the subject of the sentence. Or, the subject of the sentence is not what the sentence is about. For example, what is the subject of the following sentence?

1. Ten months after the Hartford Project began in which a team of our experts conducted close observations of management actions, our final conclusion is that the scarcity of monetary funds is at the basis of the inability of Hartford Industries to appropriate resources to essential projects that have the greatest necessity.

This sentence is difficult to read for a variety of reasons, but the most significant problem is the lack of a clear subject. What is this sentence about? The word *conclusion* is currently in the subject position, but the sentence might also be about the *experts,* the *Hartford Project,* or *scarcity of monetary funds.* Indeed, many other nouns and nounlike words also seem to be competing to be the subject of the sentence, such as *observations, management, structure, conclusion, inability,* and *company.* These nouns and nounlike words bombard the readers with potential subjects, undermining their efforts to identify what the sentence is about.

When the sentence is restructured around *experts* or *scarcity,* most readers will find it easier to understand:

1a. Ten months after the Hartford Project began, our experts have concluded through close observations of management actions that the scarcity of monetary funds is at the basis of the inability of Hartford Industries to appropriate resources to essential projects that have the greatest necessity.

1b. The scarcity of monetary funds, our experts have concluded through close observations of management actions ten months after the Hartford Project began, is at the basis of the inability of Hartford Industries to appropriate resources to essential projects that have the greatest necessity.

Both of these sentences are still rather difficult to read. Nevertheless, they are easier to read than the original because the noun occupying the subject slot is the focus of the sentence—that is, what the sentence is about. We will return to this sentence about Hartford Industries after discussing the other guidelines for plain style.

Guideline 2: Make the "doer" the subject

Guideline 3: State the action in the verb

In your opinion, which revision of sentence 1 above is easier to read? Most people would point to sentence 1a, in which *experts* is in the subject slot. Why? In sentence 1a, the experts are actually doing something. In sentence 1b, *scarcity* is an inactive noun that is not doing anything. Whereas experts take action, scarcity is merely something that happens.

Guidelines 2 and 3 reflect the tendency of readers to focus on who or what is doing something in a sentence. To illustrate, which of these sentences is easier to read?

2a. On Saturday morning, the paperwork was completed in a timely fashion by Jim.

2b. On Saturday morning, Jim completed the paperwork in a timely fashion.

Most people would say sentence 2b is easier to read because Jim, the subject of the sentence, is actually doing something, while the paperwork in sentence 2a is inactive. The active person or thing usually makes the best subject of the sentence.

Similarly, Guideline 3 states that the verb should contain the action in the sentence. Once you have determined who or what is doing something, ask yourself what that person or thing is actually doing. Find the action in the sentence and make it the verb. For example, consider these sentences:

3a. The detective investigated the loss of the payroll.
3b. The detective conducted an investigation into the loss of the payroll.
3c. The detective is the person who is conducting an investigation of the loss of the payroll.

Sentence 3a is easier to understand because the action of the sentence is expressed in the verb. Sentences 3b and 3c are increasingly more difficult to understand, because the action, *investigate,* is further removed from the verb slot of the sentence.

Guideline 4: Put the subject early in the sentence

Subconsciously, readers start every sentence looking for the subject. The subject anchors the sentence, because it tells the reader what the sentence is about. So, if the subject is buried somewhere in the middle of the sentence, the readers will have greater difficulty finding it, and the sentence will be harder to read. To illustrate, consider these two sentences:

4a. If deciduous and evergreen trees experience yet another year of drought like the one observed in 1997, the entire Sandia Mountain ecosystem will be heavily damaged.
4b. The entire Sandia Mountain ecosystem will be heavily damaged if deciduous and evergreen trees experience yet another year of drought like the one observed in 1997.

The problem with sentence 4a is that it forces the readers to hold all those details (i.e., trees, drought, 1997) in short-term memory before the sentence identifies its subject. Readers almost feel a sense of relief when they find the subject, because they cannot figure out what the sentence is about until they locate the subject. Quite differently, sentence 4b tells the readers what the sentence is about up front. With the subject early in the sentence, the readers immediately know how to connect the comment with the subject.

Of course, introductory or transitional phrases do not always signal weak style. But when these phrases are used, they should be short and to the point. Longer introductory phrases should be moved to the end of the sentence.

Guideline 5: Eliminate nominalizations

Nominalizations are perfectly good verbs and adjectives that have been turned into awkward nouns. For example, look at these sentences:

5a. Management has an expectation that the project will meet the deadline.
5b. Management expects the project to meet the deadline.

In sentence 5a *expectation* is a nominalization. Here, the perfectly good verb *expect* is being used as a noun. After turning the nominalization into a verb, sentence 5b is not only shorter than sentence 5a, it also has more energy because the verb *expect* is now an action verb.

Consider these two sentences:

6a. Our discussion about the matter allowed us to make a decision on the acquisition of the new x-ray machine.

6b. We discussed the matter and decided to acquire the new x-ray machine.

Sentence 6a includes three nominalizations *discussion, decision,* and *acquisition,* making the sentence hard to understand. Sentence 6b turns all three of these nominalizations into verbs, making the sentence much easier to understand. An additional benefit to changing nominalizations into verbs is the energy added to the sentence. Nouns tend to feel inert to the readers, while verbs tend to add action and energy.

Why do writers use nominalizations in the first place? We use nominalizations for two reasons. First, humans generally think in nouns, so our first drafts are often filled with nominalizations, which are nouns. While revising, an effective writer will turn those first-draft nominalizations into action verbs. Second, some people mistakenly believe that using nominalizations makes their writing sound more formal or important. In reality, though, nominalizations only make sentences harder to read. The best way to sound important is to write sentences that readers understand.

Guideline 6: Avoid excessive prepositional phrases

Prepositional phrases are necessary in writing, but they are often overused in ways that make writing too long and too tedious. Prepositional phrases follow prepositions (e.g., in, of, by, about, over, under) and they are used to modify nouns. For example in the sentence "Our house by the lake in Minnesota is lovely," the phrases *by the lake* and *in Minnesota* are both prepositional phrases. They modify the nouns *house* and *lake*.

Prepositional phrases are fine when used in moderation, but they are problematic when used in excess. For example, in sentence 7a the prepositions have been italicized and prepositional phrases underlined. Sentence 7b is the same sentence with fewer prepositional phrases:

7a. The decline *in* the number *of* businesses owned *by* locals *in* the town *of* Artesia is a demonstration *of* the increasing hardship faced *in* rural communities *in* the southwest.

7b. Artesia's declining number of locally owned businesses demonstrates the increased hardship faced by southwestern rural communities.

You should never feel obligated to eliminate all the prepositional phrases in a sentence. Rather, look for places where prepositional phrases are chained together in long sequences. Then, try to condense the sentence by turning some of the prepositional phrases into adjectives. In sentence 7b, for example, the phrase *in the town of Artesia* was reduced to the adjective *Artesia's*. The phrases *in rural communities in the southwest* were reduced to *by southwestern rural communities.* The resulting sentence 7b is much shorter and easier to read.

Guideline 7: Eliminate redundancy

In our efforts to stress our points, we often use redundant phrasing. For example, we might write *unruly mob,* as though some mobs are orderly, or we might talk about *active participants,* as though someone can participate without doing anything. Sometimes buzzwords and jargon lead to redundancies like, "We should collaborate together as a team" or "Empirical observations will provide a new understanding of the subject." In some cases, we might use a synonym to modify a synonym by saying something like, "We suggested important, significant changes."

Redundancies should be eliminated because they use two words to do the work of one. As a result, the readers need to work twice as hard to understand one basic idea.

Guideline 8: Make sentences "breathing length"

A sentence is a statement designed to be spoken in one breath. When a text is read out loud, the period at the end of each sentence is the reader's signal to breathe. Of course, when reading silently, we do not actually breathe when we see a period. Nevertheless, readers do take a mental pause at the end of each sentence. A sentence that runs on and on forces readers to mentally hold their breaths. By the end of an especially long sentence, readers are more concerned about getting through it than deciphering it.

The best way to think about sentence length is to imagine how long it takes to comfortably say a sentence out loud. If the written sentence is too long to say in one breath, it probably needs to be shortened or cut into two sentences. After all, you don't want to asphyxiate your readers. On the other hand, if the sentence is very short, perhaps it needs to be combined with one of its neighbors to make it a more comfortable breathing length. You also want to avoid hyperventilating the readers with a string of short sentences.

A Simple Method for Writing Plainer Sentences

To sum up the eight sentence guidelines, here is a process for writing plainer sentences. First, write out your draft as usual, not paying too much attention to the style. Then, as you revise, identify difficult sentences and apply the six steps shown in Figure 2.[1]

---

[1]In his book, *Revising Business Prose,* Richard Lanham offers a simpler technique that he calls the "paramedic method." His method is less comprehensive than the one shown here, but it works well also.

1. Identify who or what the sentence is about.
2. Turn that who or what into the subject, and then move the subject to an early place in the sentence.
3. Identify what the subject is doing, and move that action into the verb slot.
4. Eliminate prepositional phrases, where appropriate, by turning them into adjectives.
5. Eliminate unnecessary nominalizations and redundancies.
6. Shorten, lengthen, combine, or divide sentences to make them breathing length.

FIGURE 2   Six Steps to Plainer Writing

With these six steps in mind, let us revisit sentence 1, the example of weak style offered at the beginning of our discussion of plain style.

Original

1. Ten months after the Hartford Project began in which a team of our experts conducted close observations of management decisions, our final conclusion is that the scarcity of monetary funds is at the basis of the inability of Hartford Industries to appropriate resources to essential projects that have the greatest necessity.

Revision

1a. After completing the ten-month Hartford Project, our experts concluded that the Hartford Industries' budget shortfalls have limited support for priority projects.

In the revision, the subject (our experts) was moved into the subject slot, and then it was moved to an early place in the sentence. Then, the action of the sentence (concluded) was moved into the verb slot. Prepositional phrases like *to appropriate resources to essential projects* were turned into adjectives. Nominalizations like *conclusion* and *necessity* were turned into verbs or adjectives. And finally, the sentence was shortened to breathing length. The resulting sentence still offers the same content to the readers—just more plainly.

WRITING PLAIN PARAGRAPHS

As with sentences, some rather simple methods are available to help you write plainer paragraphs in proposals.

The Elements of a Paragraph

Paragraphs tend to include four kinds of sentences: a transition sentence, a topic sentence, a support sentence, and a point sentence. Each of these sentences plays a different role in the paragraph.

Transition Sentence

The purpose of a transition sentence is to make a smooth bridge from the previous paragraph to the present paragraph. For example, a transitional sentence might state "With these facts in mind, let us consider the current opportunity available." The *facts* mentioned were explained in the previous paragraph. By referring back to the previous paragraph, the transition sentence provides a smooth bridge into the new paragraph. Most paragraphs, however, do not need a transition sentence. These kinds of sentences are typically used when the new paragraph handles a significantly different topic than the previous paragraph.

Topic Sentence

The topic sentence is the claim or statement that the rest of the paragraph is going to prove or support. In a proposal, topic sentences typically appear in the first or second sentence of each paragraph. They are placed up front in each paragraph for two reasons. First, the topic sentence sets a goal for the paragraph to reach by telling the readers the claim you are trying to prove. Then, the remainder of the paragraph proves that claim with facts, examples, and reasoning. If the topic sentence appears at the end of the paragraph, the readers are forced to rethink all the details in the paragraph now that they know what the paragraph was trying to prove. For most readers, all that mental backtracking is a bit annoying.

The second reason for putting the topic sentence up front is that it is the most important sentence in any given paragraph. Since readers tend to pay the greatest attention to the beginning of a paragraph, placing the topic sentence up front guarantees they will read it closely. Likewise, scanning readers tend to concentrate on the beginning of each paragraph. If the topic sentence is buried in the middle or at the end of the paragraph, they will miss it.

Support Sentences

The support in the body of the paragraph can come in many forms. . . . Sentences that use reasoning tend to make if/then, cause/effect, better/worse, greater/lesser kinds of arguments for the readers. Meanwhile the use of examples illustrates points for the readers by showing them situations or items that support your claim in the topic sentence. For the most part, sentences that contain reasoning and examples will make up the bulk of a paragraph's support sentences. Other support will come in the form of facts, data, definitions, and descriptions. In the end, support sentences are intended to prove the claim made in the paragraph's topic sentence.

Point Sentences

Point sentences usually restate the topic sentence at the end of the paragraph. They are used to reinforce the topic sentence by restating the paragraph's original claim in new words. Point sentences are especially useful in longer paragraphs where the readers may not fully remember the claim stated at the beginning of the paragraph. These sentences often start with transitional devices like *therefore, consequently,* or *in sum* to signal to the readers that the point of the paragraph is being restated. Point sentences are optional in paragraphs, and they should be used only occasionally when a particular claim needs to be reinforced for the readers. Too many point sentences will cause your proposal to sound too repetitious and even condescending to the readers.

Of these kinds of sentences, only the topic sentence and the support sentences are needed for a good paragraph. Transitional sentences and point sentences are useful in situations where bridges need to be made between paragraphs or specific points need to be reinforced.

Here are the four kinds of sentences used in a paragraph:

8a. How can we accomplish these five goals? (transition) Universities need to study their core mission to determine whether distance education is a viable alternative to the traditional classroom (topic sentence). If universities can maintain their current standards while moving their courses online, then distance education may provide a new medium through which nontraditional students can take classes and perhaps earn a degree (support). Utah State, for example, is reporting that students enrolled in their online courses have met or exceeded the expectations of their professors (support). On the other hand, if standards cannot be maintained, we may find ourselves returning to the traditional on-campus model of education (support). In the end, the ability to meet a university's core mission is the litmus test to measure whether distance education will work (point sentence).

8b. Universities need to study their core mission to determine whether distance education is a viable alternative to the traditional classroom (topic sentence). If universities can maintain their current standards while moving their courses online, then distance education may provide a new medium through which nontraditional students can take classes and perhaps earn a degree (support). Utah State, for example, is reporting that students enrolled in their online courses have met or exceeded the expectations of their professors (support). On the other hand, if standards cannot be maintained, we may find ourselves returning to the traditional on-campus model of education (support).

As you can see in paragraph 8b, a paragraph works fine without transition and point sentences. Nevertheless, they can make texts easier to read while amplifying important points.

Aligning Sentence Subjects in a Paragraph

Have you ever needed to stop reading a paragraph because each sentence seems to go off in a new direction? Have you ever run into a paragraph that actually feels bumpy as you read it? More than likely, the problem was a lack of alignment of the paragraph's sentence subjects. To illustrate, consider this paragraph:

9. The lack of technical knowledge about the electronic components in automobiles often leads car owners to be suspicious about the honesty of car mechanics. Although they might be fairly knowledgeable about the mechanical workings of

their automobiles, <u>car owners</u> rarely understand the nature and scope of the electronic repairs needed in modern automobiles. For instance, the <u>function and importance</u> of a transmission in a car is generally well known to all car owners; but the <u>wire harnesses and printed circuit boards</u> that regulate the fuel consumption and performance of their car are rarely familiar. <u>Repairs</u> for these electronic components can often run over 400 dollars—a large amount for a customer who cannot even visualize what a wire harness or printed circuit board looks like. In contrast, a <u>400-dollar charge</u> for the transmission on the family car, though distressing, is more readily understood and accepted.

There is nothing really wrong with this paragraph—it's just hard to read. Why? It is difficult to read because the subjects of the sentences change with each new sentence. Notice the underlined subjects of the sentences in this paragraph. These subjects are different, causing each sentence to feel like it is striking off in a new direction. As a result, each new sentence forces the readers to shift focus to concentrate on something new.

To avoid this bumpy, unfocused feeling, line up the subjects so each sentence in the paragraph stresses the same things. To line up subjects, first ask yourself what the paragraph is about. Then, restructure the sentences to align with that subject. Here is a revision of paragraph 9 that focuses on the "car owners" as subjects:

9a.   Due to their lack of knowledge about electronics, some <u>car owners</u> are skeptical about the honesty of car mechanics when repairs involve electronic components. Most of our <u>customers</u> are fairly knowledgeable about the mechanical features of their automobiles, but <u>they</u> rarely understand the nature and scope of the electronic repairs needed in modern automobiles. For example, most <u>people</u> recognize the function and importance of a transmission in an automobile; but, the average <u>person</u> knows very little about the wire harnesses and printed circuit boards that regulate the fuel consumption and performance of their car. So, for most of our customers, a <u>400-dollar repair</u> for these electronic components seems like a large amount, especially when <u>these folks</u> cannot even visualize what a wire harness or printed circuit board looks like. In contrast, <u>most car owners</u> think a 400-dollar charge to fix the transmission on the family car, though distressing, is more acceptable.

In this revised paragraph, you should notice two things. First, the words *car owners* are not always the exact words used in the subject slot. Synonyms and pronouns should be used to add variety to the sentences. Second, not all the subjects need to be related to car owners. In the middle of the paragraph, for example, *400-dollar repair* is the subject of a sentence. This deviation from *car owners* is fine as long as the majority of the subjects in the paragraph are similar to each other. In other words, the paragraph will still sound focused, even though an occasional subject is not in alignment with the others.

Of course, the subjects of the paragraph could be aligned differently to stress something else in the paragraph. Here is another revision of paragraph 9 in which the focus of the paragraph is *repairs.*

9b.   <u>Repairs</u> to electronic components often lead car owners, who lack knowledge about electronics, to doubt the honesty of car mechanics. The <u>nature and scope of these</u>

repairs are usually beyond the understanding of most nonmechanics, unlike the typical mechanical repairs with which customers are more familiar. For instance, the importance of fixing the transmission in a car is readily apparent to most car owners, but adjustments to electronic components like wire harnesses and printed circuit boards are foreign to most customers—even though these electronic parts are crucial in regulating their car's fuel consumption and performance. So, a repair to these electronic components, which can cost 400 dollars, seems excessive, especially when the repair can't even be visualized by the customer. In contrast, a 400-dollar replacement of the family car's transmission, though distressing, is more readily accepted.

In this paragraph, the subjects are aligned around words associated with repairs. Even though the subjects have been changed, the paragraph should still seem more focused than the original.

One important item you should notice is that paragraph 9a is easier for most people to read than paragraph 9b. Paragraph 9a is more readable because it has "doers" in the subject slots throughout the paragraph. In paragraph 9a, the car owners are active subjects, while in paragraph 9b the car repairs are inactive subjects. Much like sentences, the best subjects in a paragraph are people or things that are doing something.

The Given/New Method

Another way to write plain paragraphs is to use the *given/new method* to weave sentences together. Developed by Susan Haviland and Herbert Clark in 1974, the given/new method is based on the assumption that readers will always try to fit new information into what they already know. Therefore, every sentence in a paragraph should contain something the readers already know (i.e., the given) and something new that the readers do not know. To illustrate, consider these two paragraphs:

10a. Santa Fe is a beautiful place with surprises around every corner. Some artists choose to strike off into the mountains to paint, while others enjoy working in local studios.

10b. Santa Fe offers many beautiful places for artists to work, with surprises around every corner. Some artists choose to strike off into the mountains to paint, while others enjoy working in local studios.

Both of these examples are readable, but paragraph 10b is easier to read because the word *artists* appears in both sentences. Example 10a is a little harder to read, because there is nothing given that carries over from the first sentence to the second sentence.

Typically, the given information should appear early in the sentence and the new information should appear later in the sentence. Placed early in the sentence, the given information will provide a familiar anchor or context for the readers. Later in the sentence, the new information builds on that familiar ground. Consider this larger paragraph:

11. Recently, an art gallery exhibited the mysterious paintings of Irwin Fleminger, a modernist artist whose vast Mars-like landscapes contain cryptic human artifacts. One of Fleminger's paintings attracted the attention of some young school children

who happened to be walking by. At first, the children laughed, pointing out some of the strange artifacts in the painting. Soon, though, the artifacts in the painting drew the students into a critical awareness of the painting, and they began to ask their bewildered teacher what the artifacts meant. Mysterious and beautiful, Fleminger's paintings have this effect on many people, not just school children.

In this paragraph, the beginning of each sentence provides something given, usually an idea, word, or phrase drawn from the previous sentence. Then, the comment of each sentence adds something new to that given information. By changing together given and new information, the paragraph builds the readers' understanding gradually, adding a little more information with each sentence.

In some cases, however, the previous sentence does not offer a suitable subject for the sentence that follows it. In these cases, transitional phrases can be used to provide the readers given information in the beginning of the sentence. To illustrate,

12.   This public relations effort will strengthen Gentec's relationship with leaders of the community. <u>With this new relationship in place</u>, the details of the project can be negotiated with terms that are fair to both parties.

In this sentence, the given information in the second sentence appears in the transitional phrase, not the subject. Transitional phrases are a good place to include given information when the subject cannot be drawn from the previous sentence.

To sum up at this point, there are two primary methods available for developing plain paragraphs: (1) aligning the subjects of the sentences and (2) using the given/new method to weave the sentences together. Both methods are useful in proposal writing and should be used interchangeably. In some cases, both methods can be employed in the same paragraph as the writer uses various techniques to weave the paragraph into a coherent whole.

When Is It Appropriate to Use Passive Voice?

Before discussing the elements of persuasive style, we should expose one important bogie monster as a fraud. Since childhood, you have probably been warned against using passive voice. In fact, you might even remember various people decreeing that passive voice was off-limits, period. It's bad for you, they said, never use it.

Indeed, passive voice can be problematic when it is misused. One problem is that passive voice removes the doer from the sentence. For example, consider this passive sentence and its active counterpart:

13a.   The door was closed to ensure privacy. (passive)
13b.   Frank Roberts closed the door to ensure privacy. (active)

Written in passive voice, sentence 13a lacks a doer. The subject of the sentence, the door, is being acted upon, but it's not really doing anything. The second reason passive voice can be problematic is the use of an extra *be* verb (i.e., is, was, were, has been). The extra verb might slow the readers down a bit.

Despite dire warnings about passive voice, it does have a place in proposals, especially highly technical or scientific proposals. Either of these conditions makes a passive sentence appropriate:

- The readers do not need to know who or what is doing something in the sentence.
- The subject of the sentence is what the sentence is about.

For example, in Sentence 13a, the person who closed the door might be unknown or irrelevant to the readers. Is it important that we know that *Frank Roberts* closed the door? Or, do we simply need to know the door was closed? If the door is what the sentence is about and who closed the door is not important, then the passive is fine.

Consider these other examples of passive sentences:

14a. The shuttle bus will be driven to local care facilities to provide seniors with shopping opportunities (passive).
14b. Jane Chavez will drive the shuttle bus to local care facilities to provide seniors with shopping opportunities (active).
15a. The telescope was moved to the Orion system to observe the newly discovered nebula (passive).
15b. Our graduate assistant, Mary Stewart, moved the telescope to the Orion system to observe the newly discovered nebula (active).

In both these sets of sentences, the passive sentence may be more appropriate, unless there is a special reason Jane Chavez or Mary Stewart need to be singled out for special consideration.

When developing a focused paragraph, passive sentences can often help you align the subjects and use given/new strategies. For example, does the use of passive voice in the following paragraph help make the paragraph more readable?

16a. The merger between Brown and Smith will be completed by May 2004. Initially, Smith's key managers will be moved into Brown's headquarters. Then, other Smith employees will be gradually worked into the Brown hierarchy to eliminate any redundancies. During the merger process, employees at both companies will be offered all possible accommodations to help them through the uncertain times created by the merger.
16b. Brown and Smith will complete their merger in May 2004. Initially, Bill's Trucking Service will move the offices of key managers at Smith into Brown's headquarters. Brown's human resources manager will then gradually move Smith's other employees into the Brown hierarchy to eliminate any redundancies. During the merger, vice presidents, human resources agents, and managers at all levels will offer accommodations to employees at both companies to help them through the uncertain times created by the merger.

Most people would find Paragraph 16a more readable because it uses passive voice to put the emphasis on *employees*. Paragraph 16b is harder to read because it includes irrelevant doers, like Bill's Trucking Service, and it keeps changing the subjects of the sentences, causing the paragraph to seem unfocused.

In scientific and technical proposals, the passive voice is often the norm because *who* will be doing *what* is not always predictable. For example, in Sentence 15b, we might not be able to predict in our proposal that Mary Stewart will actually be the person adjusting the telescope on a given evening. More than likely, all we can confidently say is that *someone* at the observatory will move the telescope on a particular day. So, the passive is used because *who* moves the telescope is not important. The fact that the telescope will be moved, on the other hand, is important.

Used properly, passive voice can be a helpful tool in your efforts to write plain sentences and paragraphs. Passive voice is misused when the readers are left wondering who or what is doing the action in the sentence. In these cases, the sentence should be restructured to put the doer in the subject slot of the sentence.

Part 5

Resumes and Other Written Materials for a Job Search

The written materials that are part of an application for a job face one of the most difficult audiences imaginable: experienced recruiting managers. Because the selections that follow in this section of *Strategies* offer extensive advice on how to prepare and write resumes and cover letters, I will limit my remarks here to some general comments.

First, there is no *one* way to write a resume or a cover letter. Slavishly following some model that an applicant mistakenly thinks represents the ideal can defeat the purpose of writing a resume and a cover letter: to get an interview. The shelves of bookstores and libraries across the country are filled with guides offering advice on how to write resumes and cover letters. These guides offer advice, not commandments. Job applicants should consult as many of these guides as they want, read the selections that follow in *Strategies,* but, in the end, let common sense be their guide when they write what are essentially advertisements for themselves.

The best advice on how to write effective resumes and cover letters quite naturally comes from the recruiting managers who read them. Several years ago, my own university's placement bureau sponsored a panel featuring the recruiting managers from the companies that traditionally have hired the greatest number of our graduates. All the managers agreed that they looked at resumes and cover letters to determine as quickly as possible—sometime merely by scanning candidate's application materials—what preparation and experience candidates had in the following skills and areas:

- written and oral communication skills
- computer skills
- interpersonal skills, as demonstrated by the ability to work as a member of a team
- self-reliance and initiative, as demonstrated by the ability to work alone
- a sense of what the world of work demands in terms of professionalism and deadlines
- specific skills in at least one business or technical area supplemented by secondary skills in a variety of related areas
- a sense of business and personal ethics
- the ability to manage time, set priorities, and work under stress.

While recruiting managers do not expect job candidates to excel in all these areas or possess all these skills from the start, the list does provide some general guidelines for communicating with recruiting managers by resumes and cover letters—or, for that matter, in person during an interview.

This section of *Strategies* begins with the late John L. Munschauer's detailed discussion of how to write a resume and a cover letter. Based on his many years of experience working at and eventually directing the Cornell University Career Center, Munschauer places the writing of resumes and cover letters within the context of how they are read by busy recruiting managers. Steven Graber then presents an additional no-nonsense discussion of the basics of a cover letter. Margaret Riley Dikel and Frances E. Roehm next offer advice on how to use the Internet to post your resume when applying for a job. And Karl Weber and Rob Kaplan remind applicants of an important, though sometimes neglected, bit of etiquette that they should practice in their job searches; they should use follow-up letters to impress their readers further with their background, their professionalism, and their personality.

John L. Munschauer

Writing Resumes and Letters in the Language of Employers

The late John L. Munschauer was Director of the Cornell University Career Center.

"I am sorry, Father O'Mega, I can't let you in."

"But, St. Peter, I did everything I was supposed to. I changed the Mass from Latin to English. I had the communicants hold their hands the new way when they took communion. Everyone, so far as I knew, genuflected properly, and even Mrs. O'Reilley's Protestant husband stood when I read the gospel. I can't think of a thing I did that was wrong."

"The message, Father O'Mega, what about the message?"

"The message? What message?"

"The message of the Lord, Father. Don't you remember? The Ten Commandments? The Beatitudes? The Golden Rule? It's the *message* that gets people in here, not the ritual. The ritual was supposed to help deliver the message, but you made the ritual the message and the meaning got lost."

Meanwhile, back on earth . . . in Tucumcari, New Mexico, and all over the United States, job hunters are at their typewriters and computers pushing words around to make their resumes look like the ones they have seen in books. Whether to have it professionally printed or not, that is the question. Does it have enough action verbs?

In Peoria, Illinois, and elsewhere, employers are scanning resumes. Some are beautiful to look at; a few are even printed commercially on expensive paper. But the resumes are laid aside. Employers are reading them with one thought in mind: What can the candidates do for us? The message they are looking for isn't there.

WHY USE A RESUME?

The purpose of a resume is to convey a message, a purpose easily forgotten in the ritual of preparing it. At every turn, you will get conflicting advice about how to conduct the ritual:

- You must have a resume to get a job.
- The purpose of a resume is to get an interview.
- Every resume must have a job objective.
- A resume should never, ever be longer than one page.
- On a resume, list experience chronologically.

On the other hand, others advise:

- Don't use a resume if you are looking for an executive position or merely seeking information.
- Resumes typecast you and narrow your options.
- Interview for information first, determine what an employer wants, and then decide whether offering a resume will suit your purposes.
- Two or more pages present no problem if your resume follows a logical, easy-to-read outline.
- List your experience by function, not chronologically.

The more opinions you get, the more confused you become, but you finally work up something. You send it out. You get little or no response. You change your resume from one page to two pages—or from two pages to one. The results are no better. Somebody says to try pink paper; *that* will get attention. You decide you should have used blue. You fiddle and fiddle with your resume, trying to find the magic formula that will get you what you want. You are caught up in the ritual, forgetting that the purpose of a resume is to send a message.

It is hard not to be distracted from the message. You concentrate on developing a format, forgetting the message, instead of concentrating on the message and *then* working on a format that will convey it best. But what is your message? To find the answer, use your imagination to step out of yourself and become an employer. Think about what the employer is trying to accomplish and the talent required to get it done. Now, from your imaginary employer's chair, look back at yourself and ponder the answer. If you can't come up with one, you will have difficulty writing a resume that will say to the employer, "I have something to offer you."

Resumes are not tickets to a job. They are just one of several ways to court employers. And, as in any courtship, sometimes it's just as well not to put everything in print so the other party can't draw the wrong conclusion. Take the case of Cheryl Fender, an alto who learned that the Springfield Opera Company was auditioning for an alto to sing *Carmen*. Cheryl was well-qualified, and her voice teacher was well-known as a coach of only the most gifted singers. On the other hand, her resume, incomplete and with gaps in her history, showed extensive experience as a secretary. Wisely, she did not send the resume, which might have established her as a secretary who wanted to sing rather than as a singer who had supported herself by being a secretary. Instead, she asked herself what was important to the director of an opera company. Voice, of course, so she outlined her training in a letter. But dramatic ability also mattered, so she included pictures of herself on stage in roles that beautifully illustrated her dramatic ability. She got an audition. Cheryl followed a good marketing rule: Don't confuse customers by flaunting things that don't speak to their needs.

GIVING YOUR MESSAGE

While the language of employment for you is "I want" and for employers it is "I need," you can create resumes and letters in your language that will be read by employers in theirs. I don't mean statements such as "I have analyzed my qualifications and feel confident that they fit your needs." If you were one of thousands of employers who read this kind of thing, would you ask what qualifications and what needs? If the answer isn't clear—and it rarely is—there won't be a message.

It isn't all that difficult to create a statement that is effective. Start with the written word, with prose. Even if you are going to be approaching employers in person, go through the exercise of writing letters. Otherwise, like most people, you can get caught up in the resume ritual and neglect to develop the words that can be more effective in telling your story. People sweat over resumes, then dash off letters without much thought. Perfect resumes arriving in the mail won't even be read if the cover letters don't impress and engage employers.

I recall being asked to review a proposed application letter for a job on the staff of a yachting magazine. The letter was beautifully written, but it left the impression of being all "I": "I want to write so very much, and I am sure I can learn if you give me a chance. . . . I got straight A's in English. . . . I love boats. . . . I am a sailor. . . . I was captain of the sailing team in college. . . ."

What do you say to a person who has written a letter like that? He had obviously spent hours composing it. In its appearance and use of language, it met the highest standards. I could not squelch the young man's hopes, so I had him read *Book Publishing*, a pamphlet Daniel Melcher wrote when he was president of R. R. Bowker Company. Although the pamphlet is concerned only with book publishing, I thought its message would be useful for someone interested in magazines as well.

Let me paraphrase a portion of Melcher's pamphlet:

> I like publishing and you might like it too. It is only fair to warn you, however, that publishing attracts a great many more people than the industry can possibly absorb. Sometimes it seems as though half of the English majors in the country besiege publishing offices for jobs each year. While we hope that you will have something to offer us, you might as well face it. Publishers are experts in the art of the gentle brush-off.
>
> The interviewer hopes that you have what he needs, but it turns out that you have never looked into any of the industry's trade journals nor read any books about the industry. You haven't even acquainted yourself with the work of your university press. You tell

the interviewer that you are willing to start anywhere, but it develops that a file clerk's job would not interest you, you do not want to type, you don't think you can sell, and you know nothing about printing.

The fact is, all you have thought about is what you want, but it is his needs that create jobs, and you must address yourself to needs.

Your problem, therefore, is to learn as much as possible about the industry before you go looking for a job. Only in this way will you be able to put yourself in the publisher's place and talk to him about his needs rather than about your wants.

After reading Melcher's advice, the fellow said he got the point, thanked me, and left, returning about three weeks later to show me his revised letter. It began in much the same way as his earlier effort—with "I's"—"I majored in English. . . . I have done considerable sailing. . . ."—but after a few short sentences, he suddenly changed his tack. A new three-sentence paragraph began like this: "With my interests, naturally I want to work for you. But more to the point is not what I want but what you need." He followed this with three words that many women refuse to use and that men almost never think to use: "I can type." Right away, he showed that he knew he must be useful.

So much for the easy part of the letter. Next came the difficult part. Although he had had no experience and had never submitted an article to a magazine or even written for a student newspaper, he still had to come up with something that would interest the editor. He found the solution. The letter continued in this manner:

> . . . In looking into the field of journalism, I visited the editor of our alumni magazine, and I talked to magazine space salesmen and to executives responsible for placing ads in magazines. I also visited a printer who has contracts with magazines. In addition, I have been reading trade journals and several books on the industry. As I looked into publishing, it occurred to me that, of all the things I have done, the one I could most closely relate to the field was, strangely enough, an experience I had as a babysitter.

Immediately, he had the editor's full attention. How could babysitting fit in with publishing? During the summer of his junior year in college, he had taken a job as a sailing instructor, tutor, and companion to the children of a wealthy family that summered on the coast of Maine. The parents often went away for a week or more at a time, leaving a governess in charge, with a cook, chauffeur, maid, and gardener to do the chores and the student to keep the children busy. While the parents were on a cruise, the governess suffered a stroke, sending the cook into a tizzy, the maid into tears, and the chauffeur and gardener to the local bar. Only the student could cope, and he took charge and managed the estate for the rest of the summer.

In his letter of application, he described the crisis and its subsequent problems and told how he had met them. Then he related those experiences to the problems that he had learned editors, advertisers, printers, and others encounter in the publishing industry. Reading his letter, you could picture the student working for a publisher. There would be no slipups with the printers. Advertisers and authors would be handled with tact, yet he would get them to turn in their copy on time. He came through as someone with ingenuity, energy, and reliability; and the letter itself testified to his writing ability.

Did he get the job? Yes and no. He got an offer, but because of the publisher's urgent need, the job had to be filled immediately, and he could not accept it. He was teaching school at the time and felt that in fairness to his pupils he should finish the school year. However, a while later, the letter surfaced again when a group of editors at a meeting chatted during lunch. The subject of good editorial help came up. It followed the usual theme: "They don't make 'em like they used to." The editor of another sailing magazine complained that he had been looking for an assistant, but despite hundreds of applicants he had found no one suitable. At this point, the editor of the yachting magazine described the young man's letter and agreed to share it. The upshot was, again, a job offer. This time the timing was right, and the young man took the job. There is nothing quite like the staying power of a well-written letter. It is remembered.

Later, I complimented the young man, telling him I had never read a better letter. "It was easy," he said. "The first letter was the tough one. I didn't have anything to say other than 'I have a good record. Please give me a chance,' but I knew everyone else was saying the same thing, so I would have to say it better. I struggled with every word, trying to make an ordinary message extraordinary, but even elegant words can't make something out of fluff. I didn't know anything about publishing, so I didn't have anything to say to publishers; nor could I be really convincing without knowing enough about the work to decide whether I wanted that kind of job or not. But publishing sounded exciting, so I thought I would give it a fling."

The Importance of Knowing What the Job Is All About

"When I looked into the field in depth," the young man continued, "I became confident that I had something to offer. Thanks to a few good high school teachers and college professors, I knew where to place a comma and a colon, so I had a technical skill to offer publishers. And I knew sailing. But the big need I saw was one I discovered when I was teaching school and again when I was an assistant manager at a McDonald's—a need for people who can get things done. Such a quality is hard to describe without an analogy. I could have alluded to any number of jobs I had held, but I chose the babysitting job because I thought it would be different and would introduce an element of surprise. Apparently, the analogy worked. I got the job. More important, I wanted it. If I had received a job offer after sending the first letter, before I had really investigated publishing, I would have taken the job with an attitude of 'teach me.' That's a passive role, an observer's role, and observers tend to be critical. The chances are fifty-fifty that I would not have known how or where to contribute and would have quit after a while. Instead of offering my employer solutions to his problems, I would have become one of his problems."

I asked the young man if he had used any other supporting documents, such as a resume, to help him get the job. "I had a resume," he replied, "but I held it back, because my task was to transfer the qualities I had demonstrated as a baby-sitter to the needs of a publisher, and I couldn't seem to do that in a resume. A resume is a good way to outline facts, but I had to use prose to develop the analogy.

"I did think about including a writing sample, but when I studied the magazine I realized that most of the articles were written by contributors rather than staff. Their job was to select and edit articles. If I presented myself as a writer, I wouldn't have received an offer; the magazine didn't hire authors. When a friend of mine saw an advertisement for an editing job, he made a list of the required qualifications, then presented his case point by point, right on target. Then the dope attached a resume that said loud and clear, 'I want to be a writer.' He either should have skipped the resume or prepared a new one."

Are these examples intended to be good arguments for not using resumes? No. They simply emphasize that it is important to determine the best way to get a message across. There are times when there is no substitute for a resume. When employers advertise and list the qualifications they seek, there is no better way to respond than to send a resume outlining qualifications.

LETTERS OF APPLICATION

Sometimes, however, it is hard to figure out how to make the letter of application you send with your resume more effective than those of the hundreds of other people who are responding to the ad. Imagine being on the receiving end of applications at General Motors or Exxon.

To find out about the effectiveness of letters and resumes, I visited corporations and asked employment managers for their comments. "Here," said one employment manager, as he picked up an 18-inch stack of letters and handed it to me. "This is my morning's mail. Read these letters and you'll have your answer."

"I can't," I protested. "I have only 2 hours, and there's a day's reading here."

"Yes, you can," he replied. "Unfortunately, you'll get through the pack in half an hour, because a glance will tell you that most are not worth reading."

It was hard to believe that the letters could be that bad, but he was right. The typical letter was an insult. Among the letters that I did not finish reading was one that was obviously a copy—the machine must have run out of toner thirty copies back. It began:

Dear Sir:

I am writing to the top companies in each industry and yours is certainly that. I want to turn my outstanding qualities of leadership and my can-do abilities to. . . .

Enough of that. Also, the applicant hadn't even bothered to type in the employment manager's name, which he could easily have found in reference books such as the CPC Annual, published by the College Placement Council, and Peterson's Job Opportunities, published by Peterson's Guides.

The next letter was written in pencil on notepaper. There may have been an Einstein behind that one, but I can't imagine anyone taking the time to find it out. Many other letters were smudged and messy. Some applicants tried to attract attention with stunts, such as putting cute cartoons on their letters. One piece of mail contained a walnut and a note that read, "Every business has a tough nut to crack. If you have a tough nut to crack and need someone to do it, crack this nut." Inside the nut, all wadded up, was a resume. Cute tricks and cleverness don't work at the General Mammoth Corporation.

At the same time, the good letters stood out like gold. Five letters, only five letters in that pile of hundreds, were worth reading. They had this in common:

- They looked like business letters. Their paragraphing, their neatness, and their crisp white 8 1/2" X 11" stationery attracted attention like good-looking clothing and good grooming.
- They were succinct.
- There were no misspellings or grammatical errors.

As I read them, I heard a voice—the voice of a fusty old high school English teacher—commanding out of the past:

- If you can't spell a word, look it up in a dictionary.
- Use a typewriter. Pen and ink are for love letters.
- Clean your typewriter. Avoid fuzzy type. You wouldn't interview in a dirty shirt, so don't send a dirty resume or letter.
- For format, use a secretarial manual. If you don't have one, borrow one from the library. What do you think libraries are for?

How that English teacher would have loved the following advice given by the late Malcolm Forbes when he was editor-in-chief of *Forbes* magazine:

Edit ruthlessly. Somebody ~~has~~ said that words are ~~a lot~~ like inflated money—the more ~~of them that~~ you use, the less each one ~~of them~~ is worth. ~~Right on.~~ Go through your entire letter ~~just~~ as many times as it takes. ~~Search out and~~ Annihilate all unnecessary words ~~and~~ sentences—even ~~entire~~ *paragraphs*.

The following letter is typical of those I saw that day. Give it the Forbes treatment, and see what you can do with it. You may need a scissors as well as a pencil.

Dear Mr. Employer:

I am writing to you because I am going to be looking for employment after I graduate which will be from Michigan where I studied Chemical Engineering and I will be getting a Bachelor of Science degree in June. The field in which I am interested and hope to pursue is process design and that is why I am writing your company to see if you have openings like that. I think I have excellent qualifications and you will find them described in the resume which I have attached to this letter.

The five letters that stood out favorably were characterized by their simplicity. Here is a letter, fictitious of course, but enough like the letter I remember to give you an idea of the ones that created a favorable impression:

February 1, 1991

Mr. Paul Boynton
Manager of Employment
The United States Oil Company
1 Chicago Plaza
Chicago, Illinois 60607

Dear Mr. Boynton:

This June I will receive a Bachelor of Science in Chemical Engineering from the University of Michigan, and I hope to work in process design or instrumentation. I saw your description in *Peterson's Job Opportunities for Engineering, Science, and Computer Graduates* soliciting applicants with my interests. Enclosed is a resume to help you evaluate my qualifications.

While I find all aspects of refining interesting, my special interest in process design and instrumentation developed while working as a laboratory assistant to Professor Juliard Smith, who teaches process design. I wrote my senior thesis on the subject of instrumentation under him, and part of what I wrote will be used in a textbook he is writing and editing.

Would it be possible to have an interview with you in Chicago during the week of March 1? To be even more specific, could it be arranged for 10 A.M. on Tuesday, March 3? I am going to be in Chicago that week, and this time and date would be best for me, but of course I would work out another time more convenient for you. In any event, I will call your office the week before to determine whether an interview is possible.

Sincerely yours,

Charles C. Thompson

Thompson's effective letter, and three of the four others, followed a similar pattern:

1. The first paragraph stated who the writer was and what he wanted.
2. The second paragraph, sometimes the third, and in one case a fourth paragraph, indicated why the writer wrote to the employer and mentioned areas of mutual interest, special talents that might be of interest to employers, or other factors relating to qualifications that could be better described in a letter than in a resume.
3. A final paragraph suggested a course of action.

The fifth letter covered the same points in a different order. I remember it because it complemented the good but not outstanding resume shown in Figure 1 [p. 286]. In that resume, a perceptive employer could see a person he might like, someone who was energetic and personable. Yet it didn't quite hang together, because the work history and activities didn't seem to support what the writer wanted to do. But look at the letter that "made" the resume:

Mr. Paul Boynton, Manager of Employment
The United States Oil Company
1 Chicago Plaza
Chicago, Illinois 60607

March 10, 1991

Dear Mr. Boynton:

This June, following my graduation from the University of Puget Sound, I want to pursue a career in sales. Between April 10 and 23, I plan to call on leading companies whose products I would like to sell. The purpose of this letter is to determine whether you would like to have me include you in my itinerary.

Let me tell you why I believe I can sell. It seems to me that I am always selling. As a camp counselor, I persuaded the director to buy a fleet of small sailboats so I could start a sailing program. When our college housing co-op needed painting, I persuaded the members to give up a vacation to do the job. In thinking about how I enjoyed selling these and other projects, I decided to look into a sales career. I persuaded several sales representatives to let me spend a day or more traveling with them to see what it was like.

While with them I realized something more about myself that further convinced me I belong in sales. The best sales representatives were well organized, had high energy levels, and used their time efficiently, qualities I feel I have. As evidence, I have enclosed a resume that outlines my accomplishments in college and during vacations.

I hope to hear from you. United States Oil is in the top group of employers on my list.

Sincerely,

Lance Zarote

Encl: Resume

Hard Work and Attention to Detail Make for a Good Letter

While only the five letters I have mentioned were effective, the rest of the correspondents could have done as well. The point is they didn't. Most people won't. Therein lies your opportunity, because, like Thompson and Zarote, you can write letters that set you apart. You don't have to create a literary masterpiece; just don't knock off a letter hastily with thoughts that wander all over the page. Write it and rewrite it, following Forbes's advice. Unless you are an exceptional typist, you are not good enough to type it yourself. Hire a professional or use a word processor, but be sure the print is letter quality. Also, get an English teacher or someone in the word business to check your spelling, punctuation, and grammar.

Don't Delegate the Job of Letter Writing

More important than style, however, is the thought process used in preparing letters and resumes. Don't shortchange yourself by delegating your thinking to someone else. Be sure it is *your* letter. Somehow, a ghost-written letter always has a phony ring to it. When you write to employers, think about their needs; then

LANCE ZAROTE

Campus Address   Permanent Address
101 Morril Hall   25 The Byway
University of Puget Sound   Provincetown, Massachusetts 05840
Tacoma, Washington 95840   (617) 555–6026
(206) 555–6206

GOAL: A SALES CAREER

EDUCATION: UNIVERSITY OF PUGET SOUND
                    Bachelor of Arts 1991
                    Philosophy Major

                    Business-Related Courses
                    Statistics   Introduction to Computers
                    Economics   Calculus
                    Accounting   English

EMPLOYMENT: UNIVERSITY OF PUGET SOUND (Work-Study Program)
                    Dining hall supervisor   1990–91
                    Kitchen helper   1987–89

                    OTHER WORK while in college
                    Ma and Pa Motel, Tacoma—night clerk   1990
                    Joe's Bar and Grill—weekend waiter
                        & bartender   1989–1991
                    Baby-sitting, gardening, house cleaning   1985–1991
                    WATCHEE OUTEY SUMMER CAMP, Nome, Alaska
                    Sailing coach and waterfront director   1990
                    Counselor   1986–1989

ACTIVITIES: CAMPUS
                    Chair, Campus Chest Drive
                    Intramural hockey, tennis, and volleyball
                    LIVING UNIT
                    House Manager, $25,000 budget
                    Secretary
                    Membership Committee
                    CIVIC
                    Reader-companion in nursing home
                    Big Brother-Sister Program, Southside Youth Center
                    CHURCH
                    Choir
                    Youth leader
                    Sunday school teacher

INTERESTS: Skiing   Music
                    Chess   Dancing

FIGURE I   An ordinary resume can become effective when attached to a powerful letter.

think about yourself and what you offer, and relate this to what you would like to do. Putting your thoughts on paper—thoughtfully—will make you sort out your ideas and interrelate them. When you see them on paper they will talk back to you, at times to suggest better ideas, at other times to tell you that you are off the mark. To organize your ideas, create an outline. In other words, prepare a resume even if you decide not to use it. *The value of a resume is frequently more in its preparation than in its use.*

RESUME PREPARATION

When you do give an employer your resume, make it a testimony to your ability to organize your thoughts. Remember, too, it must look sufficiently attractive to get an employer to read it. Unfortunately, most of the resumes I have seen on employers' desks were just as unattractive as the letters; they had sloppy, crowded margins, were poorly organized, and were badly reproduced. At least 30 percent of the resumes had been put aside with hardly a glance because their physical appearance was so awful. The rest got a 20-second scan to see if they were worth studying.

Following are two resumes that pass the appearance test with flying colors. Let's see how they fare during a 20-second scan and beyond.

Nancy Jones—A Good Resume Made Better

Nancy Jones's resume has arrived at the desk of a laboratory director who needs an assistant to help run a quality-control laboratory in a pharmaceutical company (Figure 2 [p. 288]). With candidates far outnumbering openings in biology, the advertisement for the job has brought in hundreds of applications, and the director is wearily scanning them one by one to find the few that will be of interest to him. Conditions are not favorable for Nancy. She has to catch his eye with impressive qualifications, or she is not going to get anywhere.

The director picks up Nancy's resume. Immediately, he is impressed, because it looks attractive. He thinks the resume reflects an orderly mind. Most resumes he has looked at just do not put it all together.

He begins to read. The job objective annoys him; it strikes him as being long-winded. Why couldn't she simply say she is interested in applied biology? What is this business about working with people? Is that there because she has doubts about biology?

If resumes are supposed to say only what needs to be said, what about this line?

Born January 6, 1969   5'7"   135 lbs.   Single   Excellent health

Does it say anything about her ability to do the job? The biologist doesn't think so.

His eyes move down the page:

NANCY O. JONES

Present Address
105 Belleville Place
Ames, Iowa 50011
Phone: 515-555-6674

After June 1, 1991
1212 Centerline Road
Old Westbury, New York 11568
Phone: 516-555-7664

Born January 6, 1969   5'7"   135 lbs.   Single   Excellent health

Career Objective	Research and development in most areas of applied biology, with an opportunity to work with people as well
Education	Iowa State University, Ames, Iowa
	Bachelor of Science, June 1991
	Major: Biology Concentration: Physiology
	GPA: 3.3 on a 4.0 scale

Major Subjects	Minor Subjects
Mammalian Physiology	Qualitative Analysis
Vertebrate Anatomy	Quantitative Analysis
Histology	Organic Chemistry
Genetics	Biochemistry

Scholarships and Honors	University Scholarship: $2850/year
	Iowa State Science and Research Award
	Dean's List two semesters
Activities	Volunteer Probation Officer, 1988–89
	Probation Department, Ames, Iowa
	Tutor, Chemistry and Math, 1988–91
	Central High School, Ames, Iowa
	Member, Kappa Zeta social sorority
	Women's Intercollegiate Hockey Team
Special Skills	Familiarity with Spanish; PL/C and FORTRAN computer languages; typing
Work Experience	Teaching Assistant and Laboratory Instructor
	Freshman Biology School year, 1990–91
	Waitress, Four Seasons Restaurant
	Catalina Island Summers, 1989–90

FIGURE 2   The resume of Nancy O. Jones does not communicate her career-related experience.

Education	Iowa State University, Ames, Iowa
	Bachelor of Science, June 1991
	Major: Biology Concentration: Physiology

GPA: 3.3 on a 4.0 scale

Major Subjects	Minor Subjects
Mammalian Physiology	Qualitative Analysis
Vertebrate Anatomy	Quantitative Analysis
Histology	Organic Chemistry
Genetics	Biochemistry

She has used the outline form well, so he is able to take in a great deal of information in one look. Her education impresses him.

Double spacing above and below her grade point average makes it stand out. However, if her average had not been quite as good, and if she had not wanted to feature it, she could have used single spacing to make it less conspicuous, like this:

Education	Iowa State University, Ames, Iowa
	Bachelor of Science, June 1991
	Major: Biology Concentration: Physiology
	GPA: 2.7 on a 4.0 scale

Now that she has told him about her grades and her courses she wants to drive home the point that she was no ordinary student. She flags him down with the headline "Scholarships and Honors," and he sees her financial aid and science awards, which make a favorable impression. Two times on the dean's list may not be important enough to set apart by double spacing, but it makes a modest impression.

Up until now, Nancy has made a favorable impression overall. The director is ready to take in the next batch of information:

Activities	Volunteer Probation Officer, 1988–89
	Probation Department, Ames, Iowa
	Tutor, Chemistry and Math, 1988–91
	Central High School, Ames, Iowa
	Member, Kappa Zeta social sorority
	Women's Intercollegiate Hockey Team

She almost loses him by featuring her work as a probation officer. That is not of primary interest to him, but through good spacing and placement, she draws his eye to the next item, which states that she has tutored chemistry and math. Unfortunately, Nancy now loses him permanently by ranking tutoring along with the sorority and the hockey team. He guesses she has made her major statement about biology, and so he turns to the next resume.

Good as it is, Nancy Jones's resume could be improved by reorganization. Her tutoring chemistry and her two years as a teaching assistant and laboratory instructor in biology reveal a more-than-academic interest in biology. They should be featured. Everything related to biology and any other information of possible use to the employer should be put under a new marginal headline and statement, as follows:

Career-Related Experience	Biology Lab Instructor and Teaching Assistant Freshman Biology (1990–91)
	Chemistry and Mathematics Tutor Central High School, Ames, Iowa (1988–91)
Skills and Interests	Microscopy Computer Languages: Electron Microscopy FORTRAN, PL/1, Histology COBOL Spectrum Analysis Statistics Small-Animal Surgery

Now, the director is able to see the things she can do. "Good," he says to himself, "she can use an electron microscope. We need someone with that skill."

By no means should Nancy eliminate mention of the hockey team, but since she is applying for a professional job, the first bait to throw out is credentials; they testify to her ability to do the work. After that, what may sink the hook is how the employer sees her as a person. He may have picked out bits here and there that testify to her diligence, but the picking out depends on chance reading; it would be best not to leave anything to chance. With a slightly different presentation, she might ensure that he receives an impression of her diligence.

There are other areas she could strengthen, things she has underplayed or totally neglected to mention. Sticking the waitress job in down at the bottom of the page is almost an apology for it. She may also have had other jobs, such as baby-sitting, household work, or door-to-door sales, that she belittles in her mind and has not even mentioned. Mention of such things might make a good impression on an employer looking for somebody who is not afraid to work and who is mature for her years.

If we quizzed Nancy, we might find that she could put something like this on her resume:

Scholarships and Financial Support	90% self-supporting through college as follows: University Scholarship: $2850/year Iowa State Science and Research Award Waitress, Four Seasons Restaurant Catalina Island (Summers, 1989–90) Teaching and instructing, baby-sitting, home maintenance, selling

And this, because her activities tell something about her as a person:

Activities	Volunteer Probation Officer (1988–89) Kappa Zeta social sorority Women's Intercollegiate Hockey Team Skiing, sailing, singing, tennis

Double spacing has been cut down so as not to overemphasize the less important items, yet a string of other things not terribly important in themselves has been inserted to support the impression of an active, interesting person. Sometimes you want to leave an impression, at other times you want to emphasize a qualification. For example, Nancy wanted to feature her studies, and the way she brought them out by listing them in a column was good. If she had listed them like this, they would not have stood out:

> Major Subjects: Mammalian Physiology, Vertebrate Anatomy, Histology, Genetics
> Minor Subjects: Quantitative Analysis, Qualitative Analysis, Organic Chemistry, Biochemistry

In Figure 3 [p. 292], you will see the Nancy Jones resume as she might have revised it to emphasize the strong points that would have been of interest to that director who was looking for a resourceful assistant. The added emphasis might have made the difference that would have landed the job for her.

Nancy's revised resume is pure outline, devoid of prose. It works well for her. When she states that she has studied quantitative and qualitative analysis and knows PL/1 and FORTRAN, a scientist reading her resume knows what this means.

Janet Smith—The Proper Use of Headlines

Janet Smith, whose resume is shown in Figure 4 [p. 293], has a different problem in presenting her qualifications. She needs to *describe* what she did in order to tell an employer about her qualifications, so her resume calls for a mixture of key words and prose to get her message across. Yet prose can destroy the effect of an outline. The solution lies in imitating newspaper editors, who use headlines and subheadlines to attract readers. Like a newspaper, a resume should lend itself to skimming so the reader can quickly pick up a good overview of what is important. Then the reader can select specific things of interest and read further. There is an art to using headlines, but Janet Smith hasn't mastered it, at least not in the resume she used to apply for a job with Hermann Langfelder, a hardbitten old hand with thirty years in personnel and labor relations in the machinery business. He picked up her resume and read:

| CAREER | A challenging position in personnel administration |
| OBJECTIVE | requiring organizational ability and an understanding of how people function in business and industry. |

That was pure baloney, and he choked on it. He thought of the hours he had spent in meetings, listening to a lot of hot air. "Challenging, my foot!" he muttered. Then he read the bit about her organizational ability and her understanding of how people function in business and industry. "I've been at this business for thirty years," he groused to himself, "and I still can't figure out how people function in industry. But she knows all about it."

Beware of Misleading Headlines

"Well, let's see what she's done," he said to himself, and then his eyes fell on "UNIVERSAL METHODIST CHURCH." That did it! He didn't want to bring any do-gooder into his factory to preach. He rejected her. And the fault was hers, in using the headline.

NANCY O. JONES

Present Address
105 Belleville Place
Ames, Iowa 50011
Phone: 515-555-6674

After June 1, 1991
1212 Centerline Road
Old Westbury, New York 11568
Phone: 516-555-7664

Career Objective   Research and development

Education   Iowa State University, Ames, Iowa
                        Bachelor of Science, June 1991
                        Major: Biology   Concentration: Physiology

                        GPA: 3.3 on a 4.0 scale

                        Major Subjects   Minor Subjects
                        Mammalian Physiology   Qualitative Analysis
                        Vertebrate Anatomy   Quantitative Analysis
                        Histology   Organic Chemistry
                        Genetics   Biochemistry

Career-Related   Biology Lab Instructor and Teaching Assistant
Experience   Freshman Biology (1990–91)

                        Chemistry and Mathematics Tutor
                        Central High School, Ames, Iowa (1988–91)

Skills and Interests   Microscopy   Computer Languages:
                        Electron Microscopy   FORTRAN, PL/1,
                        Histology   COBOL
                        Spectrum Analysis   Statistics
                        Small-Animal Surgery

Scholarships   90% self-supporting through college as follows:
and Financial
Support   University Scholarship: $2850/year
                        Iowa State Science and Research Award

                        Waitress, Four Seasons Restaurant
                        Catalina Island (Summers, 1989–90)

                        Teaching and instructing, baby-sitting, home
                        maintenance, selling

Activities   Volunteer Probation Officer (1988–89)
                        Kappa Zeta social sorority
                        Women's Intercollegiate Hockey Team
                        Skiing, sailing, singing, tennis

FIGURE 3   The revised resume of Nancy O. Jones effectively features her career-related experience in a separate section.

JANET V. SMITH
111 Main Street
North Hero, Vermont 05073
802-555-1234

CAREER
OBJECTIVE

A challenging position in personnel administration requiring organizational ability and an understanding of how people function in business and industry.

EDUCATION

PURDUE UNIVERSITY
Hammond, Indiana
Master of Industrial Relations   June 1991

SMITH COLLEGE
Northampton, Massachusetts
Bachelor of Arts, magna cum laude   June 1986

WORK
EXPERIENCE

UNIVERSAL METHODIST CHURCH
1 Central Square
New York, New York 10027
Assistant Personnel Officer. 1988–90
Responsible for interviewing applicants for clerical positions within the organization and for placing those who demonstrated appropriate skills, for accepting and dealing with employees' grievances, and for developing programs on career advancement.

CORTEN STEEL COMPANY
10 Lake Street
Akron, Ohio 44309
Assistant, Personnel Office. 1986–88
Responsible for all correspondence of Personnel Director and for interviewing some custodial applicants and referring them to appropriate supervisors for further interviewing.

BORG-WARNER, INC.
Ithaca, New York 14850
Assembly-line worker. Summer 1985
Assembled parts of specialized drive chains in company with thirty other men and women.

COMMUNITY
SERVICE

PLANNED PARENTHOOD
North Hero, Vermont 05073
Counselor. Summers, 1983–84
Explained various aspects of family planning and provided birth control information to clients of Planned Parenthood. Made referrals to other counselors and physicians where appropriate.

AUXILIARY
SKILLS

French: Fluent. Knowledge of office procedures.
Experience with mainframe computes and PCs.

FIGURE 4   The resume for Janet V. Smith misleads the reader with irrelevant headings.

Janet's job at the church was administrative, not ministerial, and the church didn't care whether she was Jewish, Catholic, or agnostic. Only in its ministerial work does the Methodist Church need Methodists. But people—Langfelder and the rest of us—respond to symbols and make snap judgments on the basis of symbols. Janet had put the symbol of the church—its name—in a heading, when she could have done something better.

When you lay out your resume, think of symbols. Imagine you are a newspaper editor who wants to put a story across. As an editor planning headlines, you must imagine yourself in the position of the reader. Ask yourself, "What words will catch the reader's eye? What words will put the reader off?"

Use words that fit the job in question, and play down those that can lead an employer to think of you in terms that don't relate to the job. Ask yourself, "Does this say something to the employer?" Janet Smith missed Langfelder, and some of her duties would have yielded key words to send appropriate messages. The following arrangement would have been more effective for her:

WORK EXPERIENCE	ASSISTANT PERSONNEL OFFICER, 1988–90
	Universal Methodist Church
	1 Central Square
	New York, New York 10027
	Interviewing, placement, grievances, and training of applicants and employees in the clerical and support services of the church organization. Developed programs for the career advancement of employees.
	ASSISTANT, PERSONNEL OFFICE, 1986–88
	Corten Steel Company
	10 Lake Street
	Akron, Ohio 44309
	Interviewing and referring as an assistant to Personnel Director. Responsible for all correspondence and for interviewing some custodial applicants and referring them to supervisors for further interviews.

Let's suppose that Janet has a chance to send Langfelder the revised resume shown in Figure 5. This time, she uses headlines that pinpoint the ideas she most wants to get across, so that he makes it past the Universal Methodist Church and gets down to the assembly line. "Hey, now, look at that!" he thinks. "She's worked out there on the floor. That means she's heard all the language and knows the gripes and the tedium. We have a lot of women in this factory, and it might be a good thing to have a down-to-earth, smart woman on my staff." (His sexism may have been showing, but this kind of employer is alive and kicking somewhere out there, and you may have to deal with him.)

What if Janet Smith wanted a job in the computer industry and had ten years' experience with IBM? Employers are impressed by "graduates" of companies like

JANET V. SMITH
111 Main Street
North Hero, Vermont 05073
802-555-1234

CAREER
INTERESTS   Personnel Administration and Labor Relations

EDUCATION   PURDUE UNIVERSITY   June 1991
                     Hammond, Indiana
                     Master of Industrial Relations

                     SMITH COLLEGE June 1986
                     Northampton, Massachusetts
                     Bachelor of Arts, magna cum laude

WORK   ASSISTANT PERSONNEL OFFICER. 1988–90
EXPERIENCE   Universal Methodist Church
                     1 Central Square
                     New York, New York 10027
                     Interviewing, placement, grievances, and training of
                     applicants and employees in the clerical and support services
                     of the church organization. Developed programs for the
                     career advancement of employees.

                     ASSISTANT, PERSONNEL OFFICE. 1986–88
                     Corten Steel Company
                     10 Lake Street
                     Akron, Ohio 44309
                     Interviewing and referring as an assistant to Personnel
                     Director. Responsible for all correspondence and for
                     interviewing some custodial applicants and referring them to
                     supervisors for further interviews.

                     ASSEMBLY-LINE WORKER. Summer 1985
                     Borg-Warner, Inc.
                     Ithaca, New York 14850
                     Factory work experience on an assembly line. Worked with a
                     team of thirty other men and women.

COMMUNITY   COUNSELOR. Summers, 1983–84
SERVICE   Planned Parenthood
                     North Hero, Vermont 05073
                     Explained various aspects of family planning and provided
                     birth control information to clients of Planned Parenthood.
                     Made referrals to other counselors and physicians where
                     appropriate.

AUXILIARY   French: Fluent. Knowledge of office procedures.
SKILLS   Experience with mainframe computers and PCs.

FIGURE 5   The revised resume for Janet V. Smith stresses what she did rather than the less important point of where she did it.

IBM that are known as leaders in their fields. Ten years with them is significant. It might make sense to present her experience this way:

EXPERIENCE   1981–91
                                    IBM Corporation
                                    Binghamton, New York
                                    Assistant Personnel Officer

Were she an engineer after a technical job that could use her IBM experience, highlighting the name of the company would have been a good idea. Imagine employers giving the resume a 20-second scan. What words should you use and how should they be placed to catch the eye and make employers want to read further?

Mark Meyers—The Functional Resume

Janet Smith and Nancy Jones were lucky. Their training and experience translated into satisfactory headlines to highlight their experience, but that doesn't always work. Mark Meyers, whose resume appears in Figure 6 [p. 299], adopted a different technique to help him get a job in community recreation. He got his message across by creating a resume based on functions.

When he began to write his resume, he tried time and time again to get his message across in a conventional form in which he first listed his education, then his experiences in chronological order, and finally his activities, hobbies, and interests. But writing it conventionally raised all sorts of problems. He wanted to highlight his public relations and promotion experience, some of which he had been paid for and some not. Some of it had also been secondary to a primary assignment. Dividing up this experience and placing bits and pieces of it in various parts of the resume to make it conform to a conventional style diluted its impact. Also, his athletic ability and experience would mean a great deal to an employer in his field, but how could he show it effectively? Some of it had been gained as a participant, some through training, and some as a coach. Could he expect an employer to sift through the various sections of the resume to find out all he had done in athletics? (Remember, you can only count on an employer giving a resume a quick scan before deciding whether or not to study it more fully.)

His solution, as seen in Figure 6, was to feature the functions of the job he wanted and then describe things he had done that pertained to each area. Thus, under each function he developed the equivalent of a mini-resume.

Preparing a Resume for a Specific Job

Mark stated his case well, but you can't get blood from a stone. He found he had to look outside his field, because jobs in it were virtually nonexistent due to cutbacks in government funding. In his search, he ran across the following job listing from the publisher of a magazine for parents:

EDITORIAL SECRETARY

BA. in Liberal Arts
Interested in childhood training. Well organized, outstanding language skills. Typing and
clerical skills, potential to use electronic text-
editing equipment. Reporting to Coordinating
Editor, Happy Days magazine. Assist in all editorial functions. Evidence of creativity essential. Entry-level position with career potential.

Can you put yourself in his shoes, analyze the job, and devise a resume that
speaks to the stated needs of this employer? The clue to doing it is to go through
the job description and step by step take your cue from the employer. Right off,
you will hit a bit of a snag because the employer has specified a B.A., while Mark
has a B.S. His drama minor might give him appropriate credentials, however. If
he shows his education as in the following example it might reflect the liberal
background the employer apparently prefers:

EDUCATION   HOBOKEN UNIVERSITY, B.S., 1987
                             Major: Recreation   Minor: Drama

                             Humanities courses:
                             Introduction to Dramatic Literature
                             British Drama to 1700
                             History of Theater
                             Playwriting
                             Introduction to Poetry
                             Shakespeare
                             English History

Next, the job calls for an interest in childhood training, then language skills,
and so on. Each specification suggests a headline for a resume. Mark faces
another stumbling block when it comes to demonstrating an interest in childhood training, since his experience has been with older youths only. Since a
resume makes points by stating facts, he cannot demonstrate an interest in childhood training in his resume because he lacks the appropriate experience. However, he *can* describe his interest in the letter that usually goes hand in glove with
a resume. He has a good basis for doing so, because the field of recreation certainly has much to do with the entire range of human development from childhood on. He should be able to point out correlatives in his education and
experience with the work being done in childhood training, and he could check a
library for information about childhood training to help develop the correlatives.
Above all, he should read the magazine and try to tie in as many of his experiences as possible with the purpose of the magazine.

With his interest in childhood training brought out in a letter to complement
his resume, he might then proceed to develop his outline as follows:

EXPERIENCE   HUMAN DEVELOPMENT

<u>Dover Youth Bureau.</u> Planned and implemented programs in drama, photography, athletics, health. Summer 1986

<u>Research Project.</u> Studied recreational needs of a Hoboken neighborhood. Interviewed residents, developed questionnaire. Project provided an insight into the family life and problems of parents in a neighborhood setting. Fall semester 1986

<u>Outdoor Leadership Training.</u> Practical experience in human development through a knowledge of nature and survival skills. Summer 1983.

LANGUAGE SKILLS AND CREATIVITY

<u>Writing.</u> Wrote releases, developed advertising, and prepared radio announcements for Drama Club, men's sports programs, and other events. Wrote report on Hoboken research project. 1984–87

<u>Lecturing.</u> Gave talks and led tours at Atlantic County (Delaware) Park nature trail and visitors' center. Summer 1985

<u>Audiovisual.</u> Prepared slides for audiovisual presentation for visitors' center. 1985.

<u>Promotional.</u> Created innovative techniques—such as costume parade of cast—to arouse interest in avant-garde staging of "Alice in Wonderland," which gained campuswide attention. 1984–85.

<u>Design.</u> Designed posters, fliers, and other graphics for sporting events, plays, and other campus events. 1984–87.

Constructed nature exhibits at nature center. 1985.

CLERICAL AND ADMINISTRATIVE

<u>Clerical.</u> Typing and use of office machinery. McGary's Department Store. Summer 1982.

<u>Administration.</u> Responsible for equipment, scheduling of programs, and coordinating of competitions. Hoboken University, 1984–86.

The functional resume allows you to develop a different message for each job or type of job you wish to apply for. Different functions can be highlighted, depending on what the job requires, and your specific experiences rearranged under different headings. It gives you the flexibility you need if your experience has been diverse.

Almost every resume ought to have something of the functional resume about it. With computers making it so easy to change a text, there is no reason why each resume can't be slanted to appeal to the particular employer, even if it's

---

MARK MEYERS

1414 South Harp Road
Dover, Delaware 19901
302-555-4444

CAREER INTEREST
  Community recreation.

RECREATION PROGRAMMING EXPERIENCE
  Planned and implemented programs in stagecraft and drama; assisted with pro-
    gramming in ceramics, photography, and physical fitness for Dover Youth
    Bureau summer program. Summer 1986.
  Lectured and led tours at Atlantic County (Delaware) Park nature trail and visi-
    tors' center. Prepared slides to illustrate lecture; helped in construction of
    nature exhibits. Summer 1985.
  Coordinated men's intramural sports competitions for Hoboken University.
    Had responsibility for equipment and scheduling. 1984–86.

PUBLIC RELATIONS AND PROMOTION EXPERIENCE
  Directed publicity efforts of University Drama Club for several productions.
    Developed innovative techniques—such as a costumed cast parade to arouse
    interest in an avant-garde staging of "Alice in Wonderland", which gained
    campuswide attention. 1984–85.
  Advertised in various media and became familiar with advertising methods,
    including writing news releases, taping radio announcements, designing
    graphics for posters and fliers. Drama Club, 1985–87; men's sports,
    1984–87.

LEADERSHIP AND ATHLETIC ABILITIES
  Trained in outdoor leadership and survival skills by Outward Bound.
    Summer 1983.
  Coached hockey and basketball. Dover Youth Bureau, 1986.
  Participate in hockey, swimming, backpacking.

RESEARCH AND EVALUATION ABILITIES
  Prepared college research project on the recreational needs of residents of a
    Hoboken neighborhood. Designed questionnaire to solicit residents' own
    perceptions of their needs; interviewed residents and local officials. Fall
    semester 1986.
  Reported on effectiveness of Dover Youth Bureau programming. Summer 1986.

OFFICE SKILLS
  Dealt with unhappy and irate customers at McGary's Department Store, Dover.
    Worked at service desk, tracking down problems and rectifying errors.
    Summer 1982.
  Typing and use of office machinery.

EDUCATION
  Bachelor of Science, June 1987   Major: Recreation
  Hoboken University, Hoboken, New Jersey Minor: Drama

FIGURE 6   The resume of Mark Meyers illustrates the functional style.

the resume of a generalist like Bruce Gregory Robinson, who hopes to get into a training program.

Bruce Gregory Robertson—A Resume Reflecting an Active Mind and Body

Put yourself in the chair of a merit employer, someone who is interested in candidates not so much for what they know as for what they can learn. Imagine yourself as a banker, a merchant, or a manufacturer. The position you have to fill can be learned easily on the job. What interests you are candidates' traits—their energy, intelligence, leadership qualities—the things that tell you the candidate can grow in your employ and eventually become an executive in the company. You are looking for a resume that reflects an active person with an active mind. Bruce Robertson is such a person, and his resume shown in Figure 7 has been designed accordingly.

Now make yourself a textbook publisher. Again, you are a merit employer and promote from within. Your company hires people with a good liberal education and little or no postgraduate experience. The men and women you hire will travel from college to college either soliciting manuscripts from professors or showing them texts that might fit their courses. Impressive candidates should have a high energy level and be able to show that they have used time efficiently and worked independently. How does Bruce look to you?

Now change the scenario slightly. This time you are a publisher of texts for primary and secondary school. To get your texts adopted, your representatives will have to make presentations to state regents, school superintendents, and other educational groups. Remember that experience and college major don't matter, but do you think it might help Bruce's cause if he did a bit more to highlight his interest in public speaking and his honors as a debater? He has a computer. It would be a cinch for him to change his resume slightly. Can you think how he might do it?

Next, put yourself in the chair of an employer on Main Street. You need someone in customer relations, a person to follow up on complaints and misunderstandings about credit arrangements or bills. You want to hire someone who has had experience of this kind. Still, the work easily could be learned on the job . . . if the right person came along. In response to your classified advertisement, Bruce's resume arrives accompanied by a routine cover letter. He recounts people problems he has resolved and describes his administrative experience in a way that indicates he is a team worker. What's your decision? What if his letter is more inspired?

Michelle Trio—The Curriculum Vitae

A curriculum vitae (literally, "course of life" in Latin), sometimes called a C.V. or vita, is a resume for academic positions and as such does not need a statement of goals or interest. While there is merit in keeping nonacademic resumes brief by focusing on employers' needs, a faculty tends to select colleagues not just to teach but for the prestige they will bring to the department, especially in the long run. An eminent faculty attracts eminent associates. Publications, research,

BRUCE GREGORY ROBERTSON

Home Address
105 Comstock Drive
Pierre, South Dakota 57501
605-555-5445

College Address
5 Erasmus Drive
St. Paul, Minnesota 55101
612-555-4554

CAREER
INTERESTS
— The marketing of products or services in industries such as banking, publishing, and retailing.

EDUCATIONAL
BACKGROUND
— Macalester College, St. Paul, Minnesota   A.B., May 1991
Major: English   Minor: Economics
Honors: Dean's List, 1990–91

EMPLOYMENT
— Entrepreneurial—Organized a company to paint road markings in parking lots; rented necessary equipment; employed two additional workers; had personal contact with mall managers, store owners, etc., to promote the company's services: summers, 1989–90.

— Administrative and Clerical—Temporary Help, Inc. Typing and other business machine operations, bookkeeping, clerking, complaint adjustments; short-term assignments with auto dealers, banks, real estate operations, schools, and similar employers: summer 1988; part-time 1989.

— Miscellaneous—Camp counselor, stock clerk in grocery store, baby-sitter, newspaper boy. From high school on have earned money for clothes, travel, and purchase and maintenance of an automobile.

ACTIVITIES
— Vice President, Seven-Come-Eleven investment club
— Varsity basketball
— Coach and tutor, St. Paul Concordia Boys Club
— Debate Club
— Book and Bottle literary club

SKILLS
— Typing
— PL/C and FORTRAN computer languages
— German: fluent

HONORS
— Dakota Interstate Scholarship
— Hubert H. Humphrey First Prize, Minnesota Intercollegiate Debate

INTERESTS
— Public speaking and debate
— Investments
— Writing
— Parachute jumping

FIGURE 7 A resume designed more to reflect an active, energetic personality than specific experience.

memberships, and honors all contribute to telling what a candidate is like; hence long vitae that reflect many achievements are traditional. The same candidate applying for an industrial position wouldn't list a raft of publications, for example, if those publications didn't relate to the job in question.

The C.V. that is shown in Figure 8 lays out in a logical way all the essentials of a good vita. Note the correct way to list publications. In academia, those who list them incorrectly are jeopardizing their chances of being hired.

The Job Objective

"Do I have to have a job objective?" According to my calculations, as of this writing I have been a career counselor for 2,288 weeks, and I have been asked this question at least six times a week, except for the 132 weeks when I was on vacation. When I answer, "No, I don't like the heading JOB OBJECTIVE," the sigh of relief is audible. The job seekers think I have let them off the hook for one of the most important parts of a resume. I haven't. With rare exceptions, a resume *should* open with an objective—it's the way it is stated that can be changed.

I show them the preceding Robertson, Meyers, and Smith resumes and tell them that I prefer the headline CAREER INTEREST, because it leads to a simple and direct way of stating the purpose of the resume. For example, I suggest they try rewriting the Robertson and revised Smith statements of interest as job objectives to see if they don't find it awkward. After a struggle they come up with the same kind of baloney found in the unrevised Jones and Smith resumes.

They listen politely. I may have helped them with a minor problem of phrasing, but I know that I really haven't dealt with their question. Then they tell me what I already know: They don't want to state a goal because they don't know what they want to do. I ask them to imagine themselves as the employers reading their resumes. If theirs is like the revised Nancy Jones resume, which is so obviously slanted toward biology, then employers can figure out what the resume is for without a stated objective. However, I wouldn't fool with a resume that didn't tell me at the outset what it was all about. When I have to start studying a resume to guess what the writer wants, I throw it in the wastebasket.

Next I point to the Zarote and Robertson resumes, which wouldn't make any sense without a statement of purpose. And I can't get up much enthusiasm for a letter as a substitute for an objective. A letter stating a purpose accompanied by a resume without purpose is a wasted letter.

I remember a young woman who, on short notice, got a chance to be interviewed by a recruiter from a large department store. Knowing very little about merchandising, she headed for the nearest department store, where several managers were kind enough to give her information and the loan of their trade journals. In three hours of investigation, her eyes were opened to an industry in which people pursued careers in training, employee relations, promotion, credit, public relations, merchandising, and other occupations. Several of these looked interesting, so thanks to word processing, she easily changed a few things on her resume, then listed her objective like this:

CAREER INTERESTS: Training, promotion, and public relations in a retail setting.

MICHELLE TRIO

Department of English
Athens College
Athens, NY 14850
607-555-4567

406 East Bates Street
Athens, NY 14850
607-555-7654

EDUCATION

Ph.D., 1984, Cornwall University.

Dissertation:	"The Anatomy of Sin: Violations of *Kynde* and *Trawbe* in *Cleanness*," directed by L. E. Cooper (DAI 40/09, p. 5046-A).
Major Subject:	Old and Middle English language and literature.
Minor Subject:	Medieval philology.
Courses:	Old English; *Beowulf;* seminars on the Junius Manuscript, the Exeter Book, and hagiography; Middle English literature; Chaucer; *Piers Plowman;* Medieval Latin; paleography; Old French; Middle High German; Old Icelandic; Dante.

A.M., 1979, with High Honors, Boston University.

B.A., 1977, *magna cum laude,* Honors in English, State University of New York at Stony Brook.

PROFESSIONAL EXPERIENCE

1985–present: Assistant Professor, Medieval and Renaissance Literature, Department of English, Athens College. Position includes: Engl 325 Chaucer, Engl 323 Triumph of English (a course that I instituted on the history of English), Engl 232 Medieval Literature, Engl 420 Shakespeare Seminar, Engl 231 Ancient Literature, Engl 107 Introduction to Literature: Myth, Legend, and Folktale.

1985: Assistant Professor, Department of English, Cornwall University Summer Program. One section of Practical Prose and Composition and training of a graduate teaching assistant.

1984–85: Lecturer, Department of English, Cornwall University. Two courses in Freshman Seminar Program: Shakespeare and Politics, Practical Prose Composition. I was Co-director of the latter, with responsibility for planning the syllabus, training and evaluating graduate teaching assistants, and leading staff meetings on problems and goals of teaching composition.

PUBLICATIONS

"Heroic Kingship and Just War in the Alliterative *Morte Arthure*," to be published in *Acta*, 11.

Articles on the Beatitudes, the Handwriting on the Wall, the Parable of the Marriage Feast, Sarah, and Sodom and Gomorrah in *Dictionary of Biblical Tradition in English Literature,* ed. David L. Jeffrey, to be published by Oxford University Press and W. B. Eerdmans.

"On Reading *Bede's Death Song:* Translation, Typology, and Penance in Symeon of Durham's Text of the *Epistola Cuthbert: de Obitu Bedae*." *Neuphilologische Mitteilungen,* 84 (1983), 171–81.

FIGURE 8   A Curriculum Vitae

PROFESSIONAL SERVICE AND ACTIVITIES

1. At Athens College
   1989–90: English Department Library Representative.
   1989–90: Member, Athens College Faculty Enrichment Committee.
   1988–90: English Department Personnel Committee.
   1987–89: English Department Committee on the London Center.

2. Elsewhere
   1985: Organizer and Chair, Latin Section, Northeast Modern Language Association. Topic: "Eschatology and Apocalypticism."
   1980–82: Organizer and Chair, *Quodlibet:* The Cornwall Medieval Forum.
   Memberships in MLA, Medieval Academy of America, International Arthurian Society.

HONORS

Charles A. Dana Fellowship for Excellence in Teaching, Athens College.

Goethe Prize in German Literature, Cornwall University.

George Lincoln Fellowship in Medieval Studies, Cornwall University.

Teaching Assistantship, Medieval Studies program, Cornwall University ("Medieval Literature in Translation").

Teaching Fellowship, English Department, Boston University ("Freshman Rhetoric and Composition").

LANGUAGES

Reading knowledge of Latin, Old French, Spanish, Italian, Middle High German, and Old Icelandic, in addition to the usual French, German, and Old and Middle English.

CREDENTIALS

Dossier may be obtained from the Educational Placement Bureau, Barnes Hall, Cornwall University, Athens, NY 14853.

FIGURE 8 *(Continued)*

Like most of us, her range of aptitudes was wide, so, like a chameleon, she showed the recruiter only those that seemed to match retailing.

One Page or Two?

If a resume can be kept to one page, so much the better. The length depends on the message. In reading resumes in which everything is jammed on one page with none of the white space or headings that can make them attractive and readable, I have wearied of trying to find information and have given up. On the other hand, it has never been the least bit tiring to lift a piece of paper and turn to a second page when scanning an interesting resume. A resume is an outline. It needs a white space. It needs headings that stand out. Don't sacrifice them for some arbitrary notion about a one-page maximum.

Additional Advice About Resumes

No matter how you develop your message, test it before you send it to employers. Get friends to give you a critique of your resume or vita, especially if they are in an occupation in which you hope to find a job. A word of caution, however. Unless you guide them, their critique may relate more to the ritual than to the message. One job hunter had modeled a resume after Janet Smith's. The critic took a red pencil and put all the dates of employment in the left-hand margin. That would have been a good idea if it had indicated long years of experience with a company such as 3M, signifying considerable experience with one of the best-managed companies in the country. But the dates in question referred to short-term summer jobs that were of no consequence to the message and cluttered the margin with information that distracted from the headlines.

One way to get a resume criticized is to hold it up a few feet from the reader and ask for comments on its appearance. Does it look neat? Is the layout pleasing? Does it look easy to read? Is the print good-looking?

Next, give the critics the resume to read. Let them make all the comments they want. You may pick up valuable ideas for improving its style and layout, but be careful you don't get caught up in inconsequentials. What you really want is to have your critics look at the resume as if they didn't know you. You might even show them a resume with an alias, then ask:

- What qualifications does this person have?
- What do you see this person doing with these qualifications?
- What kind of an employer would want to hire this person?
- Does the resume project an image of a certain kind of person? What kind? Aggressive? Thoughtful? Energetic? What?

In other words, ask your critics the most important question about your resume: "What message do you get about me?"

Steven Graber

The Basics of a Cover Letter

Steven Graber is the former managing editor of both the JobBank *and the* Adams Almanac *series.*

Your cover letter, like your resume, is a marketing tool. Too many cover letters are merely an additional piece of paper accompanying a resume, saying "Enclosed please find my resume." Like effective advertisements, effective cover letters attract an employer's attention by highlighting the most attractive features of the product. Begin by learning how to create an effective sales pitch. As with resumes, both the format and the content of your cover letter are important.

FORMAT

Before reading a word of your cover letter, a potential employer has already made an assessment of your organizational skills and attention to detail simply by observing its appearance. How your correspondence looks to a reader can mean the difference between serious consideration and dismissal. You can't afford to settle for a less than perfect presentation of your credentials. This chapter outlines the basic format you should follow when writing a cover letter and shows you how to put the finishing touches on a top-notch product.

The Parts of a Letter

Your cover letter may be printed on the highest-quality paper and typed on a state-of-the-art computer, but if it isn't arranged according to the proper format, you won't come across as a credible candidate. Certain guidelines apply when composing any letter.

Either of two styles may be used for cover letters: business style (sometimes called block style) or personal style. The only difference between them is that in business style, all the elements of the letter—the return address, salutation, body, and complimentary close—begin at the left margin. In personal style, the return address and complimentary close begin at the centerline of the page, and paragraphs are indented.

Return Address

Your return address should appear at the top margin, without your name, either flush left or beginning at the centerline, depending on whether you're using business style or personal style. As a rule, avoid abbreviations in the addresses of your cover letter, although abbreviating the state is acceptable. Include your phone number if you're not using letterhead that contains it or it doesn't appear in the last paragraph of the letter. The idea is to make sure contact information is on both the letter and the resume, in case they get separated in the hiring manager's office (this happens more often than you would expect!).

Date

The date appears two lines below your return address, either flush left or centered, depending on which style you're using. Write out the date; don't abbreviate. *Example:* October 12, 2000.

Inside Address

Four lines beneath the date, give the addressee's full name. On subsequent lines, give the person's title, the company's name, and the company's address. Occasionally, the person's full title or the company's name and address will be very long and can appear awkward on the usual number of lines. In this case, you can use an extra line.

The text of the letter below the date should be centered approximately vertically on the page, so if your letter is short, you can begin the inside address six or even eight lines down. If the letter is long, two lines is acceptable.

Salutation

The salutation should be typed two lines beneath the company's address. It should begin "Dear Mr." or "Dear Ms." followed by the individual's last name and a colon. Even if you've previously spoken with an addressee who has asked to be called by his or her first name, never use a first name in the salutation. In some cases, as when responding to "blind" advertisements, a general salutation may be necessary. In such circumstances, "Dear Sir or Madam" is appropriate, followed by a colon.

Length

Three or four short paragraphs on one page is ideal. A longer letter may not be read.

Enclosure

An enclosure line is used primarily in formal or official correspondence. It's not wrong to include it in a cover letter, but it's unnecessary.

Paper Size

As with your resume, use standard 8½- by 11-inch paper. A smaller size will appear more personal than professional and is easily lost in an employer's files; a larger size will look awkward and may be discarded for not fitting with other documents.

Paper Color and Quality

Use quality paper that is standard 8½ by11 inches and has weight and texture, in a conservative color like white or ivory. Good resume paper is easy to find at stores that sell stationery or office products and is even available at some drugstores. Use *matching* paper and envelopes for both your resume and cover letter. One hiring manager at a major magazine throws out all resumes that arrive on paper that differs in color from the envelope!

Do not buy paper with images of clouds and rainbows in the background or anything that looks like casual stationery you would send your favorite aunt. Do not spray perfume or cologne on your cover letter. Also, never use the stationery of your current employer.

Typing and Printing

Your best bet is to use a word processing program on a computer with a letter-quality printer. Handwritten letters are not acceptable. You will generally want to use the same typeface and size that you used on your resume. Remember that serif typefaces are generally easier to read.

Don't try the cheap and easy ways, like photocopying the body of your letter and typing in the inside address and salutation. Such letters will not be taken seriously.

Envelope

Mail your cover letter and resume in a standard, business-sized envelope that matches your stationery. Unless your handwriting is *extremely* neat and easy to read, type your envelopes. Address your envelope, by full name and title, specifically to the contact person you identified in your cover letter.

CONTENT

Personalize Each Letter

If you are *not* responding to a job posting that specifies a contact name, try to determine the appropriate person to whom you should address your cover letter. (In general, the more influential the person, the better.) Try to contact the head of the department in which you're interested. This will be easiest in mid-sized and small companies, where the head of the department is likely to have an active role in the initial screening. If you're applying to a larger corporation, your application will probably be screened by the human resources department. If you're instructed to direct your inquiry to this division, try to find out the name of the senior human resources manager. This may cut down on the number of hands through which your resume passes on its way to the final decision-maker. At any rate, be sure to include your contact's name and title on both your letter and the envelope. This way, even if a new person occupies the position, your letter should get through.

Mapping It Out

A cover letter need not be longer than three or four paragraphs. Two of them, the first and last, can be as short as one sentence. The idea of the cover letter is not to repeat what's in the resume. The idea is to give an overview of your capabilities and show why you're a good candidate for the job. The best way to distinguish yourself is to highlight one or two of your accomplishments or abilities. Stressing only one or two increases your chances of being remembered.

Be sure it's clear from your letter why you have an interest in the company— *so many candidates apply for jobs with no apparent knowledge of what the company does!* This conveys the message that they just want any job. Indicating an interest doesn't mean you should tell every employer you have a burning desire to work at that company, because these statements are easy to make and invariably sound insincere. Indicating how your qualifications or experience meet their requirements may be sufficient to show why you're applying.

First paragraph. State the position for which you're applying. If you're responding to an ad or listing, mention the source. *Example:* "I would like to apply for the position of research assistant advertised in the *Sunday Planet*" (or "listed on the Internet").

Second paragraph. Indicate what you could contribute to this company and show how your qualifications will benefit them. If you're responding to an ad or listing, discuss how your skills relate to the job's requirements. Don't talk about what you can't do. Remember, keep it brief! *Example:* "In addition to my strong background in mathematics, I offer significant business experience, having worked in a data processing firm, a bookstore, and a restaurant. I am sure that my courses in statistics and computer programming would prove particularly useful in the position of trainee."

Third paragraph. If possible, show how you not only meet but exceed their requirements—why you're not just an average candidate but a superior one. Mention any noteworthy accomplishments, high-profile projects, instances where you went above and beyond the call of duty, or awards you've received for your work. If you have testimonials, commendations or evaluations that are particularly complimentary, you may want to quote a sentence from one or two of them. *Example:* "In a letter to me, Dewayne Berry, president of NICAP Inc., said, 'Your ideas were instrumental to our success with this project.'"

Fourth paragraph. Close by saying you look forward to hearing from them. If you wish, you can also thank them for their consideration. Don't ask for an interview. If they're interested, they'll call. If not, asking won't help. Don't tell them you'll call them—many ads say "No phone calls." If you haven't heard anything in one or two weeks, a call is acceptable.

Complimentary close. The complimentary close should be two lines beneath the body of the letter, aligned with your return address and the date; Keep it simple—"Sincerely" followed by a comma, suffices. Three lines under this, type your full name as it appears on your resume. Sign above your typed name in black ink.

Don't forget to sign the letter! As silly as it sounds, people often forget this seemingly obvious detail. An oversight like this suggests you don't take care with your work. To avoid this implication if you're faxing the letter and resume directly from your computer, you can type your name directly below the complimentary close, without any intervening space. Then follow up with a hard copy of the resume and the signed letter, with your name typed in the traditional place under the signature.

TIPS FOR SUCCESSFUL COVER LETTERS

What Writing Style Is Appropriate?

Adopt a polite, formal style that balances your confidence in yourself with respect for the employer. Keep the style clear, objective, and persuasive rather than narrative. Don't waste space boasting instead of presenting relevant qualifications.

Example: "In addition to a Bachelor of Arts degree in Business Administration, I recently received a Master's, *cum laude,* in International Marketing from Brown University. This educational experience is supported by two years' part-time experience with J&D Products, where my marketing efforts resulted in increased annual product sales of 25 percent."

Tone: Reserved Confidence Is Always in Style

Think of how you'd sell your qualifications in a job interview. You'd probably think harder about what to say and how to say it than in an informal conversation. Above all, you'd want to sound polite, confident, and professional. Adopt a similar tone in your cover letter. It should immediately communicate confidence in your abilities. The trick is to sound enthusiastic without becoming melodramatic. Take, for example, the candidate who expressed his desire to enter the advertising field as "the single most important thing I have ever wanted in my entire twenty-three years of existence." The candidate who was actually offered the position began her letter as follows: "My extensive research into the industry, coupled with my internship and education, have confirmed my interest in pursuing an entry-level position in advertising."

Emphasize Concrete Examples

Your resume details the duties you've performed in your jobs. In contrast, your cover letter should highlight your most significant accomplishments. Instead of stating something like "My career is highlighted by several major achievements," use concrete examples:

"While Sales Manager at Shayko Chicken, I supervised a team that increased revenues by 35 percent in 18 months."

"I published four articles in *The Magical Bullet Newsletter.*"

"At MUFON Corporation, I advanced from telephone fundraiser to field manager to canvassing director within two years."

List tangible, relevant skills rather than personal attributes. A sentence like "I am fluent in C++, Pascal, and COBOL" is a good substitute for a vague statement like "I am a goal-oriented, highly skilled computer programmer." Avoid using "etc."—don't expect a potential employer to imagine what else you mean. Either describe it or leave it out.

Use Powerful Language

Your language should be hard-hitting and easy to understand. Your message should be expressed using the fewest words possible. As with your resume, make your letters interesting by using action verbs like "designed," "implemented," and "increased," rather than passive verbs like "was" and "did." Use simple, common language and avoid abbreviations and slang. Change "Responsible for directing" to "Directed" if appropriate. Also steer clear of language that's too technical or jargon-heavy. The first person who reads your cover letter may not possess the same breadth of knowledge as your future boss.

Avoid Catchphrases

In the course of a job search, it's tempting to use catchphrases you've picked up from advertisements or reference materials, phrases that sound as though they *should* go in a resume or cover letter. Many people are tempted to reach for expressions like "self-starter," "excellent interpersonal skills," and "work well independently or as part of a team."

Improve on these descriptions by listing actual projects and goals. For example, rephrase "Determined achiever with proven leadership skills" as follows: "Supervised staff of fifteen and increased the number of projects completed before deadline by 10 percent." Once you begin working, employers will discover your personal attributes for themselves. While you're under consideration, concrete experiences are more valuable than vague phrases or obscure promises.

Mention Personal Preferences?

Candidates often worry if, and how, they should include salary requirements and availability to travel or relocate. Refrain from offering salary information unless the advertisement you are responding to requires it. If you must include salary requirements, give a salary range rather than a number. Another option is to simply indicate that salary concerns are negotiable.

If you're applying to an out-of-state firm, indicate a willingness to relocate; otherwise, a hiring manager may question your purpose in writing and may not take the initiative to inquire.

Proof with Care

Mistakes on resumes and cover letters are not only embarrassing, they will often remove you from consideration (particularly if something obvious, like your name, is misspelled). No matter how much you paid someone else to type, write, or typeset your resume or cover letter, *you* lose if there is a mistake. So proofread it as carefully as possible. Get a friend to help you. Read your draft aloud as your friend checks the proof copy. Then have your friend read aloud while you check. Next, read it letter by letter to check spelling and punctuation.

If you're having it typed or typeset by a resume service or a printer and you don't have time to proof it, pay for it and take it home. Proof it there and bring it back later to get it corrected and printed.

If you wrote your cover letter with a word processing program, use the built-in spell checker to double-check for spelling errors. Keep in mind that a spell checker will not find errors like "to" for "two" or "wok" for "work." Many spell-check programs don't recognize missing or misused punctuation, nor are they set to check the spelling of capitalized words. It's important to still proofread your cover letter for grammatical mistakes and other problems, even after it's been spell-checked.

If you find mistakes, do not fix them with pen, pencil, or white-out! Make the changes on the computer and print out the letter again.

COVER LETTER BLUNDERS TO AVOID

The following discussion focuses on examples that have been adapted from real-life cover letters. Although some of these blunders may seem obvious, they occur far more often than one might think. Needless to say, none of the inquiries that included these mistakes met with positive results.

Unrelated Career Goals

Tailor your cover letter to the position you're applying for. A hiring manager is only interested in what you can do for the company, not what you hope to accomplish for yourself. Convey a genuine interest in the position and a long-term pledge to fulfilling its duties.

Example A (wrong way): "While my true goal is to become a professional dancer, I am exploring the option of taking on proofreading work while continuing to train for the Boston Ballet's next audition."

Example B (right way): "I am very interested in this proofreading position, and I am confident of my ability to make a long-term contribution to your capable staff."

Comparisons and Clichés

Avoid clichés and obvious comparisons. These expressions detract from your letter's purpose: to highlight your most impressive skills and accomplishments.

Examples of what not to do:

"My word processor runs like the wind."

"I am a people person."

"Teamwork is my middle name."

"Your company is known as the crème de la crème of accounting firms."

"I am as smart as a whip."

"Among the responses you receive for this position, I hope my qualifications make me leader of the pack."

Wasted Space

Since cover letters are generally four paragraphs long, every word of every sentence should be directly related to your purpose for writing. In other words, if you are applying for a position as a chemist, include only those skills and experiences most applicable to that field. Any other information weakens your application.

Examples of what not to do:

"As my enclosed resume reveals, I possess the technical experience and educational background to succeed as your newest civil engineer. In addition, I am a certified gymnastics instructor who has won several local competitions."

"I am writing in response to your advertisement for an accounting clerk. Currently, I am finishing an associate degree at Peacock Junior College. My courses have included medieval architecture, film theory, basic home surgery, and nutrition."

Form Letters

Mass mailings, in which you send a form letter to a large number of employers, are not recommended. This approach doesn't allow you to personalize each application. Every cover letter you write should be tailored to the position you're seeking and should demonstrate your commitment to a specific industry and familiarity with each employer. Mass mailings may indicate to a hiring manager that you're not truly interested in joining that organization.

Inappropriate Stationery

White and ivory are the only acceptable paper colors for a cover letter. Also, don't rely on graphics to "improve" your cover letter; let your qualifications speak for themselves. If you're a cat enthusiast, don't use stationery with images of favorite felines. If you're a musician, don't send a letter decorated with a border of musical notes and instruments.

"Amusing" Anecdotes

Imagine yourself in an interview setting. Since you don't know your interviewer, you wouldn't joke with him or her until you determined what demeanor was appropriate. Similarly, when writing, remain polite and professional.

Erroneous Company Information

If you were the employer, would you want to hire a candidate who confuses your company's products and services or misquotes recent activities? To avoid such errors, verify the accuracy of any company information you mention in your cover letter. On the other hand, if you haven't researched the company, don't bluff. Statements like "I know something about your company" or "I am familiar with your products" signal to an employer that you haven't done your homework.

Desperation

In your cover letter, sound determined, not desperate. While an employer appreciates enthusiasm, he or she may be turned off by a desperate plea for employment. However, a fine line often separates the two.

Examples of what not to do:
"I am desperately eager to start, as I have been out of work for six months."
"Please call today! I'll be waiting by the phone."
"I really, really need this job to pay off medical bills."
"I AM VERY BADLY IN NEED OF MONEY!"

Personal Photos

Unless you're seeking employment in modeling, acting, or other performing arts, it's inappropriate to send a photograph.

Confessed Shortcomings

Some job seekers mistakenly call attention to their weaknesses in their cover letters, hoping to ward off an employer's objections. This is a mistake, because the letter emphasizes your flaws rather than your strengths.

Examples of what not to do:

"Although I have no related experience, I remain very interested in the management consultant position."

"I may not be well qualified for this position, but it has always been my dream to work in the publishing field."

Misrepresentation

In any stage of the job-search process, never, *ever,* misrepresent yourself. In many companies, erroneous information contained in a cover letter or resume will be grounds for dismissal if the inaccuracy is discovered. Protect yourself by sticking to the facts. You're selling your skills and accomplishments in your cover letter. If you achieve something, say so, and put it in the best possible light. Don't hold back or be modest—no one else will. At the same time, don't exaggerate to the point of misrepresentation.

Examples of what not to do:

"In June, I graduated with honors from American University. In the course of my studies, I played two varsity sports while concurrently holding five jobs."

"Since beginning my career four years ago, I have won hundreds of competitions and awards and am considered by many the best hairstylist on the east coast."

Demanding Statements

Your cover letter should demonstrate what you can do for an employer, not what he or she can do for you. For example, instead of stating "I am looking for a unique opportunity in which I will be adequately challenged and compensated," say "I am confident I could make a significant contribution to your organization, specifically by expanding your customer base in the northwest and instituting a discount offer for new accounts." Also, since you're requesting an employer's consideration, your letter shouldn't include personal preferences or demands. Statements like "It would be an overwhelmingly smart idea for you to hire me" or "Let's meet next Wednesday at 4:00 P.M., when I will be available to discuss my candidacy further" come across as presumptuous. Job candidates' demands are rarely met with an enthusiastic response.

Missing Resume

Have you ever forgotten to enclose all the materials you refer to in your cover letter? This is a fatal oversight. No employer is going to take the time to remind you of your mistake; he or she has already moved on to the next application.

Personal Information

Do not include your age, health, physical characteristics, marital status, race, religion, political/moral beliefs, or any other personal information. List your personal interests and hobbies only if they're directly relevant to the type of job you're seeking. If you're applying to a company that greatly values teamwork, for instance, citing that you organized a community fundraiser or played on a basketball team may be advantageous. When in doubt, however, leave it out.

Choice of Pronouns

Your cover letter necessarily requires a thorough discussion of your qualifications. Although some applicants might choose the third person ("he or she") as a creative approach to presenting their qualifications, potential employers sometimes find this disconcerting. In general, using the first person ("I") is preferable.

Example A (wrong way): "Bambi Berenbeam is a highly qualified public relations executive with over seven years of relevant experience in the field. She possesses strong verbal and written communication skills, and has an extensive client base."

Example B (right way): "I am a highly qualified public relations executive with over seven years of relevant experience in the field. I possess strong verbal and written communication skills and have an extensive client base."

Tone Trouble

Tone problems are subtle and may be hard to detect. When reading your cover letter, patrol for tone problems by asking yourself, after each sentence, "Does this statement enhance my candidacy? Could a hiring manager interpret it in an unfavorable way?" Have a second reader review your letter. If the letter's wording is questionable, rewrite it. A cover letter should steer a middle course between extremely formal, which can come across as pretentious, and extremely informal, which can come across as presumptuous. Try to sound genuine, not stilted. When in doubt, err on the side of formality.

Gimmicks

Gimmicks like sending a home video or a singing telegram to replace the conventional cover letter may seem attractive. No matter how creative these ideas may sound, the majority of employers will be more impressed with a simple, well-crafted letter. In the worst-case scenario, gimmicks can even work against you, eliminating you from consideration. Examples include sending a poster-sized cover letter by courier service or a baseball hat with a note attached: "I'm throwing my hat into the ring!" Avoid such big risks; most hiring decisions are based on qualifications, not gimmicks.

Typographical Errors

It's easy to make mistakes in your letters, particularly when you're writing many in succession. But it's also easy for a hiring manager to reject any cover letter that contains errors, even those that seem minor. Don't make the mistake that one job-hunting editor made, citing his attention to detail while misspelling his own name! Here are a few common technical mistakes to watch out for when proof-reading your letter:

Misspelling the hiring contact's name or title in the address or salutation or on the envelope.

Forgetting to change the name of the organization you're applying to each time it appears in your application, especially in the body of the letter. For example, if you're applying to Boots and Bags, don't express enthusiasm for a position at Shoe City.

Indicating application for one position and mentioning a different position in the body of the letter. For instance, one candidate applying for a telemarketing position included the following statement: "I possess fifteen years experience related to the marketing analyst opening." Another mistake here is that the applicant didn't use "years" as a possessive: ". . . fifteen years' experience. . . ."

Messy Corrections

Your cover letter should contain *all* pertinent information. If, for any reason, you forget to communicate something to your addressee, retype the letter. Including a supplementary note, either typed or handwritten, will be viewed as unprofessional or, worse, lazy. For example, one candidate attached a "post-it" note to his cover letter, stating his willingness to travel and/or relocate. This and all other information must be included in your final draft. Also, avoid using correction fluid or penning in any corrections.

Omitted Signature

However obvious this may sound, don't forget to sign your name neatly in blue or black ink. Far too many letters have a typed name but no signature. Also, don't use a script font or a draw program on your word processor.

COVER LETTERS FOR SPECIAL SITUATIONS

Writing a cover letter can seem like an even more formidable task when you find yourself in what we call "special situations." Perhaps you lack paid job experience, have been out of the workplace to raise children, are concerned about possible discrimination due to age or disability, or are trying to enter a field in which you have no practical experience. The key to improving your cover letter in these special situations is to emphasize your strengths. Focus on your marketable skills (whether they were acquired in the workplace or elsewhere), and highlight impressive achievements, relevant education and training, and/or related interests. And, of course, you should take care to downplay or eliminate any information that may be construed as a weakness.

For example, if you're a "displaced homemaker" (a homemaker entering the job market for the first time), you can structure your cover letter to highlight the special skills you've acquired over the years while downplaying your lack of paid experience. If you're an older job candidate, use your age as a selling point. Emphasize the depth of your experience, your maturity, your sense of responsibility, and your positive outlook. Changing careers? Instead of focusing on your job history, emphasize the marketable skills you've acquired that are considered valuable in the position you're seeking. For example, let's say your career has been real estate and, in your spare time, you like to run marathons. Recently, you heard about an opening in the sales and marketing department at an athletic shoe manufacturer. What you need to do is emphasize the skills you have that the employer is looking for. Not only do you have strong sales experience, you're familiar with the needs of the company's market, and that's a powerful combination!

RESPONSE TO A "BLIND" ADVERTISEMENT

A form of classified advertisements, "Blind" advertisements do not list employer information and generally direct inquiries to a post office box rather than a company's address. Since you're not provided with a company name in a blind ad, your cover letter should sharply define your knowledge of the industry, position (if mentioned) and how your qualifications specifically match up to the stated requirements. In other words, tailor your letter to any information given. For example, consider a blind ad that reads:

> *Large-size law firm in need of paralegal with experience in legal research, writing briefs, and office administration.*

You need to target everything in your response: what you know of the operations of large-size firms; why you want to be and remain a paralegal; how much experience you have in legal research and writing; and exactly what office skills you have. Avoid longwinded passages that don't follow these guidelines. Without knowing your readers, you've caught their attention. They're more likely to invite you for an interview, and suddenly you're one step closer to getting the job.

COLD LETTERS

With a "cold" cover letter, you can directly contact potential employers without a referral or previous correspondence. Job-seekers most commonly use this type of letter to advertise their availability to hiring managers or personnel departments. Presumably, after researching your field, you will have devised a list of the top employers you would like to work for and gathered basic company information for each.

BROADCAST LETTERS

With a broadcast letter, well-qualified candidates can advertise their availability to top-level professionals in a particular field. The candidate attempts to entice the potential employer to consider his or her impressive qualifications for available positions. Although the broadcast letter discusses a candidate's background in detail, a resume is usually included. Since this type of letter is used primarily by seasoned executives, its tone should reflect the candidate's experience, knowledge, and confidence in his or her capabilities.

A candidate using the broadcast letter format might begin, "Are you in need of a management accountant who, in her most recent association, contributed to productivity improvements resulting in an annual savings of $20 million?" This attention-grabbing opening is effective only if the reader understands the significance of such an accomplishment. For this reason, broadcast letters are not recommended for those candidates conducting widespread job searches, where cover letters may end up in the human resources department rather than in the hands of a fellow industry executive.

LETTER TO AN EMPLOYMENT AGENCY

When searching for a job, many candidates rely on the help of employment agencies. These agencies offer services to a wide range of job-seekers, primarily for clerical or support staff positions. Letters addressed to employment agencies should focus on who you are, what type of position you are looking for and in what specific industry, and some of your strongest skills related to that field. For the agency to place you in an appropriate position, mention personal preferences, including geographic and salary requirements.

LETTER TO AN EXECUTIVE SEARCH FIRM

Although executive search firms actively recruit candidates for client companies, don't let this discourage you from writing. A well-crafted cover letter can alert an otherwise unknowing recruiter to your availability. Highlight your most impressive accomplishments and attributes and briefly summarize all relevant experience. If you have certain preferences, like geographical location, travel, and salary, mention them in your cover letter.

NETWORKING LETTERS

For the most part, networking letters refer to a third-party industry contact to garner the reader's attention and induce him or her to assist you in your job search. It is essential to achieve the right tone in your networking letters. Unless you are familiar with a contact, word your correspondence in a businesslike manner. In other words, do not use your addressee's first name or rely on an overly casual writing style. Likewise, if you have been in contact with this person recently, it could be useful to remind him or her, "It was great seeing you at the Chicago Writers' Convention last month" or "It's been several months since we bumped into each other on that flight to London. How are you?"

Many networking letters are written to an addressee whom the candidate has not met but has been referred to by a mutual acquaintance. In this case, immediately state the name of the person who referred you, such as "Jean Rawlins suggested I contact you." It is generally more effective to ask a contact with whom you are unfamiliar for assistance and names of people to contact than it is to ask for a job. Chances are, if your letter is politely persuasive, people will be interested in talking with you.

THANK YOU LETTERS

Your correspondence doesn't end with cover letters. Other types of letters, such as thank you letters, are often appropriate, even obligatory. It's acceptable to handwrite your thank you letter on a generic blank note card (but *never* a postcard). Make sure handwritten notes are neat and legible. If you're in doubt, typing your letter is always a safe bet. If you met with several people, it's fine to send each an individual thank you letter. Call the company if you need to check on the correct spelling of their names. Remember to keep the letters short, proofread them carefully, and send them *promptly*.

Margaret Riley Dikel and Frances E. Roehm

Your Resume on the Internet

Margaret Riley Dikel is the author of The Riley Guide *(www.rileyguide.com) and an internationally recognized consultant and expert on using the Internet for employment and career development.*

Frances E. Roehm is the webmaster of www.ChicagoJobs.org and manager of SkokieNet and the Citizens' Library of Illinois portal project.

When talking about the job-search process, the action of distributing a resume is rated the least effective of the four activities that make up a complete search. However, job seekers rate it as the second most productive thing they can do online. While we de-emphasize this activity, we do so because in terms of finding you a new job it is not as effective as networking, researching and targeting employers, or even reviewing job leads. Believe it or not, we actually want you to create a great resume for yourself and post it online. We just don't want you to think that this will lead you to job-search happiness.

Writing your resume, actually creating the one that employers and others will review, is one of the most important tasks in a job search, one place where you must take the time to do it right, really thinking about what you want to do before you can start pounding the pavement. Even with our statements on how ineffective it is to post your resume in the many databases online, if you are going to do it—and we think you should—you must do it right.

Your resume is your product brochure, the piece of paper that summarizes all the benefits of you and why the employer should "buy" you. A great resume can help you win the position you want. A bad one will knock you out of consideration, no matter how qualified you are. You must have a great resume, but we are not the ones to help you with that. At the end of this chapter you'll find a list of great books and online workshops offered by experts in this field. They are the ones to help you write a resume. Your local librarians, career counselors and coaches, and bookstore managers can recommend even more books and resources, so don't be afraid to ask should you not find something you like here.

What we're good at is helping you take that resume and get it online, so that is what this . . . [section] will cover. We'll also address the problems associated with posting a resume, how to format it so posting it in websites and e-mailing it to employers is fast and easy, and how to select where to post it. We encourage you to read this before you create your resume and then come back to it when you have your resume ready to go online. Consider the issues and advice we present here, and think about them as you are working on your resume. And remember, you do not need to limit yourself to just one resume. You can have several that are presented in different ways. If you have access to a computer with word-processing software, you are limited only by the available space on your disk or diskette.

Please Note: Unless indicated otherwise by the inclusion of *http://* you must add *www.* to the beginning of all URLs listed here.

THE MYTH ABOUT THE INTERNET RESUME

Many people think that the advancement of resume databases, resume-management systems, and keyword searching requires you to produce one resume for paper but an entirely different resume for online. *This is not true!*

When done correctly, your well-written, well-prepared resume will contain all of the necessary keywords to attract attention whether it is being read by a hiring manager or scanned and searched in any database, online or off. You still need only one resume, but now you want to have it in several formats, ready to produce in the proper form as needed, including these:

1. A designed or hard-copy version: a good-looking printed resume with bulleted lists, bold and italicized text, and other highlights, ready to send to contacts through the mail
2. A scannable version: a neat word-processed and printed resume without bullets, bold, italics, or other design highlights, written in a standard font and printed on white paper to send to employers who use scanning systems
3. A plain-text version: a no-frills plain-text file you keep on a diskette for copying and pasting into online forms and posting in online resume databases (See "Preparing a Perfect Plain-Text Resume" later in this [section].)
4. An e-mail version: another no-frills plain-text file you keep on disk, but one formatted to meet the length-of-line restrictions found in most e-mail systems, making it easy to copy and paste into an e-mail message and forward to an employer or recruiter in seconds (See "Preparing a Perfect Plain-Text Resume" later in this [section].)

You may also want to consider an HTML version of your resume. This can be posted on a personal website or with any site offering this kind of service. Many job seekers are creating webbed resumes in the hopes of being discovered. It's a great reference for employers who might want to see more than just that flat resume. This format works particularly well for those in a visual arts field, but if it is done correctly and for the right reasons it could serve anyone who wants to present more than what is usually found on a resume.

"Doing it correctly" means creating a simple HTML version of your designed resume, not a hip-hop page of spinning-whirling gizmos, dancing gerbils, and accompanying audio files that take more than two minutes to download over a 56K modem. "Doing it for the right reasons" means turning your resume into an employment portfolio, complete with links to former employers or projects that are already online. If you do this, you must be sure you are not violating any copyright or confidentiality clauses by putting project information online.

The main problem with HTML resumes is the too-much-information factor. Many job seekers make the resume a part of their personal website where there is often a lot of information an employer does not need to know—for example, your marital status, your ethnic background, or your personal interests. Allowing an employer to learn so much about you can lead to all kinds of problems, including discrimination against you for your physical appearance, political beliefs, religious practices, or even just the image you present. When you place your professional image online by posting your resume, it is important that you keep your presence entirely professional by never linking it to personal information of any kind. So if you decide to add an HTML resume to your campaign, post it in a location separate from your personal website, and do not create a link between the two.

RULES FOR RESPONDING ONLINE

The fastest way to respond to Internet job listings is by e-mailing your cover letter and resume to the person or organization indicated. Yes, your resume and cover letter are still your best bet for winning an interview, but if you mess up the application process, those great documents won't get you where you want to be. The rules are short and simple, so take a couple of minutes to review them before you hit the send key.

- Format your resume correctly for e-mail. If you try to copy and paste the text of your designed resume into the body of an e-mail message and then just send it without preparation, by the time it reaches the intended recipient the formatting will be such a mess it may be unreadable.
- Send your resume in the body of the e-mail message. Do not send it as an attachment unless specifically instructed by the recipient. You have only about twenty seconds to catch the eye of a recruiter or employer and to get him or her to read your resume. If you send your resume as an attachment, the recruiter has to find it and open it before he or she can read it. Your twenty seconds are over before they begin. Put the resume right in the message so the reader will see it immediately

upon opening the mail. This also helps you bypass e-mail systems that refuse attachments in this day of rampant computer viruses.

- Always include a cover letter, whether or not you are responding to an advertised opening. Make that cover letter specific to the person or organization you are contacting, and make it interesting. If you are responding to an advertisement, note where you found the advertisement and any relevant job codes. You can create and store a "standard" cover letter in text, but remember to customize it for each job listing for which you are applying, checking the format before you send it.
- Use the advertised job title or job code in the subject line of your e-mail message. This makes it easy for the recipient to sort everything coming in and route your resume and letter to the appropriate person. If you are cold calling, trying to get your resume into someone's hands without responding to an advertised job posting, put a few words stating your objective in the subject line.
- Read the application instructions included in the job announcement and follow them exactly. Sometimes employers want applications sent to a specific e-mail address according to the job location. They might require you to apply through their website using a specific code. It's even possible they will request you attach your resume in Word. Whatever they say, do it. You don't want your application to be delayed because you sent it to the wrong address or person, and you don't want to be perceived as someone who cannot follow directions.

Always remember this as you prepare to e-mail your resume to an employer: *It takes only a couple of seconds for someone to delete an e-mail message.* Don't let that happen to you. Read and think before you respond!

E-RESUMES ARE NOT JUST FOR E-MAIL

Besides the need to have a well-done plain-text resume for e-mailing, there are hundreds more reasons to take the time to create an e-resume, namely all those places online where you can post your resume. Yes, almost all sites have a copy-and-paste option for getting your resume online. Some even offer to let you build your resume right on the site, but resume expert Susan Ireland doesn't recommend using these forms for the following reasons:

1. It's very easy to have typos if you type directly into the site's form. Working first in your word-processing program (with its spell-check function) can greatly improve your chances of having a perfect resume.
2. The form may force you to use a resume format that you don't like. Most online resume builders insist on a chronological resume, a format that focuses on work history. This will put career changers at a disadvantage because the system doesn't allow you to build a functional resume, a format that focuses on skills.
3. You cannot easily save your resume for other uses because the resume bank is on a website. That means you'd have to repeat your resume-building efforts on each site where you want to post your resume.

The best way to post your resume online is to copy and paste it from a prepared copy you have already formatted to look great online. For the best results, that means transforming the hard-copy version of your resume before you copy and paste it into the website's resume form.

Job-Search Tip

Protect Your Privacy When Cutting and Pasting! Susan Joyce from Job-Hunt.org notes that people frequently sabotage their own privacy by copying the top of their resume (with all of their contact information) into the body-of-the-resume blocks on Web forms. While the job site blocks access to the contact information input into specific labeled fields, you the job seeker have accidentally revealed the information in the text block fields of the resume form. Be careful with the copy-and-paste process!

PREPARING A PERFECT PLAIN-TEXT RESUME

Preparing a resume for electronic mail is an easy process, and anyone creating a resume should take the extra few minutes needed to generate a plain-text version. Most word processors and resume-writing programs will let you save a file to plain text. The next step, altering the format, is simple. The following instructions prepared by Susan Ireland will help you take that hard-copy resume and turn it into a perfect plain-text document for posting online. Ireland even talks you through an easy way to format it for e-mailing. Why is this a separate process, and why do we suggest you have two different plain-text copies of your resume? Because e-mail has more formatting restrictions than most online resume databases, but we know what those restrictions are. To find even more complete instructions on creating resumes, cover letters, and even various formats for your e-resume, visit Ireland's website (susanireland.com).

Please note that these instructions assume that your resume is in MS Word for Windows. If your resume is in another word-processing application or on a different computer platform such as Macintosh, you may need to consult your word-processing manual for specific instructions.

Step 1: Check keywords. Be sure your resume has all the keywords that define your job qualifications.

Step 2: Save your resume as a Text Only document. A Text Only document works best for an electronic resume because you can adjust the margins and formatting to suit the database or e-mail system in which you are working. To convert your MS Word resume to Text Only, do the following:

1. Open the MS Word document that contains your resume.
2. Click File in your toolbar and select Save As.
3. Type in a new name for this document in File Name, such as "ResTextOnly."

4. Under this is the Save As Type pull-down menu. From this list, select "Text Only (*.text)." Users of Windows XP should select "Plain Text" from the pull-down menu.

5. Click Save to perform the conversion.

6. Now close the document but stay in MS Word.

7. Reopen the document you just closed by going to File in the toolbar, click Open, select the file named "ResTextOnly.txt," and click Open. *Warning:* if you exit MS Word and then open the resume document by clicking on its icon in the directory, it will be opened in Notepad, which will work for posting your resume on a website but is not what you want if you intend to use this version to prepare an e-mailable resume!

After converting your resume to Text Only, what appears in your document window is your resume stripped of any fancy formatting. You are now ready to make a few final adjustments before posting it online.

Step 3: Delete any page numbers. If your resume is more than one page, delete any indications of page breaks such as "Page 1 of 2," "Continued," or your name or header on the second page. You are making your resume appear as one continuous electronic document.

Step 4: Use all CAPS for words that need special emphasis. Since Text Only stripped your resume of all bolds, underlines, and italics used for highlighting words, use all capitalized letters to draw attention to important words, phrases, and headings. For the best overall effect, use all CAPS sparingly and judiciously.

Step 5: Replace each bullet point with a standard keyboard symbol. Special symbols such as bullet points, arrows, triangles, and check marks do not transfer well electronically. For example, bullet points sometimes transfer as "&16707," ")," or a little graphic of a thumbs-up. Therefore, you must change each to a standard keyboard symbol. Suggested replacements are:

Dashes (-)
Plus signs (+)
Single or double asterisks (*)(**)

Use the space bar to place a single space immediately after each symbol (and before the words). Do not use the tab key for spacing as you may have done in your original resume. Also, allow the lines to wrap naturally at the end of a line. Don't put a forced return (don't push the return or enter key) if it's not the end of the statement, and don't indent the second line of statement with either the tab key or space bar.

Step 6: Use straight quotes in place of curly quotes. Like bullet points and other special symbols, curly or "smart" quotes do not transfer accurately and in fact many appear as little rectangles on the recipient's screen. So you should replace curly quotes with straight quotes. To do this, select the text that includes the quotes you want to change. Click Format in your toolbar and select

AutoFormat. Click the Options button, and make sure "Replace Straight Quotes with Smart Quotes" is not selected under both the AutoFormat and AutoFormat As you Type tabs. Then click OK to exit the AutoFormat box, and your curly quotes will be changed to straight quotes.

Step 7: Rearrange text if necessary. Do a line-by-line review of your document to make sure there are no odd-looking line wraps, extra spaces, or words scrunched together in the body. Make adjustments accordingly. This may require inserting commas between items that were once in columns and are now in paragraph format because tabs and tables disappeared when the document was converted to Text Only.

Now that you have the plain-text resume for posting, it takes just a few more steps to create a perfect plain-text resume for e-mailing. Again, if you take the time to do this now, you will save yourself a lot of time later.

Step 8. Limit line lengths. Because each type or e-mail software has its own limit for the number of characters and spaces per line, your e-mail may have longer line lengths than the receiver of your e-mail allows. This can cause the employer to see line wraps in unusual places, making your resume document look odd and even illogical. To avoid this problem, limit each line to no more than sixty-five characters (including spaces). Here's an easy way to make line length changes in your document:

1. Open MS Word, click Open, select the file named "ResTextOnly.tex," and click Open. Warning: If you open the resume document by clicking on its icon in the directory, it will be opened in Notepad—not what you want right now.
2. Select the entire document and change the font to Courier, 12 point.
3. Go to File in your toolbar; select Page Setup; set the left margin at 1 inch and the right margin at 1.75 inch. (Yahoo! e-mail users should set the right margin at 2.5 inch.)

With the side margins set under these conditions, each line of your document will be no more than sixty-five characters and spaces.

Step 9: Save as Text Only with Line Breaks. To save the line-length changes you made in Step 8, you need to convert your Text Only document one more time by doing the following:

1. With your Text Only resume document open, click File in your toolbar and select Save As.
2. Type in a new name for this document in File Name, such as "ResTextBreak."
3. Directly under this is the Save As pull-down menu. From this list, select "Text Only with Line Breaks (*. text)." Windows XP users should save the document as Plain Text. When the File Conversion window appears, click "Insert line breaks" under Options, then click OK.
4. Click Save to perform the conversion.
5. Now close the document and exit MS Word.

6. Reopen the resume document (ResTextBreak.txt) by clicking on its icon in the directory. That will open it as a Notepad document.
7. Select the entire document and change the font to Times, Arial, or some other standard font you like. Don't worry that the margins automatically reset when you reopen your Text Only with Line Breaks document. Your line lengths are safely preserved by paragraph returns that were inserted by the conversion.

Step 10: Copy the entire text in your ResTextBreak.txt document that you've opened in Notepad, and paste it in the body of the e-mail message. Now that you have redone your resume in the e-mail format, e-mail it to yourself and to a friend to see how it looks after going through the Internet. This will help you identify any additional formatting problems you need to correct before you start sending it out to possible employers.

Resume Tip

Never use your current office address, e-mail address, or phone number on your resume. Employers consider the personal use of company time and resources to be stealing, and they also believe that if you'll do it to your current employer, you'll do it to the next one, too. Always use personal contact points, such as a post office box, cell phone, and a personal e-mail account. This will also help you avoid possible monitoring by your current employer.

WHERE, OH, WHERE SHOULD THAT RESUME GO?

With the hundreds, if not thousands, of possible posting sites now available online, you have ample opportunity to saturate the Internet with your resume. After all, don't you want to get your resume in front of every recruiter or employer you can, regardless of who they are? No, you don't.

Recruiters are tired of finding the same resumes for the same people in every database they search. They are even starting to ignore these resume spammers, refusing to give them any consideration for possible job openings. There is also the danger that the further your resume spreads, the less control you can exert over it. To make sure you don't encounter these problems, limit your resume exposure by limiting your postings.

- Post it on only one or two of the large online databases, preferably ones attached to popular job sites. This will give you maximum exposure to many employers and recruiters.
- Post it on one or two targeted resume databases specific to your industry, occupational group, or geographic location. This will give you a targeted exposure to employers and recruiters looking for a smaller yet more highly qualified candidate pool.

If you don't get any responses to your resume within forty-five days, remove it from its current locations and place it elsewhere.

Protect Yourself Online

Limiting the number of locations is a good way to protect your resume, but it is also important to select those few sites with care. Susan Joyce, the editor of Job-Hunt.org, encourages careful evaluation of the job sites you use, because if you aren't careful, you risk a total loss of privacy. Not only could your resume become visible to anyone who comes across it, but also your personal information might be sold to people who have products and services to sell you. Joyce's article "Choosing a Job Site" (http://job-hunt.org/choosing.shtml) outlines fourteen criteria designed to help you evaluate a site before you trust it with your resume, including the following:

1. **Does the site have a comprehensive privacy policy?** Look for a privacy policy, and *read it* before you register at a job site! The privacy policy should disclose to you the information that the site collects and what they do with it (e.g., sell or rent your e-mail address, etc.). Pay particular attention to what happens to your resume!

2. **Do you have to register a profile or resume before you can search through the jobs?** Be suspicious of a site that won't let you perform a job search before you register your profile or resume. You need to evaluate the site to determine if it has the jobs you want *before* you register.

3. **Are most of the jobs posted by employers or by agencies acting on behalf of employers?** In general, jobs posted directly by an employer are preferable because you will be dealing directly with the people who can hire you.

4. **Can you set up one or more "e-mail agents" that will send matching jobs to you when you are not at the site?** E-mail agent functions typically compare your requirements with new employer job postings and send you the results via e-mail if they find a match. You don't need to revisit the site yourself and run your search. Your agents will do the searching for you and send you the results.

5. **Who has access to the database of resumes?** The privacy policy should tell you who has access to the resumes. In addition, you can check out the "employer" side of the job site to see how easy it is to gain access to the resumes. If resume access is free, or there is only a nominal fee for access to the resumes, find another job site.

6. **Can you limit access to your personal contact information?** The best sites provide you with options to protect your contact information (name, e-mail address, street address, phone numbers, etc.). Options range from blocking access only to the contact information to keeping your resume completely out of the resume database searched by employers. Choose the option that works best for you. If you are currently employed, limiting access can help you protect your current job.

7. **Can you store more than one version of your resume so that you can customize it for different types of opportunities?** Many sites offer you the ability

to store several different resumes and apply for a job using the version of your resume you have developed for that specific kind of opportunity.

8. **Will you be able to edit your resume once you have posted it?** You shouldn't run into this very often anymore, but check to see if there is an update option for you to access your resume. You can always find ways to improve your resume, and they should allow you to do it.

9. **Will you be able to delete your resume after you have found a job?** You don't want that old resume still available for view. If your new employer finds it, he or she may be concerned that you are getting ready to leave. Good job sites provide you with the capability to delete your resume and account or to put your resume in an "inactive" mode until you are ready for your next job search.

The bottom line when posting your resume is that *you rule*. Many sites want your resume in their database. You can afford to be choosy about where you will place it and which sites you will use in your search.

Before You Post, Something to Think About

For some people, posting a resume online is a great way to find new opportunities. For others, there is a very real fear that their contact information will fall into the wrong hands, or that the wrong organizations—such as their current employer—will see their resume online and problems will arise. We are also all familiar with news reports about stolen identities. You are the only one who can say how comfortable you are with this decision and how you want to approach the idea of posting your resume. Before you begin, . . . consider the following questions very carefully:

1. **Do you want your resume public?** Once you have posted it, no matter where you place it, you should consider your resume to be public and to some extent out of your control. Anyone can look in the public databases and see what is there. Even the private resume databanks as well as those offering confidential handling of your resume may not let you dictate who can and cannot look at your resume.

2. **Are you prepared for the consequences should an electronic resume come back to haunt you?** It is a real possibility. Some job seekers fall victim to aggressive recruiters who grab their resumes from the Internet and unwittingly feed them back to the job seekers' current employers, with very bad results for the employee. Others are finding their resumes posted in places they never put them, the victims of unprofessional resume services who copy the documents from other open venues. Some employees have even been confronted by current employers brandishing copies of resumes online, not realizing that the documents were more than a year old and part of the campaign to get the current job. You should always go back and delete any resumes you posted during your search as soon as your search is over, but you might not be able to get to every electronic copy out there so you may need some strategies to help fight this. To help alleviate some of this problem put the date of posting at the very end of your resume

as a record of when it was posted or make slight changes to the wording of each copy you post, creating a code identifying where a copy originated. These small alterations will give you some ammunition should your resume float into the wrong hands at the wrong time.

We don't want to scare you away from posting your resume online, but you should be aware of problems that can occur. Susan Joyce of Job-Hunt.org has written two articles you should review before going forward with this task: "Your Cyber-Safe Resume," which offers tips on creating a confidential resume, and "Protecting Your Privacy," which is about ways to protect your identity when job searching online. Both articles and many more on this topic are available on her website (http://job-hunt.org).

RESUME BLASTERS: THE WAVE OF THE FUTURE OR A NEW FORM OF SPAM?

Resume-distribution services, sometimes called resume broadcasters, are proliferating online. While you may think this is a great way to get your resume seen, we disagree. In reference to your privacy, Susan Joyce feels that "such wide distribution may offer little, if any, control on where a copy of your resume could end up. Your name, address, and phone number, in addition to your education and work experience, could become completely public for a very long time."

There are other problems too. For one thing, not all those who are on the lists of these services actually requested they be placed on the list. Both authors have received resumes that were broadcast by these services; neither of us has requested such a service. Then there may be problems with the resumes that are sent out, problems that the job seekers may not be aware of because they were not allowed to review their resumes before they were broadcast. One hiring manager sent us the following e-mail commenting on her experience with a resume-broadcasting service.

"A recruiting site for IT and technical professionals started bombarding me with forwarded resumes two days ago (I must have received twenty or so). They told us it was a free service, which I believe to be true.

"Because I had no positions open, and because we respond to every resume we receive, this put a burden on us that we don't really have time to handle, so I asked them to stop sending the resumes. In addition, a large number of the forwarded resumes did not contain contact information for the applicants, which made it impossible for us to let them know we had received the resume, or to contact them in case we might be interested in them. Two of the resumes did not even contain the candidates' names.

"I received an e-mail later saying they had complied with the request to remove my e-mail address from their system. Well and good.

"It went on to say, however, that these were 'current resumes of candidates presently in the job market and who have paid to have their resumes reviewed by recruiters, right now.'

"That outraged me. If an organization is going to collect money from candidates to forward their resumes, then it has (I believe) an obligation to make the process of reaching a promising candidate possible. Resumes without names or contact information are useless to a recruiter, as you can well imagine. I think these candidates should know that they are paying (a) for recruiters without jobs, or (b) to have their resumes forwarded in a way that cannot be responded to. They should use their money for something else."

In this instance, resumes were being sent out without any regard as to whether or not the employer requested it. While the chances of a successful match through this activity are already low, broadcasting resumes to employers who have expressed no interest in receiving them is almost guaranteed to eliminate even the slimmest possibility those resumes will be given any consideration. Then there were problems with the contact information on some resumes. This makes it difficult, if not impossible, for the employer to contact these candidates. If you are going to cold-call an employer, you must give the employer a way to call you back. But the worst part of all this is job seekers actually paid for this service, using their hard-earned and probably limited funds to pay for a service that will provide almost no return on their investment. It is a losing proposition for both the job seeker and the employer. The only winner was the blasting service because it got the money.

HELP WITH RESUMES AND COVER LETTERS

Each of the following books and Internet services has good information and guidance for preparing your resume. Almost all will walk you through the process of translating your designed resume into the necessary scannable and e-mail formats, and a couple will even take you into Web resumes. We've also included titles covering resumes for teenagers, international variations, and resumes for positions with the U.S. government. New editions of any of these may have been released between the time we created the list and when you are reading it, so check your local library or bookstore for the most recent edition.

- Block, Jay A., and Michael Betrus. *101 Best Cover Letters.* McGraw-Hill, 1999.
- Criscito, Pat. *E-Resumes: A Guide to Successful Online Job Hunting.* Barrons, 2004.
- Ireland, Susan. *The Complete Idiot's Guide to the Perfect Cover Letter.* Alpha Books, 1997.
- ———. *The Complete Idiot's Guide to the Perfect Resume,* third edition. Alpha Books, 2003.
- Kennedy, Joyce Lain. *Cover Letters for Dummies.* For Dummies, 2000.
- ———. *Resumes for Dummies.* 4th ed. For Dummies, 2000.
- Thompson, Mary Anne. *The Global Resume and CV Guide.* John Wiley & Sons, 2000.

- Troutman, Kathryn Kraemer. *Creating Your High School Resume*, 2nd ed. Jist, 2003.
- ———. *The Federal Resume Guidebook*, 3rd ed. Jist, 2004.
- Yate, Martin. *Resumes That Knock 'em Dead*. Adams Media, 2004.

ONLINE GUIDES AND GUIDANCE

Online Writing Lab (OWL), Purdue University

http://owl.english.purdue.edu/owl

OWL was set up to help the students at Purdue with writing all types of documents. Under Handouts and Materials, the section called Professional Writing includes help with resumes, cover letters, offer acceptance and refusal letters, personal statements, references, and postinterview letters.

The Damn Good Resume

damngood.com

This is the online companion to the many resume books written by Yana Parker. Now operated under the direction of Yana's family, it still offers her outstanding samples of good resume writing along with excellent advice on preparing your own resume or helping others with theirs.

The Resume Place

resume-place.com

The Resume Place is the resume-writing service operated by Kathryn Troutman, author of *Ten Steps to a Federal Job* and *Creating Your High School Resume*. On her website you'll find free articles and advice on preparing what you'll need in order to create your resume. If you are considering applying for a job with the federal government, then you must review Troutman's information, as she is an expert in the federal resume (which is very different from the resume you need for the private sector).

Susan Ireland

susanireland.com

Susan Ireland's website has terrific information and samples for the job seeker. You will enjoy her online workshops for resume writing, e-resumes, and cover letters, along with the many samples of resumes, cover letters, and thank-you letters.

Karl Weber and Rob Kaplan

Follow-Up Letters

Karl Weber is an editor, author, and publishing consultant specializing in books dealing with business, personal finance, and current affairs.

Rob Kaplan has had a long career as a senior editor for several major publishing companies and now heads his own literary services firm.

. . . Follow-up letters provide you with an invaluable opportunity to impress your readers with your skills, your background, your professionalism, and your personality.

In this . . . [section], we'll show you why writing follow-up letters is so important. We'll help you write effective, professional letters for a variety of situations, including letters to follow-up networking and job interviews, letters for when you've been offered and accepted a position, letters for when you've been offered and turned down a position, letters thanking your interviewer even when you've not received a job offer, and letters to follow up with your networking contacts after you've started a new job.

NETWORKING INTERVIEW FOLLOW-UP LETTERS

It's always appropriate to thank someone, especially someone who doesn't know you well, for taking time out of their schedule to talk to you. And this is true whether they were considering you for a job or not. Although the individual who is actively interviewing potential job candidates obviously has a business interest in talking with you, they're still under no obligation to spend time with you. And people who *aren't* recruiting to fill a position certainly don't have to give you an hour of their valuable time. So thanking these people is courteous, considerate, and appropriate.

Michael Broderick
919 Massachusetts Avenue
Brighton, MA 07122
617-750-9086
E-mail: mikeb@bosnet.com

August 18, 2000

Ms. Corinne Blackman
Nehan, Ross & Blackman
22 Boylston Street
Boston, MA 07125

Dear Ms. Blackman:

I am writing to tell you what a pleasure it was to meet you the other day and to thank you again for taking the time to speak with me.

The information, ideas, and names you were kind enough to share with me will, I'm sure, be extremely helpful, and I appreciate it very much.

I will be in touch with you again to let you know how my job search is progressing. In the meantime, should you hear of any openings for which I might be appropriate, I would very much appreciate hearing from you.

With best wishes,

Michael Broderick

FIGURE 1   Sample Networking Interview Follow-Up Letter

More specifically, in the case of networking interviews, sending a follow-up letter [Figure 1] serves the purpose of reminding the reader who you are and the fact that you're looking for a job. Remember, the person you met with may have a job opening in the near future, and if you've sent a follow-up letter they're more likely to remember you. In addition, because executives in any given industry tend to know each other, the person you follow up with may contact you if he or she hears of a position in another company for which you would be qualified.

Like the other types of letters you've already written, the networking interview follow-up letter can be divided into three elements, or paragraphs—in this case, the opening, your comments on the meeting, and the closing.

FYI

All follow-up letters, regardless of their purpose, should be sent as soon as possible after the meeting—preferably the next day, but certainly within the next few days. It's not only courteous, it's a good way to make sure that you won't forget to do it.

The Opening

The opening should be a brief, straightforward, one- or two-sentence paragraph in which you thank your reader for taking the time to meet with you, such as:

> I'm just writing to say what a pleasure it was to meet you yesterday and to thank you again for spending some time with me.

> I'm writing to say how much I enjoyed our meeting the other day and to thank you for taking the time to help me in my efforts to find a new position.

Comment on the Meeting

The second paragraph of the follow-up letter is the appropriate place to make some comment about the discussion you had. Even if the networking meeting didn't result in your actually gaining a great deal of information, it's still appropriate for you to say something positive about it, for example:

> The information, ideas, and names you were kind enough to share with me will, I'm sure, be extremely helpful, and I appreciate it very much.

> Having the opportunity to discuss the industry with you, as well as hearing your thoughts about its future, was interesting, informative, and enjoyable.

On the other hand, the person you met with may have made some particularly interesting comment, given you an especially good idea, or provided you with several names of people to contract. In this case, it would be appropriate to be more specific about the discussion:

> Your comments about recent developments in the industry were particularly interesting and thought-provoking. I had not considered the impact of gun control legislation on the business and am now beginning to see it in a new light.

> Your suggestion that I contact the National Booksellers Association concerning entry-level positions with their member companies was an especially valuable one, and I will certainly follow up on it.

> In addition to the other valuable suggestions you made, your providing me with the names of so many people to contact will no doubt be enormously helpful, and I sincerely appreciate it.

Finally, if either you or the person you met with promised to do something in the course of the meeting, this is the appropriate place to do what you promised or (tactfully) remind the other person of his or her promise:

As I promised I would, I'm enclosing the article from the *San Francisco Examiner* about Peter Anderson's new film.

I very much appreciate your offer to contact Steven Ryan at the Ryan Company on my behalf, and I will follow up with him within the next week or so.

The Closing

The final paragraph of the follow-up letter should include both a promise and a request, for example:

I'll be in touch again to let you know how my job search progresses. In the meantime, should you hear of any positions in which you think I might be interested, I would very much appreciate hearing from you.

I will let you know how my efforts to find a new position proceed. Should you, in the meantime, learn of any positions for which you think I might be appropriate, I would appreciate your letting me know.

FYI

Remember that the ordering of the elements of the letter is less important than the inclusion of all the elements. Although having a clear opening and closing are essential, the other elements can be moved around and/or combined in a variety of ways.

JOB INTERVIEW FOLLOW-UP LETTERS

It's even more important to send follow-up letters after a job interview than after a networking interview [Figure 2, p. 338]. One reason is that, since many job candidates do *not* send them, interviewers tend to more clearly remember those who do. It's always important to find a positive way to stand out in the interviewer's mind; sending a well-crafted follow-up letter is a excellent way to do that.

Writing a follow-up letter after a job interview also gives you an opportunity to remind the interviewer of your skills and experience, as well as of their applicability to the position for which you've been interviewed.

As you'll see, there are some similarities between networking and job interview follow-up letters, but one important similarity is timing. Writing *promptly* after a job interview is essential. You have no way of knowing how quickly the interviewer is going to make a hiring decision, so you should always send a follow-up letter within 24 hours of the interview. After all, if the letter is received while the recruiter is still trying to choose a candidate, it could be the thing that tips the scales in your favor.

In certain respects, letters you send after job interviews are like the job-tailored letters you sent to request interviews. Like those letters, the job interview follow-up letter includes three elements divided into four paragraphs: the opening, the two-part pitch ("Why me?" and "Why you?"), and the closing.

Cecilia Arnold
5550 Nepperham Avenue
Los Angeles, CA 90233
206-788-8562
E-mail: cece@lanet.com

March 15, 2000

Mr. Brock Peters
Braintree Publishers
1000 Sepulveda Boulevard
Los Angeles, CA 90256

Dear Mr. Peters:

I just wanted to tell you how much I enjoyed meeting you yesterday and to thank you again for spending time with me. It's clear that Braintree's plans for the future will make it an exciting and challenging place to work, and I would very much like to be a part of it.

On the basis of our conversation, I believe that my experience in the industry, my managerial experience, and my experience and interest in acquiring and developing popular reference books would enable me to make a real contribution to your organization as a publishing director. In addition, I sense that you and I share a common philosophy about publishing and that we would accordingly be able to work well together.

I understand that you are not expecting to reach a decision about the position for some time. If, however, I haven't heard from you within the next two weeks, I'll give you a call.

Sincerely yours,

Cecilia Arnold

FIGURE 2 Sample Job Interview Follow-Up Letter

The Opening

As with the networking interview follow-up letter, the first paragraph of the job interview follow-up letter is the place to thank the reader for meeting with you, for example:

> I'm writing just to say what a pleasure it was to meet you and to thank you for taking the time to discuss with me the possibility of joining your organization.

> I'm taking this opportunity to tell you how much I enjoyed meeting you and how much I appreciate your speaking with me about the opening for an assistant manager in your department.

The Pitch: "Why Me?"

Here is where the real difference between networking and job interview follow-up letters begins. The purpose of the networking letter is basically to thank the interviewer and remind him or her of your interest in finding a new position. By contrast, the purpose of the interview follow-up letter is to make a final argument for your candidacy. In this paragraph you should reiterate the (one or two) best points that you've already made in the interview itself as well as add (one or two) others you may have neglected to mention. A couple of examples:

> On the basis of our conversation, I believe that my experience in the industry, my managerial experience, and my agreement with the goals of your organization would enable me to make a real contribution in your company as an assistant manager. The fact that I've managed a similar operation in a smaller company should also, I think, make me a good candidate for the position.

> As we discussed, the fact that I have managed a similar operation, although in a smaller firm, has provided me with the opportunity to learn and use the same skills that you are looking for in an assistant manager. In addition, the experience I gained in working directly with customers in my previous position would, I think, be very useful in dealing with the sometimes difficult suppliers you mentioned.

> One element of my background that we didn't have time to discuss is the year I spent managing the customer database for Little Industries, my previous employer. The experience I gained in managing information technology would be directly applicable to the business reorganization project you mentioned, and I think it could help me be an effective member of the team charged with implementing it.

FYI

Use the follow-up letter as a way to highlight any credential or skill you may have failed to mention in the interview itself. If, like most people, you've ever walked out of an interview and said to yourself, "I wish I'd said such-and-such," a follow-up letter is an excellent way to remedy the situation.

The Pitch: "Why You?"

The third paragraph of the job interview follow-up letter is the appropriate place for you to discuss why you're interested in working for this particular company. Your comments should be based on both whatever you learned about the company during the interview and any information you may have gathered from other sources. If there's been any (positive) "breaking news" about the company, here is a good place to mention it.

> From what you told me about your organization, it sounds like a dynamic and exciting place to work. In addition, having heard so many positive things about it from other people in the industry, I believe that it's a company in which I would not only be comfortable but also one in which I would be able to make a substantial contribution.

During our discussion I was particularly impressed with your company's ambitious goals. As we both know, ours is a particularly competitive industry, but I'm sure that with the excellent staff you mentioned you will be able to achieve those goals. On the basis of what you told me, I would very much like to be a part of that effort.

As we discussed, I'm excited about the possibility of working for Mammoth Films. I noted that two of your new releases were among the top ten box-office hits again last weekend—an impressive showing for such a young company. I hope I'll have the opportunity to help contribute to even greater achievements in the future.

The Closing

The last paragraph of the job interview follow-up letter should be short and sweet. You've made your pitch, given it your best shot, and now you just have to wait for the interviewer to make a decision. (Well, more or less.)

Toward the end of the interview, you should have asked approximately how long the interviewer expects it will take to reach a hiring decision. Thus, you'll have at least some idea of her time frame. Bear in mind, though, that many interviewers say, "We want to move quickly on this decision," and then don't. It's not that they're lying, but, rather, that things always take longer than people expect them to.

Although it's unlikely that anything you do will make her come to a decision any sooner, it doesn't hurt to prod her a bit. So the closing of your letter might read:

> Again, I appreciate you taking the time to meet with me and look forward to hearing from you soon. If I haven't heard from you within the next few weeks, I'll give you a call to follow up.

If it's true, you can subtly suggest that you might be in demand from other employers, which could increase the sense of urgency on the recruiter's part:

> Thank you again for meeting with me. As I continue to explore other potential positions, I'll look forward to hearing from you shortly. If we haven't spoken in two weeks' time, I will call you with an update on my status and to find out how your decision-making process is progressing.

FYI

Following up with a phone call to an interviewer a week or so after a meeting is acceptable, but calling him day after day is not. Not only is it unprofessional, but calling frequently is likely to hurt rather than help your cause.

JOB ACCEPTANCE FOLLOW-UP LETTERS

While sending a follow-up letter after you've been offered and accepted a position [Figure 3] may seem gratuitous (after all, you got the job!) it's still a good idea to do so. Because writing such a letter is both courteous and professional, it's likely to impress and please your new boss. It will help to confirm in his or her

---

Joshua Kriegel
224 Sansom Street
Phoenix, AZ 04405
602-890-8165
E-mail: joshk@aol.com

February 17, 2000

Ms. Marjorie Simpson
Phoenix Power & Light
2000 Arizona Avenue
Phoenix, AZ 04412

Dear Marjorie:

I'm just writing to tell you how pleased I am that we were able to come to an agreement about my joining your staff at Phoenix Power & Light as a senior planner.

I am looking forward to working with you and helping to find efficient and cost-effective ways of meeting the challenges of the increasing need for electricity in our growing community.

I will be leaving my current position a week from now, and then taking the short vacation I mentioned prior to my starting the job on March 1st.

Sincerely yours,

Joshua Kriegel

FIGURE 3 Sample Job Acceptance Follow-Up Letter

mind that they've made a good decision in hiring you, and it'll help you start your new job on a positive note.

The job acceptance follow-up letter is very different from the other types of follow-up letters we've discussed. Since you've already closed the sale, it's neither necessary nor appropriate to include any kind of sales pitch. Your letter should, however, be enthusiastic, warm, and to-the-point.

JOB TURNDOWN FOLLOW-UP LETTERS

It's entirely possible that you'll receive job offers for positions that you're not interested in accepting. It may be that they're not paying as much money as you'd like, or that you've simultaneously received an offer from another company that you'd prefer working for, or for any number of other reasons.

Whatever the reason may be, in such a situation it's likely that you'll turn down the offer over the phone (since companies extending job offers generally need a fairly prompt response). It is still, however, advisable to write a follow-up letter [Figure 4] for one very good reason.

Most industries tend to be small worlds. That is, as people move from one company to another within any given industry (and they do) they get to know a lot of other people in the industry. Moreover, they tend to run into those people again and again over the years. So the chances are that you may at some time in the future find yourself being interviewed again by, or working with, a person whose offer you rejected. When that happens, it won't hurt if they remember that you turned them down in a graceful and professional manner.

FIGURE 4   Sample Job Turndown Follow-Up Letter

JOB REJECTION FOLLOW-UP LETTERS

While it may seem odd to write a letter thanking someone for *not* hiring you, it's a good idea to send follow-up letters even when you *haven't* received a job offer [Figure 5]. Hard though it may be to do, writing such letters may well bear fruit at some later date.

For example, although you may have no way of knowing it, you may have been your interviewer's second choice, and just a tad behind his or her first. (You may even have been the first choice of some people involved in the hiring decision, who were overruled by the boss.) If they have another opening in the near future, you'll probably be the first person they'll call anyway—especially if you send a gracious, professional letter thanking them for considering you for the position.

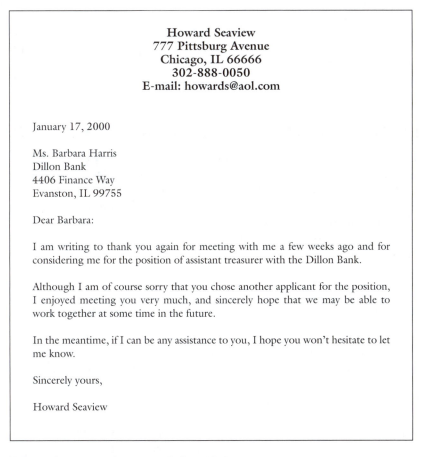

FIGURE 5   Sample Job Rejection Follow-Up Letter

In addition, as we've already mentioned, you never know when you might run into the same interviewer again, and it could well be to your advantage if they remember that you accepted their decision with maturity and class.

A rejection follow-up letter is probably the hardest kind of letter to write. But it you can bring yourself to do it, it may have very positive results at some time in the future. Like the job acceptance follow-up letter, this one should be short, to the point, professional, and infused with as much warmth as you can manage under the circumstances.

FYI

Being successful in business sometimes really is about who you know, so it's in your best interests to make an effort to stay in touch with highly placed individuals in the industry whom you've met during your job-search networking efforts. Make it your business to call your networking contacts periodically "just to catch up"; invite them for breakfast, lunch, or a cup of coffee once or twice a year; and send a quick note, a relevant clipping, or a bit of news from time to time.

FOLLOW-UP LETTERS TO NETWORKING CONTACTS

It's likely that by the time you've found a new position, you will have spoken with a good number of people, both people you already knew and those you met as a result of your networking. It's not only courteous to contact them again, especially if they either directly or indirectly led to your finding a job, it also makes good business sense [Figure 6]. Your personal network is a lifetime career tool because keeping those connections strong and positive will pay dividends for years to come.

JUST THE FACTS

- Networking interview follow-up letters are not only courteous but also help remind your contacts of who you are and that you're looking for a job.
- Sending a follow-up letter after a job interview may make the difference between an offer and a rejection.
- Writing a follow-up letter to someone from whom you've just accepted a job offer will help you get off to a good start on your new job.
- Follow-up letters after you've turned down a job, or been rejected, show style and professionalism and may pay long-term career dividends.
- Once you've started a new job, send follow-up letters to all your networking contacts; you never know when they may be in a position to help again in the future.

Dorothy Arnold
36 Pleasant Avenue
Tacoma, WA 90877
704-654-0022
E-mail: dottya@ATT.com

May 14, 2000

Mr. Howard Bean
World Communications, Inc.
666 Fifth Street
Tacoma, WA 90874

Dear Mr. Bean:

You will, I hope, remember your being kind enough to meet with me some months ago when I was in the process of seeking a position in the communications industry.

As I promised to let you know how my job search progressed, I'm writing now to tell you that I have just received and accepted an offer to become an assistant engineer with KBAL-TV here in Tacoma.

Your suggestions for who to contact were extremely helpful, and I just wanted to tell you again how much I appreciate all your help. I hope and look forward to meeting you again, and if there's ever any way I can be of help to you, please don't hesitate to ask.

Sincerely yours,

Dorothy Arnold

FIGURE 6 Networking Contact Follow-Up Letter

Part 6

And Now a Word (or Two or Three) about Ethics

Turn on the television or radio, read a newspaper, get updates on current events on your computer, and the news is discouragingly the same: corruption, ethical violations, malfeasance in public office, major corporations ruined by the greed of those who run them, high-flying—and even higher spending—executives, and bankrupt pension funds and companies. Ethical behavior seems more and more to be a concept foreign to public, business, and even, alas, academic life.

But business and technical writers have very real moral and ethical obligations, and the essays in this final section of *Strategies* discuss some of the ethical issues everyone who writes for the world of work must face. The more general topic of ethics in business and industry is beyond the scope of this section or of this entire anthology. What follows is meant to lay the groundwork for a continuing discussion in the classroom or corporate setting.

Dorothy A. Winsor begins by examining the public documents available on the explosion of the space shuttle Challenger on January 28, 1986. Her examination shows how a "history of miscommunication" contributed to this accident, which claimed the lives of all seven Challenger crew members. Winsor uses the dynamics at work in the Challenger incident to show corporate writers and managers the importance of eliminating both intentional and unintentional miscommunication in any business or technical setting.

In a timeless piece, Darrell Huff suggests—tongue-in-cheek—ways writers can misuse graphics and statistics to manipulate the truth and to inflate their prose. In separate essays, Dan Jones and Carolyn Rude follow with broader discussions of the relationship among ethics, writing style, and editing. Jones begins with a definition of ethics as it applies to technical prose, then offers "the ten commandments of computer ethics," and concludes by quoting from the ethical guidelines of the Society for Technical Communication (STC). Rude points out how necessary it is for writers and editors to work with members of product development teams and even legal experts to ensure compliance with codes of ethics regarding warnings and notices of copyright, trademarks, or patents.

Thus, the two classic essays by Winsor and Huff and the two cutting-edge pieces by Jones and Rude that end this section of *Strategies*—and indeed the entire collection itself—drive home a final important point: All business and technical writing must not only meet the needs of its intended audiences and follow a process approach, but it must also adhere to the strictest ethical and legal standards.

Dorothy A. Winsor

Communication Failures Contributing to the Challenger Accident: An Example for Technical Communicators

Dorothy A. Winsor recently retired from the faculty at Iowa State University and conducts research on the writing of engineers. Her work has won the NCTE (National Council of Teachers of English) Outstanding Research Award on three separate occasions.

A technological failure such as the explosion of the space shuttle Challenger can be puzzling in retrospect. Investigation often reveals that various people in the organization involved knew that the failure was likely and knew how to prevent it, and yet that knowledge was not shared within the organization as a whole. How does it happen that such important knowledge is not communicated? In the case of the Challenger, why did those who knew of the problem with the shuttle's solid rocket boosters not convince those in power to stop the launch?

The answer to this question lies in a complex set of factors, the most important of which seem to be (1) managers and engineers viewing the same facts from different perspectives, and (2) the general difficulty of either sending or receiving bad news, particularly when it must be passed to superiors or outsiders. An analysis of the communication failures that contributed to the Challenger accident is potentially of great interest to engineers and their managers because a large part of an engineer's job is to communicate both good and bad news upward to management for decision making. The Challenger explosion was a horrifying public event, but it resulted from factors that are probably at work more quietly in many other organizations.

The first of these factors—managers and engineers viewing the same facts from different perspectives—suggests that knowledge is not simply seeing facts but rather interpreting them, and that interpretation varies depending upon one's vantage point. Communication, then, is not just shared information; it is shared interpretation. Achieving shared interpretation within an organization is relatively easy if the sender and receiver of communication share the same corporate role and hence the same concerns and values. If sender and receiver are from different corporate subcultures, however (as they often are in technical communication), then achieving shared interpretation is more difficult [1, 2]. In an appendix to the *Report of the Presidential Commission on the Space Shuttle Accident,* for instance, Commission Member R. P. Feynman notes the difference between probability estimates of flight failure and loss of life by managers—1 in 100,000—and engineers—1 in 100 [3, V. 1, p. F-1]. This difference occurred despite the fact that the managers were almost all engineers by training. Presumably, managers and working engineers had much the same background and many of the same facts at their disposal, but they interpreted the facts quite differently because they approached them from different points of view.

By the same token, communication about the solid rocket booster joint that failed was made more difficult because it was bad news. Research has repeatedly shown that bad news is often not passed upward in organizations [4, 5]. Moreover, even when bad news is sent, people are less likely to believe it than good news [6, 7]. In the shuttle disaster, bad news moved up only slowly from engineers to management within NASA; Marshall Space Center, where the shuttle program was headquartered; and Morton Thiokol International (MTI), the contractor responsible for the solid rocket boosters. It also moved slowly among the organizations because they were in a hierarchical relationship, with MTI dependent on Marshall for the contract and Marshall dependent on NASA for funds and career opportunities.

Additionally, the three organizations seemed to view one another as outsiders despite the fact that they were working jointly on the same project and, in the case of NASA and Marshall, were nominally part of the same agency. So the taboo against airing organizational dirty linen in public was added to the general difficulties of bad news transmission [8]. Communication about O-ring problems, then, had to overcome the barriers to moving bad news between engineering and management subcultures, up through organizational hierarchies, and out to other organizations. Under these circumstances, it is hardly surprising that the communication failed.

The following paragraphs explain, first, what failed physically on the shuttle system, and then what failed in the organizations' attempts to understand those physical problems and communicate about them in the two-year period before the January 28, 1986 launch. I hope that this presentation will give engineers and engineering managers some insight into how to minimize the occurrence of events like the Challenger accident.

PHYSICAL CAUSE OF THE ACCIDENT

The physical cause of the Challenger explosion was the failure of a rubber seal in the solid rocket booster. The shuttle system consisted of three parts: the orbiter, which contained crew and experimental equipment; a large tank of liquid fuel, which was used by the orbiter's engines during liftoff; and two solid rocket boosters, which assisted at liftoff and were jettisoned to be recovered and reused on later flights. These solid rocket boosters (SRBs) were made in segments, which were stacked together at the launch site. The joints between the segments were sealed with two O-rings, which were protected from the heat of combustion by putty. The joint was pressure sealed, meaning that during rocket firing, expanding gases from burning fuel pushed the putty into the air space in the joint; this compressed air, in turn, pushed the O-ring into place and held it there. The second O-ring in each joint, in theory, provided redundancy or backup for the primary ring. During the Challenger launch, the O-rings in one of the SRB joints failed to seal, allowing hot gases to escape from the side of the SRB and burn a hole into the nearby liquid fuel tank, which exploded approximately 73 seconds into the flight.

In hindsight, the failure of the O-rings should not have been unexpected. From early 1984 on, postflight evidence increasingly showed that the joint seals were failing to meet design expectations. After every shuttle flight, the SRBs were recovered, disassembled for inspection, and readied for reuse in future flights. The inspections looked for any anomaly indicating the O-rings had not functioned as they should, and specifically for anything such as charred or eroded surfaces, which would indicate that the rings had failed to seal or had come into close contact with the heat of combustion. Before February 1984, only one O-ring anomaly had been found on the first nine flights. Beginning with the tenth shuttle flight, however, launched two years before Challenger on February 3, 1984, anomalies occurred on more than 50 percent of the flights [3, V. 1, p. 155].

We now know that the increased number of O-ring anomalies after early 1984 was probably caused by the use of increased pressure in the leak check done on the SRB joints after assembly at the launch site. The leak check involved blowing air into the joint through a hole located between the two O-rings and testing to see if the joint pressurized or sealed. Flights one through seven were tested at a pressure of 50 psi, flights eight and nine at 100 psi. From flight ten on, tests were done at 200 psi. In May 1985, MTI experiments showed that the increased pressure was likely to have blown holes through the putty that shielded the primary O-ring from the hot blast of ignition [3, V. 1, p. 156]. The holes were particularly damaging, because they not only allowed the hot gases to penetrate but actually focused them, so that they would cause maximum O-ring erosion, which is the eating away of the edge of the O-ring by the hot gases rushing past before the ring seals. Ironically, the pressure at which the leak check was conducted was increased to 200 psi because of concern that the joints were not sealing. Some officials believed that the blow holes were necessary—unless they existed, the putty rather than the O-ring could seal the joints during the test, and a defective O-ring could escape detection [3, V. 1, p. 134].

EARLY RESPONSES TO BAD NEWS: DISBELIEF AND FAILURE TO SEND UPWARD

When O-ring anomalies first began appearing in early 1984, neither engineers nor management at MTI treated them as serious problems in their communications to Marshall. They did not send a grave interpretation of the data upward and, judging by internal documents, did not believe one themselves. Marshall's reactions are more ambiguous, for they treated the O-ring situation as serious when they communicated downward to MTI but as relatively minor when they communicated up to NASA headquarters.

After MTI engineers saw erosion on the February 3, 1984 flight, they filed a problem report, and O-rings were entered into formal problem tracking systems at both Marshall and MTI. In an action that illustrates the difficulty of accepting bad news, MTI claimed (in a subsequent briefing to Marshall) that the problem was not serious because even if the primary O-ring were damaged, the second ring would provide redundancy and seal [3, V. 1, p. 128]. Tests done more than a year earlier had shown the secondary O-rings to be unreliable because of a phenomenon called joint rotation. Joint rotation means that under the pressure of launch, the two sides of the O-ring joint bent apart, widening the gap the O-ring had to seal. Joint rotation was apparently especially hard on the secondary ring, making it likely to pull completely out of its groove and never seal at all.

As a result of these tests, Marshall had changed the joint's classification from criticality 1R (a critical system with backup) to criticality 1 (a critical system without such backup) in December 1982. However, those involved in shuttle design apparently found this change hard to accept. Marshall personnel, for instance, apparently believed that they had redundancy in all but exceptional cases [3, V. 1, p. 128]. At MTI, the difficulty of belief was even more marked. Some MTI engineers and officials told the Presidential Commission that they were not notified of the criticality rating change, although their names appear on distribution lists on MTI documents [3, V. 1, p. 128]. Did these people lie about being notified? It seems more likely that they literally could not keep the bad news in mind.

Marshall's response to MTI's briefing illustrates what would be a pattern in their reaction to the O-ring difficulties. Some people at Marshall were willing to say that there was a serious problem—as long as any failure was perceived as MTI's. On February 28, for instance, John Miller, chief of the solid rocket motor branch at Marshall, wrote to his superior, George Hardy, through project engineer Keith Coates, urging that tests be done to see if the leak checks were causing problems. In an unusual recognition of the seriousness of the matter, Miller said O-ring failure could be "catastrophic" [3, V. 1, p. 245]. The next day, Coates also wrote to Hardy saying that MTI's briefing had minimized the extent of joint rotation possible and thus was too optimistic [3, V. 1, p. 128].

When the O-ring problem had to be claimed as Marshall's, however, and NASA had to be informed, Marshall, too, became optimistic. On March 8, 1984, a Flight Readiness Review for the eleventh shuttle flight was held at Marshall. These reviews were held at four levels, with each level resolving what problems it could and passing unresolved issues on to the next higher one. The March 8 meeting was Level III, meaning that it was a meeting between Marshall and its contractors and that the highest officials present were the Marshall project managers. At the meeting, MTI reported that maximum erosion on the O-rings would be 0.09 inch and that tests had shown that the rings would function with

0.095 inch of erosion. The 0.005-inch difference appears to be an extremely small safety margin. Rather than report a serious problem to a Level II meeting at Johnson Space Center, however, Marshall apparently accepted the margin, because this same information was entered in the Marshall problem assessment report with a note that future flights need therefore not be delayed. The Marshall problem tracking record reads: "Remedial action—none required" [3, V. 1, p. 128].

The 0.005-inch safety margin was also used as a rationale to justify no flight interruptions at the Level I briefing of top NASA personnel on March 27. NASA accepted Marshall's recommendation but wrote to Lawrence Mulloy, SRB project manager at Marshall, asking for further study of the O-rings. Mulloy had Marshall engineer Lawrence Wear ask MTI to identify the cause of erosion, determine its seriousness, and define any necessary changes.

Internal MTI documents show that the contractor was examining the problem but with little sense of urgency, again evidencing the tendency to see the problem in the best light possible. MTI analyzed the erosion history and test data and, on May 4, presented Marshall with a plan for studying the O-rings to produce the information NASA had asked for. The information was not actually produced, however, until a briefing on August 19, 1985—16 months later.

Despite its optimism to NASA, Marshall was apparently uncomfortable with this pace and pressed for prompter action than MTI was giving. On July 2, L. H. Sayers, MTI's Director of Engineering Design, suggested by phone to Marshall's Ben Powers that tests done to date by MTI were sufficient [3, V. 1, p. 134]. This again implies that MTI did not perceive the matter to be crucial or dangerous.

Early signs of serious O-ring problems, then, were generally not believed at MTI, were accepted at Marshall only when it was possible to see the problem as MTI's, and were not sent upward to NASA headquarters.

CONTINUED BAD NEWS REJECTION DESPITE CONTRADICTORY EVIDENCE

The optimistic view of the O-rings persisted at both MTI and Marshall over the 1984–85 period despite mounting evidence that the rings were not functioning well. This evidence had to do with the effect of cold on the rings and the amount of erosion that could occur in an O-ring.

On January 24, 1985, the fifteenth shuttle flight was launched at a temperature of 53° F, the lowest up to that time. It showed much greater O-ring erosion than any previous flight. Tests have since shown that cold reduces O-ring resiliency and increases the time to seal, thereby exacerbating the problems the joint was already experiencing. The damage on the fifteenth flight was severe enough to bring concern about the rings to the fore again. On January 31, Mulloy wrote to Wear asking him to get MTI to prepare information on O-ring erosion for a Level III Flight Readiness Review scheduled for February 8. At that review, MTI personnel mentioned the cold as a factor in the damage but labeled the risk "acceptable," mostly because they assumed the secondary ring would seal if the first one failed.

An optimistic interpretation of the data on cold was held by both managers and engineers at MTI, and one of the primary advocates of continued launch was MTI engineer Roger Boisjoly, who would later be one of the primary opponents to the launch of Challenger. The split between managerial and engineering interpretation of the data did not develop for four or five more months. MTI's claim that the secondary O-rings would seal is a further example of retaining previous theories in the face of contradictory evidence because, as noted above, the redundancy of the secondary ring had been in question since late 1982.

As they had done earlier, Marshall management accepted MTI's rationale, at least in what it told NASA. A Level I Review was held on February 21. At this meeting, the influence of temperature was not mentioned, and only a single reference was made to O-ring erosion, saying redundancy made the risk acceptable [3, V. 1, p. 136]. Although each of the next four flights experienced joint seal problems, neither MTI nor Marshall seemed unduly concerned. Perhaps the very frequency of the problem added to its acceptability because the damage kept occurring with no serious consequences.

On June 25, however, one of the joints from a flight that had been launched April 29 was examined and found to have severe erosion of not only the primary ring, but even the secondary ring, calling its redundancy into question once again. As mentioned above, MTI had predicted maximum erosion of 0.09 inch for the primary seal. The primary O-ring on this joint was eroded 0.171 inch, almost double the predicted maximum and far beyond the 0.095 inch MTI had claimed to know was safe. This was bad news indeed. It could not be ignored, since an engineer from NASA headquarters was present when the damage was discovered. The engineer wrote to Michael Weeks at NASA on June 28, reporting the damage. The joint affected was a nozzle joint—that is, the joint linking the SRB to the flared section at its base. The NASA engineer blamed the damage on the fact that, in contrast to the rest of the SRB joints, which were now pressure-tested at 200 psi, nozzle joints were still being tested at only 100 psi, which might have permitted a defective ring to escape notice [3, V. 1, pp. 137–38].

In July 1985, Mulloy placed a launch constraint on the nozzle joints. This, in theory, meant that no other flights would take place until O-ring erosion at the nozzle joint had been fixed or shown not to be a problem. By including only the nozzle joints in the constraint, Mulloy was taking the most optimistic view possible of the problem. He reasoned that the nozzle joint had failed, not because of defective design, but because of a defective ring that had escaped notice in the nozzle joint's less rigorous leak test. Thus he believed that the leak test, and not the joint, was problematic. As a consequence, although he had assigned the launch constraint, Mulloy waived it for every subsequent flight, including Challenger, believing he was justified because Marshall increased to 200 psi the leak check pressure on the nozzle joint.

The launch constraint was treated as bad news by both MTI and Marshall. MTI officials testifying before the Commission all said they did not know about it, although subsequent MTI documents refer by document number to the report imposing the constraint [3, V. 1, p. 137]. The officials apparently did not take the news in. NASA officials, on the other hand, seem genuinely not to have been informed of the constraint, although regulations required that Level II be told. Marshall seems to have kept the news to itself rather than pass it out and up to NASA [3, V. 1, p. 138].

INTERNAL VERSUS EXTERNAL COMMUNICATION OF CONCERN FROM MTI ENGINEERS

Despite MTI's ignorance of the launch constraint, the damage discovered in late June seems to have galvanized its engineers into action. Among them, at least, there seems to have been increased recognition of the problem's existence and its seriousness. MTI engineer Roger Boisjoly, for instance, became increasingly insistent about the potential danger from the O-rings. On July 22, his activity report predicted loss of the contract or flight failure if no solution was found. On July 31, he sent the following memo to R. K. Lund, MTI's Vice President of Engineering [3, V. 1, pp. 249–50]:

> SUBJECT: *SRM O-ring Erosion/Potential Failure Criticality.*
>
> This letter is written to insure that management is fully aware of the seriousness of the current O-ring erosion problem in the SRM joints from an engineering standpoint. The mistakenly accepted position on the joint problem was to fly without fear of failure and to run a series of design evaluations which would ultimately lead to a solution or at least a significant reduction of the erosion problem. This position is now drastically changed as a result of the SRM 16A nozzle joint erosion which eroded a secondary O-ring with the primary O-ring never sealing.
>
> If the same scenario should occur in a field joint (and it could), then it is a jump ball as to the success or failure of the joint because the secondary O-ring cannot respond to the clevis opening rate and may not be capable of pressurization. The result would be a catastrophe of the highest order—loss of human life.
>
> An unofficial team [a memo defining the team and its purpose was never published] with leader was formed on 19 July 1985 and was tasked with solving the problem for both the short and long term. This unofficial team is essentially nonexistent at this time. In my opinion, the team must be officially given the responsibility and the authority to execute the work that needs to be done on a non-interference basis (full time assignment until completed).
>
> It is my honest and very real fear that if we do not take immediate action to dedicate a team to solve the problem with the field joint having the number one priority, then we stand in jeopardy of losing a flight along with all the launch pad facilities.

It is evident from this memo that Boisjoly's interpretation of the data had changed and that he was trying to communicate his new interpretation to his management. This memo does not give much space to new or old factual information, but rather concentrates on what the facts mean. Boisjoly's concern is evidenced both by the way he faults his own company (particularly his own management) and by his use of emotional language unusual in engineering documents. The memo implies, for instance, that MTI management may have an inaccurate understanding of the situation. The company's previous position was "mistakenly accepted." The company had planned to solve the problem but had not actually gotten to work to do so. The situation had changed "drastically," and a "catastrophe" could result. Boisjoly's memo and subsequent similar ones evidently had some effect on their receiver, because on the night before Challenger flew, Lund did at least begin by recommending against launch. Boisjoly's concern, however, was kept within MTI. He marked his memo *COMPANY PRIVATE* at both top and bottom. Although he was sufficiently alarmed to try to reach his superiors, he still attempted to keep bad news from the prying eyes of outsiders.

Bad news went to Marshall only in response to specific questions from them. On August 9, MTI engineer Brian Russell wrote to Marshall's Jim Thomas about results of tests investigating the effect of cold on the O-rings. The tests had been initiated after the January 24 flight showed such severe damage. The tone of Russell's letter makes an instructive contrast to Boisjoly's [3, V. 5, pp. 1568–69]:

SUBJECT: *Actions Pertaining to SRM Field Joint Secondary Seal.*

Per your request, this letter contains the answers to the two questions you asked at the July Problem Review Board telecon.

1. *Question:* If the field joint secondary seal lifts off the metal mating surfaces during motor pressurization, how soon will it return to a position where contact is re-established?

    Answer: Bench test data indicate that the o-ring resiliency (its capability to follow the metal) is a function of temperature and the rate of case expansion. MTI measured the force of the o-ring against Instron plattens, which simulated the nominal squeeze on the o-ring and approximated the case expansion distance and rate.

    At 100° F the o-ring maintained contact. At 75° F the o-ring lost contact for 2.4 seconds. At 50° F the o-ring did not re-establish contact in ten minutes at which time the test was terminated.

    The conclusion is that secondary sealing capability in the SRM field joint cannot be guaranteed.

2. *Question:* If the primary o-ring does not seal, will the secondary seal seat in sufficient time to prevent joint leakage?

    Answer: MTI has no reason to suspect that the primary seal would ever fail after pressure equilibrium is reached, i.e., after the ignition transient. If the primary o-ring were to fail from 0 to 170 milliseconds, there is a very high probability that the secondary o-ring would hold pressure since the case has not expanded appreciably at this point. If the primary seal were to fail from 170 to 330 milliseconds, the probability of the secondary seal holding is reduced. From 330 to 660 milliseconds the chance of the secondary seal holding is small. This is a direct result of the o-ring's slow response compared to the metal case segments as the joint rotates.

Please call me or Mr. Roger Boisjoly if you have additional questions concerning this issue.

Russell and Boisjoly were actually of much the same opinion on the dangers of the joint, and Boisjoly helped Russell write the memo shown here, but Russell is speaking to outsiders and this affects the memo. The memo is not in any way untruthful, but it is not very communicative. In a sense, it is the opposite of Boisjoly's memo. It gives just the facts, providing little interpretation. Its tone is adamantly objective, in contrast to Boisjoly's more emotional one. In retrospect, some of the facts Russell gives should have been frightening. Note, for instance, that if there was joint rotation, the secondary O-ring never sealed when it was tested at 50° F. The conclusion reached from this information was "that secondary sealing . . . cannot be guaranteed." This is a negative wording of the finding, which was that at 50° F or below, MTI could pretty well guarantee no secondary seal.

That this memo did not communicate its intent is shown by the fact that the people who read it were uncertain about what it meant. Thomas copied the memo to be sent to NASA headquarters, but when the memo went through Mulloy's office for his signature, Mulloy returned it to Thomas saying it sounded like old news. The NASA official to whom Thomas was sending it has since said that even had he received the memo, he might not have understood it. "I don't know if anybody at that time understood the joint well enough to realize that the data was crucial," he said. When Mulloy was asked why he had not treated the temperature data as more important, he said he had not realized its significance, adding, "There were a whole lot of people who weren't smart enough to look behind the veil and say, 'Gee, I wonder what this means'" [9]. As can be seen, the urgency in Boisjoly's memo had not been conveyed to Marshall. Marshall officials had been given the facts about the effects of cold on the O-rings. They did not, however, interpret those facts in the same alarmed manner Boisjoly and Russell did, and Russell's memo did not attempt to communicate the more pessimistic interpretation.

Thus, MTI engineers concluded that the O-ring problems were serious before their management did. However, in their written communication, they varied the extent to which they voiced that seriousness, depending on whether their audience was internal or external.

THE SPLIT BETWEEN MANAGERS AND ENGINEERS

As is suggested by Boisjoly's memo above, MTI managers and engineers were beginning to disagree over the seriousness of the O-ring problem, and engineers had a difficult time communicating their view upward. Support from the mini-culture of the task force probably made it easier than it had been previously for engineers to recognize the problem and speak up about it. On October 1, MTI engineers Roger Ebeling and S. R. Stein complained in separate internal MTI memos to management that the O-ring task force was being slowed by administrative delays and lack of cooperation [3, V. 1, pp. 252–53]. Both men complained that, although the task force members regarded their work as urgent, administrators required that all testing and design be done according to routines established for more leisurely long-term development.

On October 3, the team met with Joe Kilminster, MTI's Vice President of Space Programs, to discuss these administrative difficulties, but apparently the members were not successful in convincing him of the gravity of the situation. On October 4, Roger Boisjoly's activity report complained bitterly that "upper management apparently feels" MTI has the SRB contract "for sure and the customer be damned" [3, V. 1, p. 255]. In December, Ebeling actually told fellow task force members that MTI should not ship any more SRBs to Marshall until the problem was solved. However, consistent with patterns of bad news transmittal, he did not tell this to any of his superiors [3, V. 1, p. 142].

In January 1986, final preparation for Challenger's flight began. A launch scheduled for January 27 was cancelled and rescheduled for the next day. The temperature at launch time was 36° F, 17° colder than it had been for any previous launch. When MTI engineers, including Ebeling, Russell, and Boisjoly, heard of the predicted low temperatures, they became alarmed enough to convince Lund, their Vice President of Engineering, to recommend that the launch be delayed until the temperature of the joints reached 53° F, the previous lowest launch temperature. In their argument, they cited the information from Russell's memo and the severe erosion from the 53° F launch the previous January.

In a teleconference involving numerous MTI and Marshall managers and engineers, Lund, MTI's Vice President of Engineering, did recommend delaying launch. Marshall was apparently surprised by MTI's action and, refusing to accept the bad news, they resisted the recommendation. George Hardy, who headed engineering at Marshall, said he was "appalled" [3, V. 1, p. 94]. He later testified to the Presidential Commission that he meant he was appalled at MTI's data, but MTI personnel believed he said he was appalled at their recommendation. Mulloy apparently asked if they expected him to wait until April to launch.

In general, Marshall challenged, not MTI's facts, but the conclusions drawn from them. Mulloy and Hardy conceded that the primary ring might be slower to seal in the cold and that the increased time to seal would allow more erosion to take place. But they argued that the ring would seal eventually and that the joint could sustain three times the worst erosion they had seen to date and still have a good seal. In the time before the primary ring sealed, joint rotation would not yet have taken place; therefore, the secondary ring would still be good. Moreover, they believed that temperature could not be the deciding factor because, although they had had severe erosion at 53° F, they had also had it at 75° F. (It is true that erosion occurred at various temperatures. However, the Commission later pointed out that 15 percent of the flights launched above 65° F had O-ring anomalies, while 100 percent of those launched below that temperature had them.) In the face of Marshall's opposition, Joe Kilminster, MTI's Vice-President of Space Programs, asked that MTI have a private caucus off the phone line.

During the caucus, it became obvious that MTI was split along role lines. The engineers continued to argue against launch. Boisjoly says that, in the caucus, "there was never one comment in favor . . . of launching by any engineer or other nonmanagement person in the room" [3, V. 1, p. 93]. And in words that describe his listeners shutting out bad news, he says he himself argued until it became clear "that no one wanted to hear what I had to say" [3, V. 4, p. 697].

At this point, Jerald Mason, MTI's Senior Vice President, said it was obvious that all present would not reach agreement and that a management decision would have to be made. He polled the other three vice presidents in the room, first asking Lund, who had presented the recommendation not to launch, to take off his "engineering hat" and put on his "management hat" [3, V. 1, p. 94]. When Lund changed his role, he changed his position, and the four managers voted unanimously to launch. Engineer Brian Russell describes an atmosphere in which it was difficult to maintain opposition to the launch, wondering "whether I would have the courage, if asked, . . . to stand up and say no" [3, V. 4, p. 822]. As it happened, no engineer was asked to vote, and MTI went back on the conference line and reversed its earlier warning.

During the time MTI was caucusing, Marshall engineer Ben Powers also told his immediate supervisors that he agreed with MTI's recommendation not to launch, but his supervisors did not pass his view upward to Hardy and Mulloy. Similarly, Hardy and Mulloy did not pass on to NASA officials the fact that MTI engineers were opposed to the launch. At 11:38 the next morning, Challenger took off.

CONCLUSION

Looking at prelaunch miscommunications, then, several factors are apparent. First, no one at MTI or Marshall wanted to believe the growing evidence of O-ring problems. Second, even when MTI engineers came to believe that a problem existed, they had a difficult time convincing their management, with its different perspective on operations, to interpret the facts in the same light. In turn, on the night before launch, MTI personnel were unable to convince Marshall of the situation's gravity, even though they looked at the same facts, because Marshall, too, saw things differently. Finally, both engineers and managers at MTI were especially reluctant to communicate bad news to those outside the company.

All of this suggests a number of precautions engineers and their managers might take in the face of the same kind of pressure-induced miscommunication. From a manager's point of view, one of the most important precautions is to establish an atmosphere in which engineers feel free to communicate bad news as well as good. A number of writers have offered advice on how this can be done [8, 10, 11, 12, 13]. Establishing an open atmosphere takes time and a concerted effort from the whole organization. It cannot be done on short notice when emergencies arise.

In addition, pressures for holding back bad news should be anticipated to be especially strong when contractors are involved. Encouraging bad news transmittal is difficult when the bearer of bad tidings is afraid of losing a contract. Contracts can, perhaps, be designed to lessen this fear, but those issuing the contracts should be alert for any sign of problems, since full disclosure of bad news is unlikely in this situation.

Lastly, managers and engineers alike should anticipate that they are probably erring on the side of optimism in interpreting data bearing on already established designs and programs. Failure to believe bad news is probably caused by a number of factors, including reluctance to admit that one was wrong, fear of practical consequences such as expensive redesign, and a kind of intellectual inertia that makes it easier to persist in an already established belief than to change it. Such optimism can, however, have disastrous consequences, especially when coupled with other forces, as the Challenger accident demonstrates.

REFERENCES

1. Gregory, K. L., "Native-View Paradigms: Multiple Cultures and Culture Conflicts in Organizations," *Administrative Science Quarterly* 28 (1983), 359–76.
2. Riley, P., "A Structurationist Account of Political Culture," *Administrative Science Quarterly* 28 (1983), 414–37.
3. Presidential Commission on the Space Shuttle Challenger Accident, *Report of the Presidential Commission on the Space Shuttle Accident,* 5 vols, Washington: GPO, 1986.
4. Housel, T. J., and Davis, W. E., "The Reduction of Upward Communication Distortion," *Journal of Business Communication* 14 (1977), 49–65.
5. Frank, A. D., "Can Water Flow Uphill?" *Training and Development Journal* 38 (1984), 118–28.
6. Rasberry, R. W., and Lemoine, L. F., *Effective Managerial Communication,* Boston: Kent, 1986, pp. 63–97.
7. Applebaum, R. L., and Anatol, K. W. E., *Effective Oral Communication for Business and the Professions,* Chicago: SRA, 1982, pp. 259–262.
8. Steele, F., *The Open Organization: The Impact of Secrecy and Disclosure on People and Organizations,* Reading, MA: Addison-Wesley, 1975.
9. Zaldivar, R. A., "Rocket Chief Was Warned and Failed to Act," *Detroit Free Press,* May 23, 1986.
10. Vance, C. C., "How to Encourage Upward Communication from Your Employees," *Association Management* 28 (1976), 56–59.
11. Driver, R. W., "Opening the Channels of Upward Communication," *Supervisory Management* 25 (1980), 24–29.
12. Macleod, J. S., "How to Unmuzzle Employees," *Employment Relations Today* 11 (1984), 49–54.
13. Thomas, P., "Plugging the Communication Channel—How Managers Stop Upward Communication," *Supervisory Management* 30 (1985), 7–10.

Darrell Huff

How to Lie with Statistics

Darrell Huff, a freelance writer, expanded this article into a book with the same title (Norton, 1954).

"The average Yaleman, Class of '24," *Time* magazine reported last year after reading something in the New York *Sun*, a newspaper published in those days, "makes $25,111 a year."

Well, good for him!

But, come to think of it, what does this improbably precise and salubrious figure mean? Is it, as it appears to be, evidence that if you send your boy to Yale you won't have to work in your old age and neither will he? Is this average a mean or is it a median? What kind of sample is it based on? You could lump one Texas oilman with two hundred hungry freelance writers and report *their* average income as $25,000-odd a year. The arithmetic is impeccable, the figure is convincingly precise, and the amount of meaning there is in it you could put in your eye.

In just such ways is the secret language of statistics, so appealing in a fact-minded culture, being used to sensationalize, inflate, confuse, and oversimplify. Statistical terms are necessary in reporting the mass data of social and economic trends, business conditions, "opinion" polls, this year's census. But without writers who use the words with honesty and understanding and readers who know what they mean, the result can only be semantic nonsense.

In popular writing on scientific research, the abused statistic is almost crowding out the picture of the white-jacketed hero laboring overtime without time-and-a-half in an ill-lit laboratory. Like the "little dash of powder, little pot of paint," statistics are making many an important fact "look like what she ain't." Here are some of the ways it is done.

The sample with the built-in bias. Our Yale men—or Yalemen, as they say in the Time-Life building—belong to this flourishing group. The exaggerated estimate of their income is not based on all members of the class nor on a random or representative sample of them. At least two interesting categories of 1924-model Yale men have been excluded.

First there are those whose present addresses are unknown to their classmates. Wouldn't you bet that these lost sheep are earning less than the boys from prominent families and the others who can be handily reached from a Wall Street office?

There are those who chucked the questionnaire into the nearest wastebasket. Maybe they didn't answer because they were not making enough money to brag about. Like the fellow who found a note clipped to his first pay check suggesting that he consider the amount of his salary confidential: "Don't worry," he told the boss. "I'm just as ashamed of it as you are."

Omitted from our sample then are just the two groups most likely to depress the average. The $25,111 figure is beginning to account for itself. It may indeed be a true figure for those of the Class of '24 whose addresses are known and who are willing to stand up and tell how much they earn. But even that requires a possibly dangerous assumption that the gentlemen are telling the truth.

To be dependable to any useful degree at all, a sampling study must use a representative sample (which can lead to trouble too) or a truly random one. If *all* the Class of '24 is included, that's all right. If every tenth name on a complete list is used, that is all right too, and so is drawing an adequate number of names out of a hat. The test is this: Does every name in the group have an equal chance to be in the sample?

You'll recall that ignoring this requirement was what produced the *Literary Digest*'s famed fiasco.* When names for polling were taken only from telephone books and subscription lists, people who did not have telephones or *Literary Digest* subscriptions had no chance to be in the sample. They possibly did not mind this underprivilege a bit, but their absence was in the end very hard on the magazine that relied on the figures.

This leads to a moral: You can prove about anything you want to by letting your sample bias itself. As a consumer of statistical data—a reader, for example, of a news magazine—remember that no statistical conclusion can rise above the quality of the sample it is based upon. In the absence of information about the procedures behind it, you are not warranted in giving any credence at all to the result.

The truncated, or gee-whiz, graph. If you want to show some statistical information quickly and clearly, draw a picture of it. Graphic presentation is the thing today. If you don't mind misleading the hasty looker, or if you quite clearly *want* to deceive him, you can save some space by chopping the bottom off many kinds of graph.

---

*Editor's note: The *Literary Digest* predicted that Alfred Landon would defeat Franklin Roosevelt in the 1936 presidential election. Landon carried only two states.

Suppose you are showing the upward trend of national income month by month for a year. The total rise, as in one recent year, is 7 percent. It looks like this:

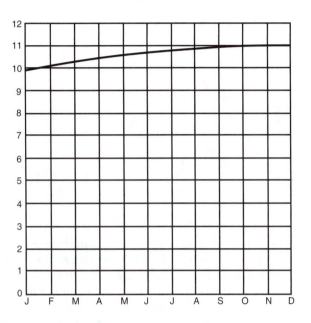

That is clear enough. Anybody can see that the trend is slightly upward. You are showing a 7 percent increase, and that is exactly what it looks like.

But it lacks schmaltz. So you chop off the bottom, this way:

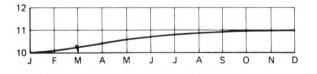

The figures are the same. It is the same graph and nothing has been falsified—except the impression that it gives. Anyone looking at it can just feel prosperity throbbing in the arteries of the country. It is a subtler equivalent of editing "National income rose 7 percent" into ". . . climbed a whopping 7 percent."

It is vastly more effective, however, because of that illusion of objectivity.

The souped-up graph. Sometimes truncating is not enough. The trifling rise in something or other still looks almost as insignficant as it is. You can make that 7 percent look livelier than 100 percent ordinarily does. Simply change the proportion between the ordinate and the abscissa. There's no rule against it, and it does give your graph a prettier shape.

But it exaggerates, to say the least, something awful:

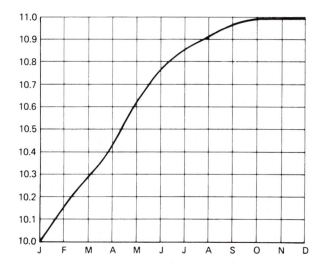

The well-chosen average. I live near a country neighborhood for which I can report an average income of $15,000. I could also report it as $3,500.

If I should want to sell real estate hereabouts to people having a high snobbery content, the first figure would be handy. The second figure, however, is the one to use in an argument against raising taxes, or the local bus fare.

Both are legitimate averages, legally arrived at. Yet it is obvious that at least one of them must be as misleading as an out-and-out lie. The $15,000-figure is a mean, the arithmetic average of the incomes of all the families in the community. The smaller figure is a median; it might be called the income of the average family in the group. It indicates that half the families have less than $3,500 a year and half have more.

Here is where some of the confusion about averages comes from. Many human characteristics have the grace to fall into what is called the "normal" distribution. If you draw a picture of it, you get a curve that is shaped like a bell. Mean and median fall at about the same point, so it doesn't make very much difference which you use.

But some things refuse to follow this neat curve. Income is one of them. Incomes for most large areas will range from under $1,000 a year to upward of $50,000. Almost everybody will be under $10,000, way over on the lefthand side of that curve.

One of the things that made the income figure for the "average Yaleman" meaningless is that we are not told whether it is a mean or a median. It is not that one type of average is invariably better than the other; it depends upon what you are talking about. But neither gives you any real information—and either may be highly misleading—unless you know which of those two kinds of average it is.

In the country neighborhood I mentioned, almost everyone has less than the average—the mean, that is—of $10,500. These people are all small farmers, except for a trio of millionaire week-enders who bring up the mean enormously.

You can be pretty sure that when an income average is given in the form of a mean nearly everybody has less than that.

The insignificant difference or the elusive error. Your two children Peter and Linda (we might as well give them modish names while we're about it) take intelligence tests. Peter's IQ, you learn, is 98 and Linda's is 101. Aha! Linda is your brighter child.

Is she? An intelligence test is, or purports to be, a sampling of intellect. An IQ, like other products of sampling, is a figure with a statistical error, which expresses the precison or reliability of the figure. The size of this probable error can be calculated. For their test the makers of the much-used Revised Stanford-Binet have found it to be about 3 percent. So Peter's indicated IQ of 98 really means only that there is an even chance that it falls between 95 and 101. There is an equal probability that it falls somewhere else—below 95 or above 101. Similarly, Linda's has no better than a fifty-fifty chance of being within the fairly sizeable range of 98 to 104.

You can work out some comparisons from that. One is that there is rather better than one chance in four that Peter, with his lower IQ rating, is really at least three points smarter than Linda. A statistician doesn't like to consider a difference significant unless you can hand him odds a lot longer than that.

Ignoring the error in a sampling study leads to all kinds of silly conclusions. There are magazine editors to whom readership surveys are gospel; with a 40 percent readership reported for one article and a 35 percent for another, they demand more like the first. I've seen even smaller differences given tremendous weight, because statistics are a mystery and numbers are impressive. The same thing goes for market surveys and so-called public opinion polls. The rule is that you cannot make a valid comparison between two such figures unless you know the deviations. And unless the difference between the figures is many times greater than the probable error of each, you have only a guess that the one appearing greater really is.

Otherwise you are like the man choosing a camp site from a report of mean temperature alone. One place in California with a mean annual temperature of 61 is San Nicolas Island on the south coast, where it always stays in the comfortable range between 47 and 87. Another with a mean of 61 is in the inland desert, where the thermometer hops around from 15 to 104. The deviation from the mean marks the difference, and you can freeze or roast if you ignore it.

The one-dimensional picture. Suppose you have just two or three figures to compare—say the average weekly wage of carpenters in the United States and another country. The sums might be $60 and $30. An ordinary bar chart makes the difference graphic.

That is an honest picture. It looks good for American carpenters, but perhaps it does not have quite the oomph you are after. Can't you make that difference appear overwhelming and at the same time give it what I am afraid is known as eye-appeal? Of course you can. Following tradition, you represent these sums by pictures of money bags. If the $30 bag is one inch high, you draw the $60 bag two inches high. That's in proportion, isn't it?

The catch is, of course, that the American's money bag, being twice as tall as that of the $30 man, covers an area on your page four times as great. And since your two-dimensional picture represents an object that would in fact have three dimensions, the money bags actually would differ much more than that. The volumes of any two similar solids vary as the cubes of their heights. If the unfortunate foreigner's bag holds $30 worth of dimes, the American's would hold not $60 but a neat $240.

You didn't say that, though, did you? And you can't be blamed, you're only doing it the way practically everybody else does.

The ever-impressive decimal. For a spurious air of precision that will lend all kinds of weight to the must disreputable statistics, consider the decimal.

Ask a hundred citizens how many hours they slept last night. Come out with a total of, say, 781.3. Your data are far from precise to begin with. Most people will miss their guess by fifteen minutes or more and some will recall five sleepless minutes as half a night of tossing insomnia.

But go ahead, do your arithmetic, announce that people sleep an average of 7.813 hours a night. You will sound as if you knew precisely what you are talking about. If you were foolish enough to say 7.8 (or "almost" 8) hours it would sound like what it was—an approximation.

The semiattached figure. If you can't prove what you want to prove, demonstrate something else and pretend that they are the same thing. In the daze that follows the collision of statistics with the human mind, hardly anybody will notice the difference. The semiattached figure is a durable device guaranteed to stand you in good stead. It always has.

If you can't prove that your nostrum cures colds, publish a sworn laboratory report that the stuff killed 31,108 germs in a test tube in eleven seconds. There may be no connection at all between assorted germs in a test tube and the whatever-it-is that produces colds, but people aren't going to reason that sharply, especially while sniffling.

Maybe that one is too obvious and people are beginning to catch on. Here is a trickier version.

Let us say that in a period when race prejudice is growing it is to your advantage to "prove" otherwise. You will not find it a difficult assignment.

Ask that usual cross section of the population if they think . . . [Blacks] have as good a chance as white people to get jobs. Ask again a few months later. As Princeton's Office of Public Opinion Research has found out, people who are most unsympathetic to . . . [Blacks] are the ones most likely to answer yes to this question.

As prejudice increases in a country, the percentage of affirmative answers you will get to this question will become larger. What looks on the face of it like growing opportunity for . . . [Blacks] actually is mounting prejudice and nothing else. You have achieved something rather remarkable: the worse things get, the better your survey makes them look.

The unwarranted assumption, or *post hoc* rides again. The interrelation of cause and effect, so often obscure anyway, can be most neatly hidden in statistical data.

Somebody once went to a good deal of trouble to find out if cigarette smokers make lower college grades than non-smokers. They did. This naturally pleased many people, and they made much of it.

The unwarranted assumption, of course, was that smoking had produced dull minds. It seemed vaguely reasonable on the face of it, so it was quite widely accepted. But it really proved nothing of the sort, any more than it proved that poor grades drive students to the solace of tobacco. Maybe the relationship worked in one direction, maybe in the other. And maybe all this is only an indication that the sociable sort of fellow who is likely to take his books less than seriously is also likely to sit around and smoke many cigarettes.

Permitting statistical treatment to befog casual relationships is little better than superstition. It is like the conviction among the people of the Hebrides that body lice produce good health. Observation over the centuries had taught them that people in good health had lice and sick people often did not. *Ergo,* lice made a man healthy. Everybody should have them.

Scantier evidence, treated statistically at the expense of common sense, has made many a medical fortune and many a medical article in magazines, including professional ones. More sophisticated observers finally got things straightened out in the Hebrides. As it turned out, almost everybody in those circles had lice most of the time. But when a man took a fever (quite possibly carried to him by those same lice) and his body became hot, the lice left.

Here you have cause and effect not only reversed, but intermingled.

There you have a primer in some ways to use statistics to deceive. A well-wrapped statistic is better than Hitler's "big lie": it misleads, yet it can't be pinned onto you.

Is this little list altogether too much like a manual for swindlers? Perhaps I can justify it in the manner of the retired burglar whose published reminiscences amounted to a graduate course in how to pick a lock and muffle a footfall: The crooks already know these tricks. Honest men must learn them in self-defense.

Dan Jones

Determining the Ethics of Style

Dan Jones is Professor of English at the University of Central Florida and the author of four books on technical communication.

Doublespeak is not the product of carelessness or sloppy thinking. Indeed, most doublespeak is the product of clear thinking and is carefully designed and constructed to appear to communicate when in fact it doesn't. It is language designed not to lead but mislead. It is language designed to distort reality and corrupt thought.[1]

—William Lutz

Sometimes we want to be unclear. We don't know what we're talking about, and we don't want anyone else to know that. Or we do know what we're talking about, and we don't want anyone else to know what we know. On those occasions, we write unclearly deliberately, and if what we write gets the job done, then we say the writing is "good." But "good" has two meanings. Assassins can be "good" at their jobs but not be "good" people. In the same way, writing can be "good" if it gets the job done, but if the job is ethically questionable, then the writing may be bad just because it is so good.[2]

—Joseph Williams

What kind of behavior is "prose behavior"? Prose is usually described in a moral vocabulary—"sincere," "open" or "devious," and "hypocritical"—but is this vocabulary justified? Why, for that matter, has it been so moralistic? Why do so many people feel that bad prose threatens the foundations of civilization? And why, in fact, do we think "bad" the right word to use for it?[3]

—Richard Lanham

WHAT IS ETHICS?

Simply defined, "Ethics is the study of right and wrong conduct."[4] More broadly, ethics is "the discipline dealing with what is good and bad and with moral duty and obligation."[5] Ethics also means "a set of moral principles or values" or "a theory or system of moral values."[6] More broadly still, ethics may be defined as a guiding philosophy.

Vincent Ruggiero observes that "the focus of ethics is moral situations—that is, *those situations in which there is a choice of behavior involving human values* (those qualities that are regarded as good and desirable). . . ."[7]

ETHICS AND TECHNICAL PROSE

Just as people must make many ethical decisions throughout their lives, you must make many ethical decisions concerning what you write throughout your career. For example, are you doing your best to document a product honestly and accurately? Are you knowingly omitting any essential information? If you are unclear or imprecise, and if your poor instructions cause injury to someone, are you morally responsible? If you are in a company's marketing department, and you are asked to exaggerate a product's features, are you guilty of lying? If you know you are promising more than you can deliver in a proposal written in response to a request for a proposal (RFP), are you unethical or just keenly competitive? If you fail to point out numerous known bugs in a program in your software manual, are you a good software documentation writer or are you unethical? If you exaggerate your qualifications on your resume or in your cover letter to gain an advantage over the competition, are you unethical? These are just some of many possible scenarios faced by writers of technical prose everywhere.

Why isn't it possible just to make a list of ethical language choices that everyone could agree on and everyone could abide by? Unfortunately, it's not that easy. Suppose that you wrote the clearest instructions you could write, but someone neglected to provide some essential information and a customer is injured following your instructions. Are you unethical in this instance? Suppose because of unreasonable deadlines, you have to cut corners and you just don't have time to document some important features of a software program. Are you a bad person? Suppose you have been told by your boss in marketing that you're expected to hype the product just to keep up with the competition. Are you unethical for trying to keep your job and doing what you are told to do? As these possible scenarios and many others show, it's not always easy to determine which writing decisions are right or wrong, moral or immoral, ethical or unethical.

ETHICS AND THE PROFESSIONS

Of course, ethical concerns are not new to those who must write technical prose. Most professions and professional organizations have published ethical guidelines. The Computer Ethics Institute, for example, published the following guidelines:

The Ten Commandments of Computer Ethics

1. Thou shalt not use a computer to harm other people.
2. Thou shalt not interfere with other people's computer work.
3. Thou shalt not snoop around in other people's computer files.
4. Thou shalt not use a computer to steal.

5. Thou shalt not use a computer to bear false witness.
6. Thou shalt not copy or use proprietary software for which you have not paid.
7. Thou shalt not use other people's computer resources without authorization or proper compensation.
8. Thou shalt not appropriate other people's intellectual output.
9. Thou shalt think about the social consequences of the program you are writing or the system you are designing.
10. Thou shalt always use a computer in ways that insure consideration and respect for your fellow humans.[8]

Search for information on *ethics* using any search engine on the World Wide Web, and you'll see all kinds of databases offering other codes of ethics and all kinds of information on ethics. You'll see medical ethics, business ethics, computer ethics, military ethics, media ethics, journalism ethics, and so on. You'll see links to many professional societies and their published codes of ethics. You'll see course syllabi, papers, online journals, and much more.

It seems as though almost everyone is concerned about ethics in one way or another. Developing ethical guidelines for technical communicators parallels the challenges of doing so for computer professionals and engineers. All three groups have obligations to society, to their employers, to their clients, and to co-professionals and even professional organizations.

In *Computer Ethics*, Tom Forester and Perry Morrison list many ethical questions faced by computer professionals:

- Is copying software really a form of stealing? What sort of intellectual property rights should software developers have?
- Are so-called "victimless" crimes (against, e.g., banks) more acceptable than crimes with human victims? Should computer professionals be sued for lax computer security?
- Is hacking merely a bit of harmless fun or is it a crime equivalent to burglary, forgery and/or theft? Or are hackers to be seen as guardians of our civil liberties?
- Should the creation of viruses be considered deliberate sabotage and be punished accordingly?
- Does information on individuals stored in a computer constitute an intolerable invasion of privacy? How much protection are individuals entitled to?
- Who is responsible for computer malfunctions or errors in computer programs? Should computer companies be made to provide a warranty on software?
- Is "artificial intelligence" a realistic and a proper goal for computer science? Should we trust our lives to allegedly artificially intelligent "expert" systems?
- Should we allow the workplace to be computerized if it de-skills the workforce and/or increases depersonalization, fatigue and boredom?
- Is it OK for computer professionals to make false claims about the capabilities of computers when selling systems or representing computers to the general public? Is it ethical for computer companies to "lock-in" customers to their products?
- Should, indeed, computer professionals be bound by a Code of Conduct and if so, what should it include?[9]

As you would expect, the Society for Technical Communication also has ethical guidelines:

STC Ethical Guidelines for Technical Communicators

Introduction. As technical communicators, we observe the following ethical guidelines in our professional activities. Their purpose is to help us maintain ethical practices.

Legality. We observe the laws and regulations governing our professional activities in the workplace. We meet the terms and obligations of contracts that we undertake. We ensure that all terms of our contractual agreements are consistent with STC Ethical Guidelines.

Honesty. We seek to promote the public good in our activities. To the best of our ability, we provide truthful and accurate communications. We dedicate ourselves to conciseness, clarity, and creativity, striving to address the needs of those who use our products. We alert our clients and employers when we believe material is ambiguous. Before using another person's work, we obtain permission. In cases where individuals are credited, we attribute authorship only to those who have made an original, substantive contribution. We do not perform work outside our job scope during hours compensated by clients or employers, except with their permission; nor do we use their facilities, equipment or supplies without their approval. When we advertize our services, we do so truthfully.

Confidentiality. Respecting the confidentiality of our clients, employers, and professional organizations, we release business-sensitive information only with their consent or when legally required. We acquire releases from clients and employers before including their business-sensitive information in our portfolios or before using such material for a different client or employer or for demo purposes.

Quality. With the goal of producing high-quality work, we negotiate realistic, candid agreement on the schedule, budget, and deliverables with clients and employers in the initial project planning stage. When working on the project, we fulfill our negotiated roles in a timely and responsible manner and meet the stated expectations.

Fairness. We respect cultural variety and other aspects of diversity in our clients, employers, development teams, and audiences. We serve the business interest of our clients and employers, as long as such loyalty does not require us to violate the public good. We avoid conflicts of interest in the fulfillment of our responsibilities and activities. If we are aware of a conflict of interest, we disclose it to those concerned and obtain their approval before proceeding.

Professionalism. We seek candid evaluations of our professional performance from clients and employers. We also provide candid evaluations of communication products and services. We advance the technical communication profession through our integrity, standards, and performance.[10]

Codes of conduct are valuable because they establish ideals and help define the character of a profession. These codes help to establish an atmosphere of professionalism, and they help to encourage members of a profession to act ethically even in the most difficult of circumstances. . . .

NOTES

1. William Lutz, *Doublespeak* (New York: Harper & Row, 1989) 18–19.
2. Joseph Williams, *Style,* 4th ed. (New York: HarperCollins, 1994) 134.
3. Richard Lanham, *Revising Prose,* 3rd ed. (New York: Macmillan, 1992) 96–97.
4. Vincent Ryan Ruggiero, *Thinking Critically about Ethical Issues,* 3rd ed. (Mountain View, CA: Mayfield, 1992) 4.
5. *Merriam-Webster's Tenth Collegiate Dictionary and Thesaurus.* Electronic Edition. CD-ROM, 1995.
6. *Merriam-Webster's Tenth Collegiate Dictionary and Thesaurus.* Electronic Edition. CD-ROM, 1995.
7. Ruggiero, p. 5.
8. Computer Ethics Institute Home Page, http://www.cpsr.org:80/dox/cei.html
9. Tom Forester and Perry Morrison, *Computer Ethics: Cautionary Tales and Ethical Dilemmas in Computing* (Cambridge, MA: MIT P, 1992) 4–5.
10. *STC Membership Directory 1995–96* 42:3A (September 1995) xi.

Carolyn D. Rude

Legal and Ethical Issues in Editing

Carolyn D. Rude was a member of the faculty at Texas Tech University from 1989 until 2003 and a past President of the Association of Teachers of Technical Writing.

Laws and codes of ethics aim to protect the good of a society as well as to protect individual rights. In technical publication, intellectual property laws govern copyright, trade secrets, and trademarks. Laws also require organizations to accept responsibility for product safety. Editors work with other members of product development teams and legal experts to verify adherence to these laws and the representation of them in the text through warnings and notices of copyright, trademarks, or patents.

Codes of ethics are less formally encoded than laws, as are the sanctions for violating them, but, like laws, they aim to protect individuals and groups from harm and to provide opportunities by creating a work environment in which individuals can achieve. They may be encoded as statements of professional responsibility, such as the "Ethical Guidelines" by the Society for Technical Communication, or derive from cultural values of right and wrong, such as the values that individuals should not harm other persons and that they have responsibility for maintaining the quality of the environment.

Legal and ethical issues pertain both to individuals and to organizations. Editors can be most effective as individuals if corporate policies establish commitment to legal and ethical behavior and if corporate procedures allow for review of products and documents by a variety of knowledgeable people, not just the editor or even just the legal department.

This . . . [selection] reviews legal and ethical issues that pertain to technical publication and suggests corporate policies to encourage ethical behavior.

LEGAL ISSUES IN EDITING

Editors share responsibility with writers, researchers, and managers for protecting documents legally and for ensuring that they do not violate intellectual property, product safety, and libel laws. Editors verify that permissions to reprint portions of other people's publications are obtained, and they review instructions and warnings for potentially dangerous products. Using the symbols ©, ®, and ™ indicates copyrights, registered patents, or trademarks.

The editor can fulfill legal responsibilities more readily and thoroughly if the organization's policy manual and style manual include policies and guidelines about intellectual property, product safety, and misrepresentation.

Intellectual Property: Copyright, Trademarks, Patents, Trade Secrets

The law of many countries, the United States among them, recognizes that one can own intellectual property just as one can own land or a business. These protections are intended as incentives to develop ideas or products that may improve the quality of life for the community. Intellectual property includes original works of fiction or nonfiction, artwork and photographs, recordings, computer programs, and any other expression that is fixed in some form—printed, recorded, or posted on the World Wide Web. These expressions are protected by copyright law. Other intellectual property includes work protected as trademarks, patents, and trade secrets. The owner of intellectual property has the legal right to determine if and where the work may be reproduced or used.

Copyright

The United States Copyright Act of 1976 protects authors of "original works of authorship," whether or not the works are published. The Copyright Act gives the owner of a copyright the right to reproduce and distribute the work and to prepare derivative works based on the copyrighted work. No one else has these rights, and reproducing work copyrighted by someone else violates the copyright law. As editor, you verify permission to use copyrighted material and take steps to protect documents published by your company.

Ownership

Copyrights belong to the author who created the work unless the author wrote the work to meet responsibilities of employment. In the case of "work for hire," the employer owns the copyright. Most technical writers and editors work for hire, and the manuals and reports they write are the intellectual property of their employers. Some publishers require authors to surrender the copyright to them. Works by the U.S. government are not eligible for copyright protection; they are in the "public domain" and can be used by other people without permission. Collections with contributions by multiple authors are generally protected by a single copyright, but the sections may also be copyrighted individually.

The copyright extends 70 years beyond the copyright owner's death. Works written for hire are protected 95 years beyond publication. When the copyright expires, works are in the public domain and may be reproduced and distributed by others unless a copyright is renewed.

Copyright Notice, Registration, and Deposit

Copyright is automatic in the United States as soon as a work exists in fixed form, and protection does not require a notice or registration. However, registration with the Copyright Office gives maximum legal protection. Registration requires sending an application form (Form TX for most technical documents), a fee ($30 in the year 2001), and two copies of the document to the Register of Copyrights in the Copyright Office.

For the best protection, a published work should include a notice of copyright. In a book, the notice usually appears on the verso page of the spread that contains the title page and is sometimes called the copyright page. The notice includes the symbol ©, abbreviation "Copr.," or the word "Copyright"; the year of publication; and the owner's name. Here is an example: © 2001 Longman.

You can get more detailed information on U.S. copyrights from the Copyright Office, either in print or on the World Wide Web (see Further Reading). Particularly useful publications are Circular 1, "Copyright Basics," and Circular 3, "Copyright Notice." You can download application forms from the website.

Copyrightable work published in the United States is subject to mandatory deposit for use in the Library of Congress. Some publications are exempt from this requirement. Circular 7d gives more information on what must be deposited and what is exempt.

International Copyright Protection

Copyright in one country does not automatically extend to another. Use of works in a particular country depends on the laws of that country. Countries that have signed one or both of two multilateral treaties, the Universal Copyright Convention (UCC) and the Berne Convention, offer some protection to work published in member countries. Circular 38a lists countries that maintain copyright relations with the United States. Some countries offer little or no copyright protection for work published in other countries.

Permissions and "Fair Use"

Because work is protected by copyright, permission must be obtained to reproduce sections of someone else's work. Usually the writer requests permission from the copyright holder, but editors verify that permissions have been acquired before the document goes to print. The permission ought to exist in writing, whether in a letter or on a form. The correspondence will establish exactly what will be reprinted and where. The copyright owner may charge a fee for use of the material. If permission is denied, the material cannot be used. Permission must be acquired for each use. Permission to use copyrighted material in one publication does not give one the right to use it elsewhere.

A request for permission to reprint copyrighted material should include the following information:

- Title, author, and edition of the materials to be reprinted.
- Exact material to be used: include page numbers and/or a photocopy.
- How it will be used: nature of the document in which it will be reprinted, author, intended audience, publisher, where the material will appear (for example, quoted in a chapter, cited in a footnote).

Fair use allows some copying for educational or other noncommercial purposes. For example, you may photocopy an article in a journal to study for your research paper. But your professor cannot copy whole sections of a book for the class so that the class won't have to buy the book because doing so would deprive the book's publisher of sales of its intellectual property. In a work setting, it is best to be cautious about copying, especially if you will distribute the work widely and profit from this distribution.

Copyright prohibits duplication of software for multiple users unless an organization has purchased a site license. The terms of use are often listed on the software package.

Copyright and Online Publication

The internet and World Wide Web have provided the public with wonderful access to information. An ethic of openness, sharing, and access has encouraged individuals and organizations to offer documents, statistics, and other information to be freely used. This ethic and the ease of copying material distributed through the internet or web makes some people think that it is all right to do so, but material on the web and even email messages are protected by the same copyright laws that protect print, especially when the information has commercial value. Because publishing on the web makes material available without purchase, you are not depriving an author of sales if you copy material from the web for your personal use and knowledge. Such use is probably fair use. But if you distribute something you found on the web to make money or to people who would otherwise read it on the web and see the advertisements that accompany it, or if you copy the words and use them as your own without citation, then you are violating copyright law. . . .

Cyberspace law is a new area that is still developing. For the present it is prudent to assume that the laws for print publication apply to online text. You need to get permission to use material from the web in other contexts. You may even need permission to link from your website to another site, especially if you wish to link to pages deep in the site, bypassing the home page where the advertisements may appear.

Trademarks, Patents, and Trade Secrets

Trademarks are brand names, phrases, graphics, or logos that identify products. The rainbow-colored apple is a trademark of Apple Corporation, as is the name Macintosh. If the marks are registered with the United States Patent and Trademark Office (PTO), no one else can use those particular marks to represent their own products. Editors look for proper representation of trademarks that may be referred to in the text. The symbol ® next to the mark means that the mark is formally registered and certified with the PTO. The symbol ™ indicates a mark registered on a state basis only or one that has not been officially placed on the Principal Register in the PTO. Trademarks are capitalized in print. Dictionaries identify words that are registered trademarks. Product literature is a good way to find out about whether to use the ® or ™. A typical procedure is to use the symbol on first use of each trademark or to list all trademarks used in a publication in the front matter. Constant repetition of the symbol in the text may become distracting, and the law does not require it.

Patents protect inventions in the way that copyright protects expressions. Their significance to editors is that the text records patent registration with the symbol ® just as it does for trademarks.

Law also protects trade secrets, such as the specifications for a new product or a customer list. Employees owe a "duty of trust" to current and former employers. It is illegal for a company to hire you to find out what a competitor is planning, and it is illegal for you to give trade secrets of a former employer to a new employer or of your current employer to anyone else. Documents produced under government contract projects may be classified and require secrecy. For either private or government work, you may have to use special protections to keep information in your computer files secure. The duty of trust represents one exception to the First Amendment protection of free speech.

Product Safety and Liability

According to U.S. law, companies and individuals must assume responsibility for safety of the products as they are used or even misused by consumers. The products themselves must be designed to be as safe as possible, but because design itself cannot ensure safety, instructions for use and warnings must be complete and clear. The instructions and warnings must cover use, anticipated misuse, storage, and disposal. Manufacturers cannot avoid responsibility with disclaimers (statements that the manufacturer is not responsible for misuse or accidents).

Instructions, Safety Labels, and the Duty to Warn

The first strategy of documenting safe use of a product is to write clear and complete instructions. Clarifying procedures where safety is an issue should be part of document planning. Editors may wish to request a summary of hazards so that they don't have to rely on the instructions alone to identify them. In reviewing a draft, editors rely on the principles of organization, style, visual design, and illustrations . . . , as well as an attitude of vigilance and care. Ambiguity must be clarified. For example, if a procedure calls for "adequate ventilation," an editor may query whether "adequate" means "fresh" air or if circulating air will suffice. Taking the perspective of the reader, the editor may note some gaps that a writer misses.

If there are hazards of using products, manufacturers and suppliers must warn of the risks unless the product is common and its hazards well known. Instructions do not constitute warnings, nor can a warning substitute for instructions. A warning calls attention to a particular procedure verbally, visually, or both. Some standards for identifying different categories of risks and for symbols and colors to identify them have been developed by the American National Standards Institute (ANSI), the Occupational Safety and Health Administration (OSHA), and the International Organization for Standardization (ISO). For example, the word "danger" and the color red are used only when serious injury or death may result. "Warning" (orange) and "caution" (yellow) warn of less serious risks.

Safety labels should be attached to products where users will see them before and as they use the product. This principle may seem obvious but is not always followed. For example, a warning to thaw a turkey slowly in the refrigerator to avoid food poisoning was placed inside the frozen cavity of the turkey where it could not be found until the turkey was already thawed!

The Editor's Legal Responsibility

Like everyone else who is involved in product distribution, writers and editors have some responsibility in the eyes of the law for safe use of a product. They can be named in product liability lawsuits, though the usual practice is to name the company. A sense of professional responsibility as much as respect for the law encourages editors to be careful in the legal edit.

Libel, Fraud, and Misrepresentation

Libel is a defamatory statement without basis in fact that shames or lowers the public reputation of an identifiable person. People who can prove libel may win damages from a publisher. The possibility of being sued for libel worries editors of fiction and periodicals more than it does technical editors because people are more likely to be discussed, referred to, or otherwise cited in works of fiction and in periodicals than in technical documents. Nevertheless, all editors should read alertly for facts to verify the accuracy of any negative statements that may be made about individuals.

Fraud and misrepresentation deceive the public. Misrepresentation may occur in labeling products or making claims about them or in claims about credentials of individuals or about data.

FURTHER READING

"Copyright Basics" (Circular 1) and "Copyright Notice" (Circular 3). Copyright Office, Publications Section, LM–455, Library of Congress, Washington, DC 20559. Also available at www.loc.gov/copyright/.

Dombrowski, Paul (1999). *Ethics in technical communication.* New York: Allyn & Bacon.

Dragga, Sam (1996). Is it ethical? A survey of opinion on principles and practices of document design. *Technical Communication,* 43.3, 255–265.

Johannesen, Richard L. (1996). *Ethics in human communication* (4th ed.). Prospect Heights, IL: Waveland.

Lessig, Larry, Post, David, & Volokh, Eugene (1996). Cyberspace law for non-lawyers. Alan Lewine (Ed.). www.ssrn.com/cyberlaw/.

Johannesen, Richard L. (1996). *Ethics in human communication* (4th ed.). Prospect Heights, IL: Waveland.

Phillips, Jerry J. (1998). *Products liability in a nutshell* (5th ed.). St. Paul, MN: West Publishing Company. See especially "Warnings, Instructions, and Misrepresentations," pp. 210–237.

Product liability (1984). *The guide to American law: Everyone's legal encyclopedia, 8* (pp. 318–324). St. Paul, MN: West Publishing Company.

Strong, William S. (1999). *The copyright book: A practical guide.* (5th ed.). Cambridge, MA: MIT Press.

Velotta, Christopher (1987). Safety labels: What to put in them, how to write them, and where to place them. *IEEE Transactions on Professional Communication, PC, 30.3,* 121–126.

Acknowledgments

Part 1

USING PAFEO PLANNING. By John Keenan. Reprinted from *Feel Free to Write* by John Keenan. Copyright © 1982 by John Wiley & Sons, Inc. This material is used by permission of John Wiley & Sons, Inc.

THE WRITING PROCESS. By Michael E. Adelstein. Reprinted from *Contemporary Business Writing* (New York: Random House 1971). Copyright © 1971 by Michael E. Adelstein. Reprinted by kind permission of Michael E. Adelstein.

THE DIRECT WRITING PROCESS FOR GETTING WORDS ON PAPER. By Peter Elbow. Reprinted from *Writing with Power*, Second Edition by Peter Elbow, copyright © 1981 by Oxford University Press. Used by permission of Oxford University Press, Inc.

EVALUATING AND TESTING AS YOU REVISE. By Linda Flower and John Ackerman. Reprinted from *Writers at Work, Strategies for Communicating in Business and Professional Settings*, 1st edition by LINDA FLOWER © 1994. Reprinted with permission of Heinle, a division of Thomson Learning: www.thomsonrights.com. FAX 800 730-2215.

THE PROJECT WORKSHEET FOR EFFICIENT WRITING MANAGEMENT. By John S. Harris. Reprinted from *Publications Management: Essays for Professional Communications*. Edited by O. Jane Allen and Lynn H. Deming (Amityville, New York: Baywood Publishing Company, Inc., 1994). Copyright © 1994 by Baywood Publishing Company, Inc. Reprinted by kind permission of John S. Harris and of the publisher, Baywood Publishing Company, Inc.

Part 2

GOBBLEDYGOOK. By Stuart Chase. Reprinted from *Power of Words*, copyright 1954, 1953 and renewed 1982, 1981 by Stuart Chase, reprinted by permission of Harcourt, Inc.

WRITING IN YOUR JOB. By William Zinsser. Reprinted from *On Writing Well* by William Zinsser, 6th Edition (HarperCollins, 1998). Copyright © 1976, 1980, 1985, 1988, 1990, 1994, 1998, 2001, 2006 by William K. Zinsser. Reprinted by permission of the author.

THE PLAIN ENGLISH REVOLUTION. By Alan Siegel. Reprinted from *Across the Board* 18 (February 1981) by kind permission of Alan Siegel.

A CRITIC OF PLAIN ENGLISH MISSES THE MARK. By Mark Mathewson. Reprinted from *The Scribes Journal of Legal Writing*, Vol. 8 (2001–2002), pp. 147–151, by kind permission of the author and the journal's editorial board.

A GUIDE TO NONSEXIST LANGUAGE. By the University of Wisconsin-Extension Equal Opportunities Program Office and Department of Agricultural Journalism. Reprinted with the kind permission of the Board of Regents of the University of Wisconsin System and the University of Wisconsin-Extension.

INTERNATIONAL COMMUNICATION AND LANGUAGE. By Gwyneth Olofsson. Reprinted with permission from *When in Rome or Rio or Riyadh . . . Cultural Q & As for Successful Business Behavior Around the World* by Gwyneth Olofsson, Intercultural Press, A Nicholas Brealey Company, 2004. © 2004 by Gwyneth Olofsson.

Part 3

MAKING YOUR CORRESPONDENCE GET RESULTS. By David V. Lewis. From *Secrets of Successful Writing, Speaking, and Listening* by David V. Lewis (AMACOM a division of the American Management Association, New York, 1982). Reprinted with the kind permission of the author.

"I HAVE SOME BAD NEWS FOR YOU." By Allan A. Glatthorn. From *Writing for Success* by Allan A. Glatthorn (Scott, Foresman and Company, 1985). Reprinted with the kind permission of the author.

HOW TO WRITE BETTER MEMOS. By Harold K. Mintz. Copyright © 1970 by Chemical Week Associates, 110 Williams Street, New York, NY 10038-3901. Reprinted by special permission from *Chemical Engineering* January 26, 1970, pp. 136–139.

HOW TO USE BOTTOM-LINE WRITING IN CORPORATE COMMUNICATIONS. By John S. Fielden and Ronald E. Dulek. Copyright © 1984 by the Trustees at Indiana University, Kelley School of Business. Reprinted with permission from *Business Horizons* July–August 1984, pp. 24–30, and from Elsevier Ltd.

E-MAIL: PRESENT A PROFESSIONAL IMAGE. By Janis Fisher Chan. Reprinted from *E-Mail: A Write It Well Guide—How to Write and Manage E-Mail in the Workplace,* by Janis Fisher Chan, published in 2005 by Write It Well, www.writeitwell.com.

Part 4

AUDIENCE ANALYSIS: THE PROBLEM AND A SOLUTION. By J. C. Mathes and Dwight W. Stevenson. Reprinted by kind permission of the authors from *Designing Technical Reports,* 2nd ed. Allyn & Bacon, 1991, pp. 27–48.

WHAT TO REPORT. By Richard W. Dodge. Reprinted from *Westinghouse Engineer,* 22 (July–September, 1962), by permission of the Westinghouse Electric Corporation.

THE WRITING OF ABSTRACTS. By Christian K. Arnold. Copyright © 2003 by IEEE, The Institute of Electrical and Electronics Engineering. Reprinted with permission from *IRE Transactions on Engineering Writing & Speech* December 1961, pp. 80–82.

TEN REPORT WRITING PITFALLS: HOW TO AVOID THEM. By Vincent Vinci. Copyright © 1975 by Chemical Week Associates, 110 Williams Street, New York, NY 10038-3901. Reprinted by special permission from *Chemical Engineering* December 22, 1975, pp. 45–48.

CREATING VISUALS. By Walter E. Oliu, Charles T. Brusaw, and Gerald J. Alred. From *Writing That Works,* 9th edition, by Walter E. Oliu, Charles T. Brusaw, and Gerald J. Alred. Copyright © 2007 by Bedford/St. Martin's. Reprinted with permission of the Publisher.

STRATEGIES OF PERSUASION. By David W. Ewing. Copyright © 1979 by John Wiley & Sons, Inc. Reprinted from *Writing for Results,* 2nd ed. by David W. Ewing. Used by permission of John Wiley & Sons, Inc.

PROPOSALS. By Philip C. Kolin. Reprinted from Kolin, Philip C., *Successful Writing at Work,* Fourth Edition, Copyright © 1994 by D. C. Heath and Company. Used by permission of Houghton Mifflin Company.

Index